AF412874

## Three-Tier Shared Spectrum, Shared Infrastructure, and a Path to 5G

Written by a leading expert in the field, this unique book describes the technical requirements for three-tier shared spectrum, the key policy rationale, the opportunity it creates for shared infrastructure, such as neutral host networks, and the likely impact for 5G.

Detail is provided on the inception of the concept and its implementation in the US Citizens Broadband Radio Service (CBRS), along with descriptions of standards for deployment, algorithms required for implementation, and the broader consequences for wireless network and service architectures. The economic and innovation incentives offered by three-tier spectrum are described, along with potential outcomes such as widely deployed neutral host networks. There is also detailed technical analysis of the unique challenges introduced by three-tier spectrum, such as coexistence among non-cooperating networks.

Covering a wide range of spectrum bands, ITU international allocations, and rule structures that can be adapted for different spectrum-sharing regimes and nations, this is ideal for an international readership of communications engineers, policy-makers, regulators, and industry strategic planners.

**Preston Marshall** led the Google and Alphabet Inc. efforts to establish the technical, commercial, and regulatory ecosystem to make three-tier spectrum a success in the USA. He was a major participant in the USA PCAST study that was the basis of the three-tier concept. He chairs the Wireless Innovation Forum Spectrum Sharing Committee, and is founding vice chair of the CBRS Alliance. Previously he was Deputy Director of the Information Sciences Institute (ISI) at the University of Southern California (USC), Research Professor at USC's Electrical Engineering Department, and Program Manager for wireless, cognitive radio, and networking programs at the US Defense Advanced Research Projects Agency (DARPA).

# Three-Tier Shared Spectrum, Shared Infrastructure, and a Path to 5G

PRESTON MARSHALL

CAMBRIDGE
UNIVERSITY PRESS

University Printing House, Cambridge CB2 8BS, United Kingdom

One Liberty Plaza, 20th Floor, New York, NY 10006, USA

477 Williamstown Road, Port Melbourne, VIC 3207, Australia

4843/24, 2nd Floor, Ansari Road, Daryaganj, Delhi – 110002, India

79 Anson Road, #06-04/06, Singapore 079906

Cambridge University Press is part of the University of Cambridge.

It furthers the University's mission by disseminating knowledge in the pursuit of
education, learning, and research at the highest international levels of excellence.

www.cambridge.org
Information on this title: www.cambridge.org/9781107196964
DOI: 10.1017/9781108165020

© Cambridge University Press 2017



First published 2017

Printed in the United Kingdom by Clays, St Ives plc

*A catalogue record for this publication is available from the British Library*

ISBN 978-1-107-19696-4 Hardback

# Contents

# Acknowledgments

The development of the three-tier spectrum concept has created new opportunities for innovation, not just in spectrum management, but in the wireless service architectures, and wireless technologies themselves. Flexible, scalable, rapid, and low cost access to the spectrum offers the opportunity to create whole new industries that could not exist with more static spectrum models. Neither this book, nor this concept, could exist without the very significant efforts and initiative by a large number of individuals and groups.

First and foremost, I thank the two co-chairs of the President's Council of Advisors on Science and Technology (PCAST) study. Mark Gorenberg, was instrumental in facilitating the development of the principles, and in creating acceptance of this concept within the US executive offices. Eric Schmidt (Alphabet Inc., then Google Inc.) contributed the concept of micro-transactions to reduce transaction costs and create the dynamic market concept that underlies the three-tier concept. It would have been hard to imagine the Department of Defense (DoD) supporting highly flexible sharing of its spectrum prior to this study. Mark worked tirelessly not only to run this study, but to ensure that its recommendations would be supported and acted on by the US Government. Another significant contributor to this initiative was Milo Medin (then Google Inc., now Alphabet Inc.) who introduced many of the commercial and competitive considerations that were central to the recommendations. I also thank the staff of the White House Office of Science and Technology Policy (OSTP) for supporting the study recommendations, and turning them into action by the Administration.

The development of the concept, and its acceptance, would not have occurred without the extensive research performed by the Dynamic Spectrum Access (DSA) and Cognitive Radio (CR) communities. Although many of the more progressive concepts developed in these research areas have not yet been incorporated into the proposed three-tier frameworks, their effort did socialize and demonstrate the value of shared and dynamic spectrum, which was instrumental in its ultimate acceptance.

Additionally, although regulators are often considered to be an obstacle to innovation, the US Federal Communications Commission (FCC) was extremely aggressive in pursuing this opportunity, and in particular, John Leibowitz (Deputy Chief of the FCC Wireless Bureau and Special Advisor to the FCC Chairman for Spectrum Policy at the time) was instrumental in initiating the regulatory process, despite much opposition from those that did not see the vision.

I personally thank the many people at Alphabet and Google, and the participants in the Wireless Innovation Forum (WinnForum) Spectrum Sharing Committee (SSC) for their many contributions and further refinement, development, and implementation of these concepts. Also, I thank the Information Sciences Institute (ISI) at the University of Southern California (USC) Viterbi School of Engineering for support during my involvement in development of the PCAST report. It has been an extreme pleasure to follow this work from its early inception in academia and the Defense Advanced Research Projects Agency (DARPA), through to its deployment in real world wireless systems serving users.

# **Preface**

The United States of America (USA) is embarking on the development and application of a new model for the management of spectrum. This model has evolved through more than a decade of academic, policy, and economic concept development and analysis. It has now been embraced by sufficient major participants in the wireless ecosystem to achieve the critical mass to succeed. Although this model is focused on spectrum management, it has significant implications on the broader questions of wireless architecture, the structure of wireless service models, and even the nature of the wireless industry itself.

As the title suggests, this book is about the intersection of spectrum and innovation. There is a fundamental linkage between the flexible availability of spectrum, and the ability of entrepreneurs and innovators to create and deploy new technology, business models, and services. Therefore, we consider spectrum policy from the perspective that it should effectively provide spectrum to the services that require it, and at the same time, ensure that the operation of the spectrum mechanism does not fossilize the industries and services it supports. Spectrum policy that focuses primarily on the technical protection of incumbents will inevitably end up primarily protecting the business and technical models of those same incumbents.

While many topics in spectrum management and wireless have a rich literature, the rapid growth of the three-tier concept has not yet created an extensive base of either rationale or descriptive material. I believe there was a need to create a single work that is inclusive of the ideas that had been developed, and the actions that had been taken in the concept's short history. Second, the available documentation, mostly consisting of government regulations, reports, and legal filings, is deficient in establishing the rationale for many aspects of, and objections to, three-tier spectrum, and do not address the link between this regime and the creation of innovation-friendly spectrum policies. From an implementation perspective, there was no source of context describing what the real-world implementations of these concepts were, and could be.

Today, the community is focused on a very small set of use cases; one country, one primary incumbent, and a limited set of bands. A general framework that can be applied internationally, with many incumbent user classes, and many bands, is necessary for a scalable introduction of the concept worldwide. Lastly, although the subject is spectrum management policy, it is important to look beyond the lower layers, and consider the impact of three-tier spectrum policies that create a band accessible by a much wider range of participants and technical models. There are many opportunities for different network and service architectures that are enabled by this flexibility in spectrum usage.

The opportunity to share spectrum through bands that are common to operators and non-operators has led to the consideration of large-scale neutral host networks that leverage shared infrastructure to provide localized capacity to multiple wireless operators. In the three-tier framework, carrier accessible deployments can be performed by a wide range of organizations, such as venue owners and enterprises that are able to perform these deployments at a lower cost through integration with their own information technology, power and management infrastructures, and where the costs would be shared among them, and the carriers using the offload service. Shared, or common, spectrum, is an enabler for these neutral host deployments, and at least one industry organization, the CBRS Alliance in the USA, is aggressively pursuing the technical standards for the integration of private capacity into carrier networks.

In this book, we will first examine the traditional methods of allocating spectrum to users and services. These mechanisms have been highly effective, enabling a vast ecosystem of wireless services and sensing to be deployed, and it is certainly the baseline from which all other concepts must evolve. This book will then examine more flexible ideas of spectrum management that have emerged, driven by the need to enable innovation, as well as the need to more flexibly utilize spectrum.

This discussion will start with the concepts developed in two successive US policy efforts, the initial PCAST report, and the subsequent US regulations that established the three-tier Citizens Broadband Radio Service (CBRS) band. These are not the complete models of three-tier spectrum, but they are valuable points of departure. The technical constraints are common for all the world, while regulatory law, economics, policy, and of course, politics will vary nation by nation and region by region; and all of these impact three-tier spectrum. That said, the US efforts are a valuable point of departure, and their development and implementation are highly instructive about the challenges all nations will face in developing appropriate three-tier policies.

To understand these challenges fully, a description of the USA's CBRS regime is provided in order to examine specific technical approaches and implementations that would enable three-tier regimes to be established in bands occupied by a much wider range of protected incumbents or mid-tier users. Lastly, in keeping with the desire for innovation, this book will examine the impact of three-tier spectrum on the ability of innovators to deploy their ideas in a scalable and effective manner. The growth of Internet services and applications was fostered because it could thrive in an incremental, scalable, funding regime, such as practiced by the venture capital community. We will consider how spectrum policy can enable wireless services to avail themselves of the same evolution path.

At the same time we consider these new objectives of a spectrum regime, we need to consider the changing needs for spectrum. Whereas the traditional wireless deployment was for Wide Area Network (WAN) base stations to provide reliable coverage, the need for spectrum is increasingly focused on small cells, serving dense populations in nomadic usage. It is not clear that the deployment model for this capacity is monolithic carrier owned, operated and deployed devices. Instead neutral host hybrid networks might evolve more in the model of Wireless Fidelity (Wi-Fi) deployment than operator deployments, but still integrated into operator network services. If this evolves, this model requires a very different method of spectrum management.

At its core, three-tier spectrum sharing addresses the following two, apparently conflicting, objectives:

1. How do we provide a spectrum regime that can support the intensive capital investment and deployments that are essential to high capacity modern wireless; and
2. Still provide a regime in which all spectrum can be utilized for some useful purpose (*Use it or share it principle*)?

An example of this objective is a management framework in which one spectrum usage pattern might emerge in Times Square, New York and Piccadilly Circus, London; another distinct one in a suburb fifty miles from a major city center; and yet a third, completely different one, far away in agricultural fields, where the major application would be low density, soil condition instrumentation. Local demand in each of these locales would drive the spectrum allocation and usage, not the regulators.

This is the fundamental challenge of three-tier spectrum; to create not the optimal spectrum regime, but create a framework that can adapt itself to meet the local demand, market conditions, economics, and system needs. It is not an argument between licensed and unlicensed, or exclusive and non-exclusive spectrum access. Instead, it is a hybrid that is inclusive of all of these models, and adapts locally to demand, rather than be dictated by "one size fits all" regulation.

The concept of three-tier spectrum management is very much in its infancy. The development of three-tier spectrum sharing is a logical consequence of a series of developments in wireless networking, and the concomitant requirement to provide these networks with the needed spectrum. If this book was written three years ago, the concept would have been, at best, a proposal from an unknown study group. Several years ago, it was only a regulatory action in the USA, almost universally opposed by the entire commercial wireless industry. Even a year ago, it was an interesting US initiative, but not of worldwide significance. Today, it is fully embraced by major players in the wireless ecosystem.

Early consideration in the USA has certainly demonstrated that both traditional and non-traditional users have shown significant interest in using the first three-tier spectrum band. Among other benefits, it appears to solve some of the intractable problems in economically deploying indoor cellular coverage sufficient to meet the escalating demand. It is reasonable to assume that, as additional spectrum is made available through this mechanism, it will be used effectively, if differently, in each local situation.

Some of the concepts that underlay three-tier spectrum were described in a previous book by the author.[1] This previous work explored the unique challenges of dense and scalable capacity networks. The concept of treating the impact of emissions footprint and receiver protection symmetrically was the basis for the quantitative model of the

---

[1] P. Marshall (2012), *Scalability, Density, and Decision Making in Cognitive Wireless Networks*, Cambridge University Press.

spectrum that was provided in the PCAST report spectrum utilization model, and was the basis for several of the recommendations. This concept considers spectrum usage as the capacity that was denied to the other users of the spectrum, regardless of whether it was the impact of the emission causing interference to other users, or limitations on the use of the band, due to receiver performance. The effective spectrum usage was considered to be the "opportunity cost" lost to other spectrum users. Readers of this previous book should recognize many of its more theoretical principles applied to practice in three-tier spectrum management.

One final note. Three-tier spectrum is very much a work in progress. Trials in the USA have just started; tailoring of existing technology standards is now only at its infancy; and no systems have yet been developed that are native to, and fully leverage, the unique opportunities provided by three-tier spectrum. And it has not yet expanded beyond the USA. There are still many waves of innovation to come.

# Part I

## Spectrum Sharing Background

# 1 Introduction to Three-Tier Spectrum Sharing

## 1.1 Overview

This book will develop the general principles of three-tier shared spectrum and shared infrastructure. Then it will explore what the impact of these principles will be on the deployment of future wireless generations, such as Fifth-Generation Wireless Systems (5G). These two highly related technology topics can fundamentally change the nature of wireless architectures as these architectures evolve from a focus on ensuring area coverage to high local capacity, and the transition from 4G to 5G technology and capability.

- Spectrum sharing, which creates the opportunity for access to common spectrum and technology by both traditional wireless operators and non-traditional operators, such as enterprises, and venue operators, and enables new sources of investment in high bandwidth services.
- Infrastructure sharing, which greatly expands the range of entities that can deploy wireless capacity, both reduces and shares costs, and creates bandwidth abundance.
- Extending the shared infrastructure model to 5G will require partnerships between operators and building and venue owners, operators and enterprises with access to siting, power, and backhaul.

Spectrum sharing is the enabler of infrastructure sharing. Spectrum sharing requires overcoming many technical, regulatory, and cultural obstacles, and much of this book will focus on overcoming these challenges. Infrastructure sharing requires these solutions for shared spectrum challenges, as well as significant changes in the concept and technology of a wireless operator, and the model of infrastructure deployment and control.

This emphasis represents the fusion of many concepts that have emerged in wireless practice, spectrum management, and an understanding of the conditions needed to encourage and promote innovation in technology and services. These considerations run the gamut from very technical considerations of path loss, link analysis and device performance, to national economic policy, economic and market principles, and fiscal policies. Creating these new opportunities that embrace and support modern wireless practice, together with policies that embrace and promote these new opportunities, is the challenge.

Individually, each of the underlying concepts is an evolutionary continuation of current thrusts. However, collectively, they offer a unique opportunity to utilize new technology approaches to solve fundamental challenges in the level of innovation in wireless ecosystems. This fusion is just in the process of being incepted in the United States of America (USA), but has already attracted massive industry interest, and it is inevitable that it will be viewed favorably worldwide.

## 1.2 The Problem of Spectrum Scarcity, Densification, and Wireless Growth

The growth of wireless demand has created an almost linear demand for spectrum in which to operate these systems. In most of the world, the opportunities to readily allocate spectrum to meet these needs has long since been exhausted, and even the opportunity to easily reallocate spectrum by relocating or terminating existing services, has also been effectively exhausted. Each successive reallocation has been more expensive, more disruptive to the uses of the spectrum being reallocated, and often results in constraints on the use of the band for the eventual (new) incumbent. The "Low Hanging Fruit" becomes rapidly exhausted to depletion!

The drive for more spectrum has resulted in massive investments by industry, governments, militaries, and other users of the spectrum. Additionally, it has driven significant compromises in wireless systems, in order to accommodate the increasing demand, and concomitant scarcity of spectrum. Spectrum has become an investment, on the principle that no more spectrum can be created, so it must become more scarce, and thus more valuable. This has had the perverse effect of actually taking spectrum out of use and "warehouse" it, in order to maximize its market timing and value.

Yet, some in the spectrum community have argued that the issue facing all of the spectrum-dependent activities is not the lack of spectrum, but is the lack of spectrum access. This argument is that there is unused spectrum available in most, if not all, locations and times, and that the methods of managing spectrum are the limiting factors in accessing this available spectrum, not its implicit scarcity.

It is certainly true that if the spectrum in any specific location is measured with receiving equipment, the spectrum will appear to be very empty. An example of spectrum measurements is shown in Figure 1.1. These measurements [1] were collected in the center of downtown Chicago, IL by Illinois Institute of Technology (IIT) from the highest building in the vicinity.

Much of the apparently unused spectrum in this measurement is, in fact, used by one or another service, or, at the very least, its use would cause interference to a usage in the vicinity. For example, short pulse width radars are hard to detect with long integration time receivers, and unlikely to be detected (very often) by scanning receivers. Satellite uplinks can be interfered with by emitters that can be anywhere in hundreds of millions of square kilometers. Fixed Satellite Service (FSS) receivers are very sensitive to interference, and operate at such low signal levels that a directional antenna must be pointed upwards directly at the satellite's orbital position. Airborne users may be receiving a ground terminal that is not in line of sight of the sensor, but where the

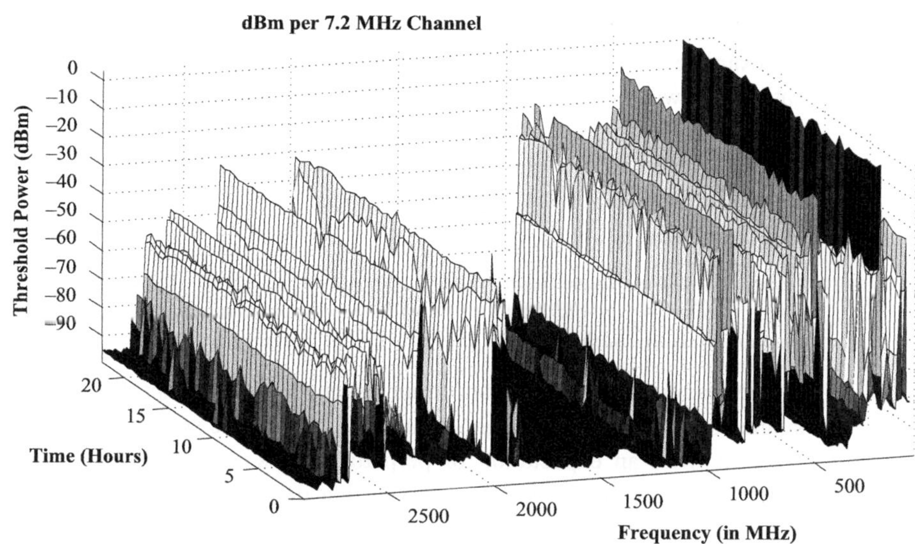


**Figure 1.1** Spectrum occupancy overview (from [2]).

aircraft would be interfered with by the usage. However, no matter how many special cases are considered, it is obvious that there is a considerable resource of spectrum that could, and should, be made available for use.

Although there is general agreement that some spectrum is available for sharing at times; the technical issues involved are significant. They include how to determine what device would interfere with what device, how to know whether there is even someone present to interfere with, or how to resolve conflicts that arise due to this usage. These issues have limited spectrum sharing to very few applications, despite the opportunity it would appear to provide.

The pressure to achieve spectrum sharing is driven not only by a general desire to deploy more access points, but to diversify the participation in densifying the networks. Spectrum sharing is less relevant if deployment is homogeneous and linear. However, if one of the adaptations needed to achieve densification is to empower many more organizations to participate in deploying wireless capacity in dense proximity, then new principles of spectrum and system management are essential.

But, before discussing new methods of spectrum sharing, it is important to examine some of the more systemic weaknesses in the current approaches to spectrum allocation, assignment, and management.

## 1.3 Inadequacy of Traditional Approaches to Spectrum Management and Allocation

Some spectrum regulators recognize that the very evolved framework of spectrum management actually hinders the effective use of spectrum, and creates artificial partitions among the various classes of spectrum users. This partitioning not only impacts spectrum policy, it forces structures onto the industries that use this spectrum. For example, a nationwide Wireless Fidelity (Wi-Fi) supplier is in a similar business as a nationwide wireless carrier, yet their business and technical models are differentiated by the impact of the spectrum they use, rather than any fundamental difference in mission.

An example of this recognition is found in the introduction to the US Citizens Broadband Radio Service (CBRS) final rule-making [3], which succinctly stated the purpose of this first three-tier band as:

*"The Citizens Broadband Radio Service takes advantage of advances in technology and spectrum policy to dissolve age-old regulatory divisions between commercial and federal users, exclusive and non-exclusive authorizations, and private and carrier networks."*

The traditional model of spectrum usage has been to provide a single usage, or single user, with exclusive rights to use a portion of the spectrum, in a specified region, typically with an expectation of perpetual rights. It is the lost opportunity cost of spectrum that is inherent in this model that drives the consideration of multiple tier approaches. Some of the weaknesses in this approach include:

**Geographic Exclusivity** An allocation of spectrum across a large region runs the risk that the permitted users of the spectrum will not have equally deep needs at all locations. For example, a broadband offload service may be starved of spectrum in a very densely populated portion of a city, but may have much less need of spectrum in the sparsely populated outlying areas. Conversely, other users might have a different focus in these less dense areas, such as wireless Internet services, and could use the spectrum, with no loss of utility to the primary network. Or, the offload network could have limited ability to deploy in some locations, such as building interiors, but the spectrum could not be used, even in those areas where the primary licensee could not deploy.

**Usage Exclusivity** An allocation of spectrum exclusively to a single purpose, and typically to a large number of users of that purpose. Broadcast, satellite up and downlinks, Land Mobile Radio (LMR) are all examples of this allocation. However, this spectrum is unavailable for use by other services that could be compatible with this protected usage. Attempts to add services in underutilized bands have generally met stiff resistance from incumbents, who view the spectrum as "theirs."

**Obsolete Allocations** Even then brilliant decisions fade, and appear less brilliant with time. The user or service receiving the exclusive spectrum rights are likely to be supplanted by different technologies and modalities. For example, fiber optic communications have displaced microwave in many applications, video is increasingly distributed by the Internet, and not by satellite, and broadcast television is viewed over cable, if at all. At the same time, telephony services and Internet access are increasingly migrating from wired connections to wireless connections, with concomitant spectrum requirements. Other than generally ineffective secondary markets,[1] there is no process to "refresh" spectrum allocations or assignments.

---

[1] The reasons secondary markets have not addressed these fundamental challenges will be discussed in Section 2.7.

Complicating the issues described above is that reclaiming spectrum is extremely difficult. Systems designed to operate in exclusive spectrum are typically incapable of accepting any forms of interference.[2] The expectation of exclusive access is built into the design of the radios, the operating procedures and business planning. And few services just disappear, unless forced out by regulators. They develop a constituency that resists any change in the spectrum allocation, despite the opportunity cost of the spectrum to other users. Since existing users are typically more organized than potential, future users, the process is highly resistant to "re-farming" of spectrum.

Another complication is that radio designers typically examine the existing radio allocations and usage when designing receivers. One of the cost and power consumption drivers of receivers is the worst case amplitude of the signals in adjacent channels (or bands) that must be rejected. A radio designed to operate next to a low power satellite downlink will not be perceived to need a high degree of robustness to strong adjacent signals. This impact is referred to as "blocking" because it blocks the receiver from being able to effectively operate, rather than directly interfering with the signal.

However, when that spectrum is "re-farmed" to a new, higher-powered application, these lower receiver performance devices will be at risk of failing in the new environment. This situation has occurred several times in the USA. When NEXTEL installed high power cellular systems, they interfered with the lower performance public safety systems that had assumed no high power adjacent band usage [4]. Similarly, when LightSquared proposed to place high power Long-Term Evolution (LTE) systems in a band previously occupied by low power INMARSAT links, the Global Positioning System (GPS) community was able to effectively block this use, based on the claim that it would cause interference to the GPS receivers that had not been designed to handle strong adjacent signals [5]. Spectrum policy is not just about allocating the emission frequency and associated rights, it is much more of an ecosystem management problem, where all users of the spectrum interact in some way.[3]

One approach has been to utilize financial pressures and opportunities to remove existing allocations that are perceived to be less beneficial than possible new uses. The US Incentive Auction is offering television broadcasters with a portion of the auction revenues if they agree to vacate their broadcast channel, or to share one with another broadcaster. While this attempt to share the Government revenue, in order to incentivize broadcast spectrum licensees to relinquish their spectrum was successful, it had to overcome considerable resistance to the concept of rewarding private license holders to give up spectrum for which they paid nothing. And the financial model has limited applicability to government-allocated spectrum. Any spectrum revenue a government organization received would likely be offset in their budgets, resulting in zero net benefit to the organization. They thus have limited incentive to take the risk and complexity of voluntarily sharing their spectrum.

---

[2] Or, at the very least, this is what incumbents claim is true in order to minimize the possibility that "their" spectrum will be forced to be shared with others.

[3] As if that was not complex enough, it is often the case that users of the spectrum do not really understand these interactions until they become an issue.

## 1.4        Spectrum Challenges to Transition from Exclusive to Shared Infrastructure

The previous discussion addressed the shortfalls in current exclusive shared spectrum allocation concept. An equivalent set of shortfalls exist in our current infrastructure concepts. The default of exclusive infrastructure was a good engineering solution when base station sites were sparse, and fully loaded with individual carrier demand. However, in the future, deployments will be made on a building-by-building, and, in 5G, a room-by-room basis, rather than over an order of tens of square kilometers. The spectrum under consideration for three-tier spectrum management typically fits in between the current 4G and future 5G frequency ranges.

Some of the spectrum allocation attributes that enable shared infrastructure include:

- The spectrum available to wireless operators and to non-operators is common to both, so that networks and devices intended for either community are available to both of the communities.
- The access to spectrum in a common band is available to all mobile operators (Mobile Network Operators (MNO), Mobile Virtual Network Operator (MVNO), and Mobile System Operator (MSO)), as well as private network users.
- The capacity available is sufficient to support the demand from multiple operators, so that sharing resources is acceptable in terms of the customer Quality of Service (QoS) commitments that must be assured.
- The cost of third-party deployment is comparable, or less than the equivalent cost of traditional deployment of equivalent capacity.

Very clearly, today's partition of carriers into exclusively licensed spectrum, and non-carriers into unlicensed, does not meet the first two of these conditions. A building owner cannot independently deploy capacity in a carrier's exclusive band except with the support of the carrier, and even if they did, only that one carrier could access the specific Radio Frequency (RF) channel.

## 1.5        Challenges in Achieving Significant Levels of Spectrum Sharing

Sharing existing spectrum has become a technically attractive opportunity to address some of the spectrum shortfalls. It enables spectrum to be utilized by new services without the necessity to remove the existing ones. It enables national spectrum regulators to say *"Yes"* to both the incumbent and the new users. It has the appearance of creating literally something from nothing. The technical community has thus invested considerable time and effort in investigating the technology and management of these regimes. Some of the specific methods and approaches are discussed in the next chapter.

Spectrum sharing has significant issues from the perspective of both national regulators, and of potential users. These perceived issues include:

**Revenue** Licensing spectrum is a significant source of revenue for many nations, and opportunistically shared spectrum appears to lack an equivalent government revenue model that would create income for these nations.

**Interference Assurance** Users of existing spectrum have essentially absolute assurance of no interference, or at least predictable interference. Shared spectrum commons appear to have no such assurances. It would be difficult for traditional operators to justify the significant investments in using such spectrum.

**Localization** Spectrum sharing opportunities are localized in place, and may be temporary in time. Therefore, the benefits of shared spectrum would be localized and possibly transient; this also makes it difficult to justify the required investment.

**Transaction Costs** The process of determining if sharing would be possible is a complex process. Today, it is often a legalistic, regulatory, and often a political process that has a very high transaction cost. The results of the process could be obsolete by the time they could address specific spectrum sharing opportunities. Performing this process on a large scale would appear to be impractical, as the transaction costs would be too high to justify the investment in localized spectrum opportunities.

**Incumbent Protection** Despite the best technical protection methods, there is always a possibility of interference to protected incumbents, due to errors in the spectrum sharing regime, device performance issues, rogue operation, or incorrect deployment data. A method to ensure that admission control is provided, and that sharing nodes can be controlled in cases of interference, is critical to the incumbent's perception of protection, but does not exist and would be difficult for governments to provide.

**Vague Rights** Most spectrum allocations have no generally accepted, established, engineering-level, defined spectrum protection rights. The lack of specific rights pertains to the conditions in their, and/or the adjacent, spectrum. Incumbents typically design their systems to be effective in the context of the adjacent usage at the time of deployment; therefore, they are at risk in any subsequent change in adjacent usage. These issues are typically resolved in regulatory, litigation, or political venues, such as in the USA when NEXTEL deployed high power cellular base stations adjacent to public safety, or when LightSquared similarly proposed to deploy high power LTE macrocells in close proximity to the GPS spectrum [4, 5].

In subsequent chapters, we will use this list of considerations as a "checklist" to evaluate the impact of advanced spectrum regimes and the extent to which they resolve these spectrum sharing issues.

## 1.6 The Concepts Behind Three-Tier Spectrum Sharing

The US President's Council of Advisors on Science and Technology (PCAST) recently provided a framework and approach to addressing these challenges, through three-tier spectrum sharing [6]. This concept has been further iterated by US regulatory action [3, 7, 8, 9, 10]. It is not inherently inconsistent with the two-tier models being developed in Europe [11], but extends the regime to enable more extensive sharing of underutilized spectrum, and provides a mechanism for shared infrastructure deployment by non-operators.

Three-tier spectrum concepts have evolved to address many of the perceived weaknesses of the traditional approaches to spectrum allocation. Later sections will provide specific descriptions of different three-tier regimes, but proposed regimes share some common characteristics, including:

1. There is more than one unique tier in which users of the band have some expectations of protection levels, which presumably are hierarchical in their rights structure.[4]
2. There is a lowest tier that can share any spectrum that is not "used" by any higher-tier user.
3. The lowest-tier users have no expectations of any interference protection.
4. The allocation of spectrum is performed in small geographic and time units to promote flexibility to maximize lower-tier entry opportunities.
5. Decisions regarding the admission of users into the three-tier band must be automated to reduce the transaction cost and delay, and to provide assurance of access.

This is not to say that there cannot be useful spectrum regimes that do not include all of these attributes. However, for the purposes of this book, we will consider this most complete vision of three-tier spectrum sharing as the baseline from which others may depart.

It is worth examining the impact of each of these considerations in depth. Table 1.1 shows the impact of each consideration on the usability, utility, and flexibility of the resulting spectrum management regime. Each of the five considerations above provides significant benefits. Note that the first principle (multiple protected tiers) could be waived if dedicated spectrum was available, but the general approach in this work is to address the more complex problem of sharing this tier with a protected incumbent, since this is likely to be the most typical deployment path.

The next chapter will examine several of the current spectrum management regimes in the context of these principles. Of course, the devil is in the details, and defining what spectrum is in use by higher tiers, and what "small" units of space and time are, are complex problems. We will address how to define them, and the consequences of these definitions, in later chapters.

## 1.7      Outline of This Book's Treatment

Part I (Spectrum Sharing Background) of this book introduces the general problem statement of spectrum management, including the concepts behind both traditional

---

4 Because of the current depth of spectrum usage, we assume that the spectrum that is used for three tier has incumbents that are automatically granted protection. If these incumbents are not present, then this tier is empty.

**Table 1.1** Consequences of three-tier spectrum regimes.

| | Consideration | Consequences |
| --- | --- | --- |
| 1 | Multiple protected tiers | Enables three-tier spectrum to be implemented in bands with an existing incumbent that continues to receive protection, while still providing a high level of assurance to secondary users of their right to receive significant protection. The middle tier(s) (Priority Access (PA)) have an explicit "Right to Receive," but no "Right to Exclude." |
| 2 | Lower tier that can share unused spectrum | A key principle is that all spectrum should be usable by someone at all times and places. This default spectrum sharing right provides the flexibility for deployments that can operate around any of the protected tiers. This provides an inherent elasticity in that high demands for protected usage cannot put lower-tier investment at risk by precluding its access to spectrum. The lower tier also provides a disincentive to spectrum warehousing and creation of barriers to entry, since the spectrum is usable until an actual local deployment occurs. |
| 3 | Lowest tier with no protection | The lowest tier not providing protection ensures it is open to all users, and cannot be blocked or have arbitrary impediments to its use implied by a protection regime. This provision assures spectrum access, although with no assurance of the quality of that access. |
| 4 | Small units of allocation | Small units of allocation (time and space) ensure that the band can support unique uses from both a locality perspective, and as the best use of the spectrum evolves over time. For example, in one section of a metropolitan area, outdoor small cells may be the best use, in another, backhaul, and in a third, interior local area networks. Small allocation units enable this mixed usage. Similarly, use of personal wireless may move to other bands, and this spectrum will become more useful for other applications, such as Internet of Things (IoT). |
| 5 | Automated transactions | Low cost and time delay is essential to the spectrum ecosystem. The number of access points will increase by orders of magnitude (from tens of thousands to millions) as bandwidth provision shifts from macrocell to small/femtocell architectures. This scaling can only be provided by an automated process that does not involve regulatory decision-making, politics, or litigation as elements of the process. |

approaches and the three-tier framework. The material on existing spectrum practices is provided only at the very high level of detail, and for the reader with detailed interest in the subject, the recommended readings provide more specific detail. Chapter 1 (Introduction to Three-Tier Spectrum Sharing), this chapter, provides a general summary of the issues that must be addressed in any new spectrum regime.

Chapter 2 (Prior Dynamic Spectrum-Sharing Regimes) discusses spectrum sharing models. Although these models do not have all of the features desired of the three-tier model, they build confidence in more complex and situational spectrum management models. Experience with each of these models provided essential input to the three-tier model, so this chapter will also discuss how the experience of each of these approaches drove aspects of the three-tier spectrum approach. This chapter discusses specific spectrum sharing approaches; some which are in use, and some which have been proposed. It provides a short summary of these approaches, and is geared to understand the concerns raised, or encountered, with each of the approaches, more than their details. These concerns serve as criteria for sharing regimes in later chapters. Each of the sharing proposals is evaluated against the criteria of Table 1.1.

Part II (Three-Tier Dynamic Spectrum Models) discusses the specifics of the three-tier model from the perspective of the initial policy and regulatory definitions. Chapter 3 (Key Aspects of PCAST Recommendations) describes the principles and arguments of the PCAST report [6] that established the initial framework of three-tier spectrum management. This report led to a US Executive Memorandum that established that it was the Government's policy to share Federal spectrum widely. The study defined a three-tier framework to implement spectrum sharing of either Federal or non-Federal spectrum. This chapter also discusses the concerns that were raised in opposition to this approach. These are highly instructive in understanding the issues facing broader application of such spectrum sharing. Some of the objections were valid, and had to be reflected in the structure of the sharing regime.

Chapter 4 (US FCC CBRS Regulations, and Other International Activity) provides the details on the first regulatory implementation of many of the PCAST report principles in establishing the rules for the US CBRS band. As in all regulatory topics, the "devil is in the details," so this section will describe the significant regulatory conditions in some detail. Many of the concerns addressed in the US adoption of these rules will be common to other spectrum management administrations, so the specific rules may be of interest outside of the US, as either technical or process examples.

Part III (Components of a Three-Tier Architecture) describes the components of a three-tier spectrum sharing ecosystem. These are the basic building blocks of all of the proposed sharing regimes. This part describes these components broadly, so that the discussion could apply to a wide range of three-tier implementations and applications. Part V (Example Use of Three-Tier Spectrum: Use of the 3.5 GHz CBRS Band in the USA) will provide the specifics of the US CBRS design specifics.

Chapter 5 (Three-Tier Spectrum Admission Control Systems) specifies the characteristics of the central control structure, or Spectrum Access System (SAS) needed to implement three-tier spectrum management. This section also includes a discussion of more decentralized approaches that could supplement or replace the functions of a central controller.

Chapter 6 (Admission Control Management of Access Point Devices and Their Clients) describes the unique requirements of devices that can support three-tier spectrum, beyond those that would be required for traditional spectrum occupancy.

Devices addressed include peer devices, such as in point-to-point links, or access point and client, as in Local Area Networks (LAN) architectures.

Chapter 7 (Taxonomy of Protection Methods to be Provided Across and Within Tiers) focuses on the specific algorithmic components that may be included in various protection modalities that three-tier regimes may require. This chapter provides a set of algorithms that will be later assembled to address different types of protections of incumbent or tiered users, and to provide coexistence between devices within a tier. The control system discussion is separate from the protection discussion in order to isolate the purely technical aspects of the control system from the policy-driven protection criteria discussion.

Part IV (Protection Processes for Incumbents and Peers) describes the specific protection methodologies for a wide range of potential incumbent systems that might share spectrum in one or more tiers of a three-tier spectrum access regime. Chapter 8 (General Process for Database System Operation) describes the basic algorithmic elements that are common to all protection situations. These are the basic elements that all control systems will likely have to implement. Chapter 9 (Protection of Incumbent Satellite Operations) discusses the unique aspects of protecting satellite operations from interference from other tiers. There is a considerable amount of spectrum allocated to satellite operation, so sharing of this spectrum is a particularly significant opportunity, although protection of the highly sensitive satellite receivers is an equally significant challenge.

We then consider specific protection cases. Chapter 10 (Protection to and From Terrestrial Access Points and Point-to-Point Networks) discusses the protection of terrestrial networks. Terrestrial networks typically include a number of mobile clients which are not explicitly managed by the admission control process. This chapter addresses the interference from lower-tier, and other protected network, users, and the converse problem of protecting other users from terrestrial networks. This is the case between a Priority Access (PA) and another PA, and a PA and a General Authorized Access (GAA) in PCAST terminology.

Similarly, Chapter 11 (Protection to and From Radar Systems) discusses the specific protection of radar systems from any of the lower tiers. Because radar protection is one of the most complex considerations, this chapter also provides a detailed description of the algorithms that are applicable to the other protection cases. Chapter 12 (Coexistence Between Unprotected Peer Devices) discusses maximizing the degree of coexistence between peer terrestrial networks, where the networks are not entitled to protection from other peer tier networks, such as between devices in the PCAST GAA status. Although this tier may not be afforded protection, or require admission control, it is still desirable that the control structure provides the maximal degree of coexistence among users in this status.

Part V (Example Use of Three-Tier Spectrum: Use of the 3.5 GHz CBRS Band in the USA) provides a specific description of the implementation of the three-tier regime in the USA CBRS band. This is included to provide context for the more general discussion in earlier portions of this book, as an example of how one country applied these principles and implemented them in one band. The basis for this discussion is the industry consensus designs being developed through the Wireless Innovation Forum (WinnForum) Spectrum Sharing Committee (SSC) process. These designs can serve as a point of departure for other implementations of the three-tier principles. The two chapters discuss the additional functionality that is required to transition designs intended to operate in conventional, single-tier spectrum to the United States (US) CBRS framework. The intent of these chapters is not to provide a detailed description of how to design devices for deployment in the US CBRS band, but to illustrate some of the challenges in developing for three-tier, and how one community chooses to address them.

Chapter 13 (Device and Network Interaction with the Spectrum Access System) described the design of the interface from devices to the SAS. This discussion includes the SAS interface protocols and interaction that is required to have devices operate in a three-tier framework. This chapter also discusses the implementation of these controls in intermediate proxies, such as the LTE Extended Packet Core (EPC) components. Chapter 14 (CBRS SAS Requirements) describes the specific requirements and functions that are allocated to the SAS in the CBSD regime. This chapter also addresses some of the constraints on the implementation of a SAS service from an operational perspective, as well as the purely technical considerations.

The previous chapters focused on the lower levels of the networks that are supported or potentially impacted by three-tier operation. Part VI (Future Bands, Network Services, Business Models, and Technology) examines the future of three-tier spectrum, and the ecosystems that it impacts. This Part looks to the future, both to how three-tier may be conceptually enhanced, and to alternatives that could supplant this model, and provide even more extensive access to the spectrum.

Chapter 15 (Potential Services and Business Models Enabled by Three-Tier Spectrum) discusses the implications for new kinds of services and disruptive business models that could emerge from the flexibility of the three-tier spectrum. The chapter discusses the creation of opportunities for innovation in the bands where spectrum is made available. These disruptive innovations require a three-tier spectrum management regime in order to be achievable, or economically realistic. This chapter will examine the scalable entry of new concepts, their funding and incremental deployment, and hypothesize some networks models that could emerge. Chapter 16 (Candidate Incumbent Bands for Three-Tier Spectrum Sharing) develops a set of criteria that can be used to examine the validity of a three-tier regime in the context of the current incumbent usage. It then considers compatibility of three-tier regimes with a wide range of incumbent systems categories, independent of the specifics of the band.

Chapter 17 (From Shared Spectrum, to Shared Infrastructure, to a New Model of 5G) discusses the implications of shared spectrum on enabling a shared 5G infrastructure. This chapter discusses the traditional TELCO model of infrastructure, and contrasts it with the emerging model of shared infrastructure in the Internet. It then develops the requirements for shared infrastructure, and the obstacles to its emergence, and then develops the issues specific to the likely deployment of 5G unique systems above 20 GHz.

Chapter 18 (Future Actions to Deploy More Three-Tier Spectrum Bands and Nations) discusses the path to future adoption of three-tier approaches more broadly in terms of spectrum and nations. This discussion considers the process of "top-down" international adoption in contrast to a nation-by-nation approach, driven by local needs and economies. It also introduces the concept of "virtualized regulation," which enables national administrators to tailor regulations without fragmenting the device of access control system ecosystem. Of course, realistically, there are many considerations (political, economic, diplomatic) that influence spectrum sharing decisions. However, since establishing secondary allocations is not directly controlled by the International Telecommunication Union (ITU), the considerations in this chapter may be applied on a nation-by-nation basis, even absent worldwide consensus in the ITU World Radio Conference (WRC) level.

Chapter 19 (Alternatives to Three-Tier Operation) discusses the possible alternatives to three-tier spectrum that could emerge. The mere existence of three-tier spectrum regimes will make it more practical for even more radical concepts of spectrum management to emerge and be adopted. As the focus of the three-tier model is geared to protection of "dumb" or legacy incumbent devices, it is reasonable to assume that these principles can be adapted, or even deconstructed, as incumbent systems become more sophisticated, cognitive, and, at least, more interference tolerant.

Lastly, Chapter 20 (Conclusions and a Look Ahead) will review what has been developed in the previous chapters, and assess what has been learned, what needs to be learned, and what are the possible outcomes, as these lessons are integrated into innovative spectrum management and wireless technology. This chapter ends with a discussion of concepts that could follow, and perhaps displace many of the three-tier principles, or adopt them to be applicable to more spectrum. Some concepts developed by the author to completely replace the protection metaphor with one focused on device self-protection and recourse will be introduced.

Each chapter includes a set of suggested readings. The specific issues associated with three-tier spectrum management have not yet appeared in the classic academic literature, so many of the suggested readings are from discussion of the initial PCAST recommendations, the regulatory positions filed during the US CBRS proceedings, or explanations of how various positions were adjudicated by the Federal Communications Commission (FCC). Hopefully, as this regime is both better understood, and some experience is gained in its operation, a deep treatment in the academic and professional literature will emerge.

To simplify locating many of the government and industry references, I have organized copies and links to key documents on my web page:

www.prestonfmarshall.com/threetierdocuments.

This site also archives documents that might be lost in changes of US Presidential Administrations, but still provide useful context on the US decision-making. As documents are developed by the various standards groups active in three-tier spectrum I will be adding them to this site to keep it current.

## 1.8    Suggested Reading

There are a number of books that address various aspects of cognitive wireless radio and networks that provide an overview of the field, and its constituent technologies, as follows:

- Martin Cave and William Webb [12] have a recent book that is an outstanding treatment of the overall issues in spectrum management and allocation, and addresses some of the traditional spectrum sharing methods.
- Understanding the inherent weaknesses in the current spectrum management regime is central to appreciating any new regime. The "no winners" conflicts previously mentioned regarding the NEXTEL/Public Safety and the LightSquared LTE/GPS conflicts are instructive, both philosophically, and in the engineering details [4, 5]. The LightSquared/GPS conflict was contemporary with the development of the PCAST report, and was influential in the development of a symmetric treatment of receiver protection.
- A somewhat cynical look at the handling of spectrum issues (primarily in the USA) is provided by a former FCC engineer, Mike Marcus, in his *SpectrumTalk* blog. It demonstrates that the current spectrum management process is highly stressed, even to manage fixed allocations. Available at: marcus-spectrum.com/Blog/Blog.html.
- A significant body of literature has been developed around the concept of Cognitive Radio (CR) and Dynamic Spectrum Access (DSA) by the academic community. Some examples include: work by the author that developed concepts regarding coexistence management, particularly as it relates to adjacent band effects [13, 2]; Fette edited a collection [14] of early work in the field of DSA and CR; and a survey of the field by Doyle [15], among others.

## References


1 M. McHenry, P. Tenhula, D. McCloskey, D. Roberson, and C. Hood, Chicago spectrum occupancy measurements & analysis and a long-term studies proposal. *Proceedings of the First International Workshop on Technology and Policy for Accessing Spectrum (TAPAS)* (2006).

2 P. F. Marshall, *Scaling, Density, and Decision-Making in Cognitive Wireless Networks* (Cambridge University Press, 2012).

3 Federal Communications Commission, *Amendment of the Commission's Rules with Regard to Commercial Operation in the 3550–3650 MHz Band, GN Docket 12-354*, Order on Reconsideration and Second Report and Order (2016). https://apps.fcc.gov/edocs_public/attachmatch/FCC-16-55A1.pdf.

4 L. Luna, NEXTEL interference debate rages on. *Mobile Radio Technology*, 21/8 (2003), 26.

5 National Space-Based Positioning, Navigation, and Timing Systems Engineering Forum (NPEF), *Assessment of LightSquared Terrestrial Broadband System Effects on GPS Receivers and GPS-dependent Applications* (2011). www.gps.gov/spectrum/lightsquared/docs/2011-06-NPEF-lightsquared-report.pdf.

6 President's Council of Advisors on Science and Technology, *Report to the President: Realizing the Full Potential of Government-Held Spectrum to Spur Economic Growth* (Executive Office of the President (EOP), Office of Science and Technology Policy (OSTP), 2012. http://obamawhitehouse.archives.gov/sites/default/files/microsites/ostp/pcast-stem-ed-final.pdf.

7 Federal Communications Commission, *In the Matter of Amendment of the Commission's Rules with Regard to Commercial Operations in the 3550–3650 MHz Band: Further Notice of Proposed Rulemaking*, no. Docket 12-354 (2014). http://apps.fcc.gov/edocs_public/attachmatch/FCC-14-49A1.pdf.

8 ——, *Amendment of the Commission's Rules with Regard to Commercial Operation in the 3550–3650 MHz Band, GN Docket 12-354*, Report and Order and Second Further Notice of Proposed Rulemaking and Order (2015). http://apps.fcc.gov/edocs_public/attachmatch/FCC-15-47A1.pdf.

9 ——, *Wireless Telecommunications Bureau and Office of Engineering and Technology Establish Procedure and Deadline for Filing Spectrum Access System (SAS) Administrator(s) and Environmental Sensing Capability (ESC) Operator(s) Applications* (2015). http://apps.fcc.gov/edocs_public/attachmatch/DA-15-1426A1.pdf.

10 ——, *Wireless Telecommunications Bureau and Office of Engineering and Technology announce Methodology for Determining the Protected Contours for Grandfathered 3650–3700 MHz Band Licenses GN Docket No. 12-354*. Public notice (2016).

11 European Telecommunications Standards Institute, *Reconfigurable Radio Systems (RRS); System Architecture and High Level Procedures for Operation of Licensed Shared Access (LSA) in the 2,300 MHz–2,400 MHz Band*. Technical report (2015).

12 M. Cave and W. Webb, *Spectrum Management Using the Airwaves for Maximum Social and Economic Benefit* (Cambridge University Press, 2016).

13 P. F. Marshall, *Quantitative Analysis of Cognitive Radio and Network Performance* (Norwood, MA: Artech House, 2010).

14 B. A. Fette, *Cognitive Radio Technology*, 2nd edn. (Amsterdam: Academic Press/Elsevier, 2009).

15 L. Doyle, *Essentials of Cognitive Radio* (Cambridge University Press, 2009).

# 2 Prior Dynamic Spectrum-Sharing Regimes

## 2.1 Introduction

This chapter discusses traditional exclusive licensing of spectrum, and some of the current proposals that have emerged recently to address spectrum access shortfalls and to address the improved utilization of underused spectrum. In general, these approaches address many of the same weaknesses in the spectrum models that were discussed in the last chapter. In this chapter, we will compare each of these approaches to the five principles developed and justified in Chapter 1, Table 1.1.

This chapter will also discuss how each of the spectrum sharing models influenced the development of the three-tier model. Each of these existing models contributed experience that was central to the refinement of the principles of the three-tier model. While many commentators consider the three-tier model as a radical departure from current practices, it was, in fact, a somewhat evolutionary response to experience with the models described in this chapter.

## 2.2 Exclusive-Usage Models

Traditionally, spectrum has had to be licensed for a single usage, or a single user. There are a number of works that provide the details of this method of spectrum management, and these details will not be included in this chapter, except as required to develop the concept of three-tier spectrum regimes. This discussion is thus very simplistic, and admittedly skips over many subtle aspects of traditional spectrum management.

The spectrum that is considered for three-tier implementation is generally assigned and managed in one of these two methods, which are significant from the perspective of how this regime can be instantiated.

**Exclusive Usage** Exclusive usage licensing dedicates spectrum to a single purpose; broadcasting, satellite uplink or downlink, radars, etc. Users are licensed to use this band for the exclusive purpose, but have no exclusivity, in that any other user can use the spectrum for the same purpose. Traditionally, it has been hard for other uses to share this spectrum, since all of the systems deployed in it were developed on the assumption that all of the uses would be identical. In most cases, this spectrum does not cost the user any significant payments, and only requires some form of

**Table 2.1** Comparison of three-tier spectrum regime with an exclusive spectrum regime.

|   | Consideration | Consequences |
|---|---|---|
| 1 | Multiple protected tiers. | Typically only one tier. |
| 2 | Lower tier that can share unused spectrum. | No. |
| 3 | Lowest tier with no protection. | No lower tier. |
| 4 | Small units of allocation. | Metropolitan area to National scope. |
| 5 | Automated transactions. | None. |

registration and perhaps coordination with other pre-existing users. The fact that this spectrum was provided by the public without significant cost makes such spectrum more (politically) available for introduction of new sharing regimes.

**Exclusive User** Exclusive use spectrum is assigned to a single user, in a region. That user has the right to exclusive use of this spectrum, and need not coordinate, or consider any other user, except possibly at the boundary of their exclusivity. Since the advent of auctions as a mechanism for assigning exclusive spectrum rights, these rights have generally been acquired at national auctions [1], and at relatively high cost. Although legal treatment of this spectrum is complex, license holders consider that they have property rights as equity in the spectrum. Since there was a considerable investment in this spectrum, most holders of auctioned, exclusive spectrum would consider that they had all of the rights to any secondary usage. This does not preclude them from establishing such frameworks, but most governments would be unwilling to challenge these rights. A possible opportunity is that although these licenses have an "expectation of renewal," their terms could be adjusted at these renewal time points.

In the rest of this chapter, we will compare the spectrum management principles of shared spectrum with the criteria from Chapter 1. Table 2.1 compares exclusive rights spectrum to the three-tier spectrum model.

## 2.3    Unlicensed Spectrum

Unlicensed spectrum has been one of the great wireless successes of the last decade. The growth of Wireless Fidelity (Wi-Fi) has been central to meeting the increasing demand for wireless services, which have grown beyond the scale that traditional Mobile Network Operators (MNO) deployments can provide. Several other services also share this class of spectrum, although Wi-Fi Local Area Networks (LAN) and Bluetooth Personal Area Networks (PAN) are the primary applications. The hypothesized Internet of Things (IoT) may be highly dependent on this spectrum model.

From the perspective of the three-tier model, unlicensed spectrum access has demonstrated that spectrum that has no management, no admission control, and no assurance of interference-free operation, has massive social, societal and economic

**Table 2.2** Comparison of three-tier spectrum regime with an unlicensed spectrum regime.

|   | Consideration | Consequences |
| --- | --- | --- |
| 1 | Multiple protected tiers. | Unlicensed spectrum typically only provides a single tier, which provides no protection. In the discussion of DFS in the next section, this model will be extended slightly. |
| 2 | Lower tier that can share unused spectrum. | Yes. The one tier provides spectrum that all users can use. |
| 3 | Lowest tier with no protection. | Yes. The one tier is usable, regardless of interference to other uses. |
| 4 | Small units of allocation. | Not applicable – there is no allocation. |
| 5 | Automated transactions. | Not applicable – there is no admission control. |

value. Without the success of unlicensed spectrum, it is unlikely that the community would ever consider the lower-tier spectrum rights in the three-tier system to have utility or value.

Table 2.2 illustrates how unlicensed spectrum relates to the criteria we established for three-tier regimes. The difference between a three-tier regime and an unlicensed one comes down to:

- Unlicensed spectrum has no concept of any protected levels within the band and area of operation.
- Unlicensed spectrum has no admission control process, thereby precluding any protection mechanisms. If a device meets the certification conditions, it has a right to use the spectrum, which is inherently incompatible with admission control. In unlicensed spectrum, devices have a right to emit, so long as they comply with the regulations. Unlicensed operation must therefore be the sole use of the band, since no protection can be provided.[1]

The absence of a protected tier in the unlicensed band has certainly had its impact on the nature of the deployments there. The deployment model of Wi-Fi has shown great appeal, even to the carrier-driven, MNO community. This usage model has been almost universally driven to low cost devices, due to the interference potential, where the low device cost, and non-existent spectrum cost, overcame any concern that interference would cause loss of efficiency to access. Even if highly congested, the low cost nature of Wi-Fi has made it highly cost competitive with traditional, higher cost, systems in protected spectrum.

---

[1] The inevitable exception to this statement is the Dynamic Frequency Selection (DFS) sensing in 5 GHz, which protects certain radar operations. This protection mode is addressed in the next section.

## 2.4    Dynamic Frequency Selection for 5 GHz

The DFS model is a spectrum sharing technology that was jointly developed by the Wi-Fi industry, and a number of radar users, including the US Department of Defense (DoD). Incumbent usage in this band includes devices operated for government weather, civilian navigation, and maritime and military users. It is intended to protect certain radar operations in the 5 GHz band, in order to expand the operation and power of unlicensed devices in the range of 5250–5730 MHz.

The DFS Access Point (AP) is required to check on the detectability of a radar signal, when first initiating operation and periodically thereafter. If it detects a radar, it is required to disassociate any connected devices, and move to a frequency on which no radar is detected. It is a very specific implementation of the more general *"sense and avoid"* principle of Dynamic Spectrum Access (DSA). Table 2.3 assesses DFS in the context of the five principles developed and justified in Chapter 1, Table 1.1.

DFS has a long history. It initially had an industry standard that was incomplete in the number of protected radar signals that the DFS implementations recognized. In order to be permitted to operate, a major effort was made by industry, the users, and the governments to make the DFS regime effective. There are a number of issues that arose in developing an effective sensing regime for such a wide range of victim radar characteristics. Some details on the implementation are provided by the Wi-Fi Alliance best practices document [2].

Table 2.3 compares DFS sharing with the three-tier regime.

The differences between a three-tier regime and the DFS one come down to:

1. It has a single protected tier, but does not have a secondary protected tier. Therefore, from the perspective on any spectrum sharing device, it is essentially unlicensed.
2. There is no admission control, thereby limiting the protection to the incumbent. The decision about band occupancy is solely up to the device, and even if interference is detected, the protected user has no recourse to resolve it.
3. The range of protected users is limited. The "sense and avoid" mode can only protect devices with high Power Spectral Density (PSD) levels so that sensing detection can be reliable. Low power, high sensitivity, and receive only devices cannot be protected.

The lack of an admission control process, and of methods, or recourse, to resolve interference from DFS controlled devices was a major concern of protected users. In developing the President's Council of Advisors on Science and Technology (PCAST) three-tier proposal, the concerns of these users had to be addressed in order to obtain support for a significant expansion of shared spectrum principles. These concerns included:

1. The existence of the protected radar was determined solely by the device, and it may or may not have good placement, performance, or the software integrity sufficient to accomplish this task correctly.

**Table 2.3** Comparison of three-tier spectrum regime with the DFS regime.

|   | Consideration | Consequences |
|---|---|---|
| 1 | Multiple protected tiers. | There is only a single protected tier. |
| 2 | Lower tier that can share unused spectrum. | Yes. The lower tier provides spectrum that all users can access. |
| 3 | Lowest tier with no protection | Yes. The lower tier is usable, regardless of interference to other users in the same tier. |
| 4 | Small units of allocation. | Not applicable – there is no allocation. |
| 5 | Automated transactions. | Not applicable – sensing based; there is no admission control. |

2. Even if a node was identified as having caused interference, there was no mechanism to preclude its entry into the band, as long as it implemented the designed detection method. Even if the device has a failed or tampered protection mechanism,[2] it would still have access to the band.

3. The secondary-tier users had no incentive to make the protection of higher-tier users effective. In fact, they had a disincentive to make sensing effective, as ineffective sensing effectively created additional spectrum opportunities for the lower tier. DFS was not viewed as a part of a larger cooperative effort, where both partners had an incentive to make the protection regime effective, in order to expand future spectrum sharing opportunities.

## 2.5    Television "White Space" Spectrum Sharing

The worldwide acceptance of the principles of Television White Space (TVWS) sharing is the first example of international acceptance of a two-tier spectrum management regime, which is mediated by an automated admission control system. The USA Federal Communications Commission (FCC) and the UK Office of Communications (OFCOM) led the efforts to establish the framework for TVWS sharing. As such, it was an important milestone in advancing new concepts of spectrum sharing, and certainly influenced both the development and acceptance of the more extensive and complex proposals provided in the PCAST report.

The United States (US) FCC was the first national administration to provide regulations for the sharing of the Television (TV) band with wireless services. The process involved a Senate Bill to require the FCC to consider TVWS, a suit in Federal court, numerous FCC decisions, and several versions of the technical requirements to operate in the band.

---

[2] There was no requirement that the devices have software protection of the DFS functionality, so DFS access points could have their software replaced with software that did not implement the DFS protections. This became an issue that had to be addressed in three-tier spectrum regimes to assure primary user confidence in the protection mechanism.

The advocates of sharing of TV bands had initially proposed that the TVWS should be managed through sensing by devices themselves. It might seem obvious that if a device could not detect a signal, then a TV receiver could not receive a signal, and therefore, there could be no harmful interference. The process of developing a proof of the sensing ability targeted a detection level of $-114$ dBm. A sensor device provided by Microsoft showed this was possible in FCC testing [3, 4].

However, even though the sensor was shown to be technically possible, the practical protection of TV broadcasts showed the difficulty in using the "sense and avoid" methodology with systems that had highly asymmetric deployments. The TV broadcast sensor might be located in a basement, with poor propagation to the TV broadcast signal. On the other hand, a TV receiver may be located up on a hill, and using an antenna on a high tower, and that antenna may have high gain compared to the omnidirectional antenna at the sensor. All of these possibilities provided issues that could be raised to resist and discredit the concept of self-sensing in the TVWS regime. The argument that was made that there were placements in which the sensor would not detect the TV broadcast, but where the transmissions of the TVWS could reach the receiver's antenna and therefore, cause interfere to the reception of the broadcast stream.

The final FCC decision recognized these issues. TVWS devices were required to access a database server prior to entering the band. The database server uses the device's location and a database of broadcaster antenna locations to determine what channels are available for use by the TVWS device. Interference is addressed as a one-on-one consideration, so it is stateless. Therefore, the admission of a node is not a function of the admission of any nodes previously. The protection contours are static. This simplifies the development of the database systems, but means that the limits on power must reflect the possibility that a large number of nodes might be located in the vicinity of any one receiver, regardless of where they actually were located.

An additional protected class in TVWS is wireless microphones that were already operating on unused TV channels. Users can register these microphones and achieve protection that is essentially equivalent to that of the broadcasters within the locale of the wireless microphone.

The model of DSA that emerged from the academic[3] and Defense Advanced Research Projects Agency (DARPA) funded research was largely based on sensing of the RF environment. In environments where the sensor and victim receiver have similar placement and antenna performance, the assertion that the sensor could detect the signal at signal levels lower than the signal level that could be reliably demodulated is a reasonable one. The difficulty that the TVWS advocates had in showing that self-sensing was a reliable mechanism arose from the vast asymmetry in possible placement and antenna performance between the self-sensing device and victim receiver. In the use cases, such as Land Mobile Radio (LMR), this was much less of an issue, and node placement and antenna performance were much more symmetric.

---

[3] For example, the IEEE Dynamic Spectrum Access Networks Conference (DYSPAN) conferences have had a significant academic involvement from the start of spectrum sharing research.

**Table 2.4** Comparison of a three-tier spectrum regime with TV white space spectrum sharing.

|   | Consideration | Consequences |
|---|---|---|
| 1 | Multiple protected tiers. | There is only a single protected tier, the TV Broadcaster's, and no process for any other type of usage to acquire protected status. |
| 2 | Lower tier that can share unused spectrum. | Yes. The one shared tier provides spectrum that all users can access equally. |
| 3 | Lowest tier with no protection. | Yes. The one tier is usable, regardless of interference to other same tier uses. |
| 4 | Small units of allocation. | Not applicable – there is no allocation. |
| 5 | Automated transactions. | Admission control provided through the database system based on the declared location of the TV broadcaster, and regulatory coverage contours. |

The TVWS regime has many similarities to the notional three-tier regime. There is a positive admission control process, a protected tier, and unlimited rights to use the lowest level. One difference from the three-tier model is that there is only one protected tier. This tier is restricted to a single incumbent use. There is no opportunity for users other than TV broadcasters or wireless microphone users to achieve protected status. Another difference is that the TVWS admission is stateless; that is, a node entering does not change the rights of future nodes to enter.

This is a fair regime, but it has the problem that the admission control system cannot assure a maximum level of interference, since there could be no nodes, one node, or any number of nodes in close proximity to a victim, all adding the interference level at a victim. So, although there is an admission control system in this regime, the admission decisions are based on the application of static contours. The admission control system does not control actual interference, since it cannot control. To provide a high level of assurance to protected users, the maximum power per device must be limited based on an estimate of worst-case density. This is unsatisfactory to both sides. The power for most users is reduced unnecessarily, and the protected users cannot be assured of absolute protection, since there is no control over the density of admitted devices. The solution to this concern, interference aggregation protection, will be discussed in more detail in the next chapter.

An example of the spectrum availability created by this framework is shown in Figure 2.1. The Washington DC metropolitan area is to the right of the figure, and relatively sparsely populated areas in Virginia to the left. Although there are a large number of channels in the rural areas, the spectrum availability in the densely populated metro area is very restricted. This graphic points to the necessity to examine spectrum sharing opportunities in terms of how the spectrum availability correlates with the likely demand. In this case, spectrum sharing opportunities are inversely correlated with the likely consumer spectrum demand.

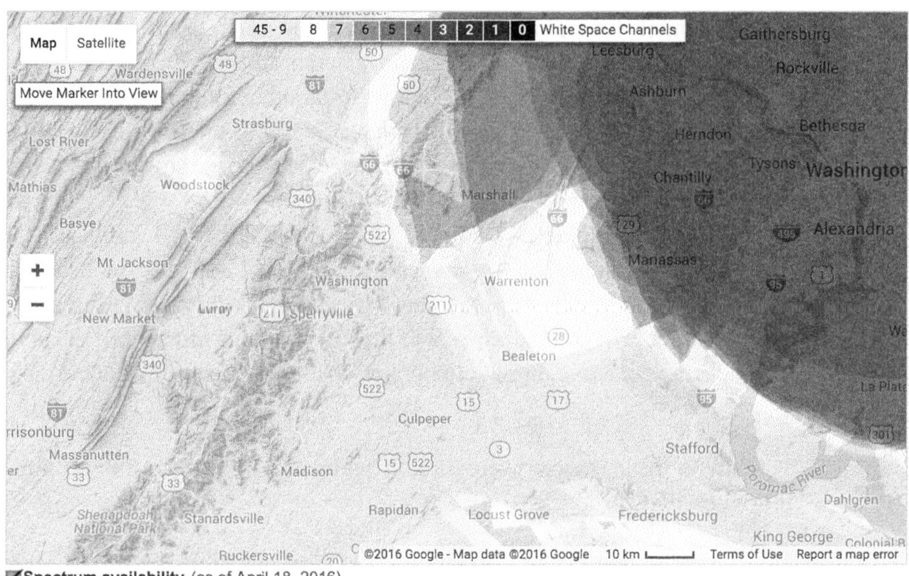


✓Spectrum availability (as of April 18, 2016)

**Figure 2.1**  TVWS spectrum availability.

The TVWS opportunity has had varied penetration in relevant markets, depending on the availability of channels, alternative technologies, user density, and, as always, economics. In the developed world, the deployment of TVWS devices was far below what had been anticipated. Some of the reasons this occurred includes a number of interlocking causes:

1. The volume of channels available in dense environments was very restricted by the broadcast protection policies. Many potential users investigating the use of the band through the white space database services discovered that the channel availability was so limited that reasonably interference-free operation was unlikely. The dense use of broadcast, plus the protection from adjacent band operation were too restrictive to provide a viable sharing opportunity.
2. The advantage of longer operating ranges also limited the density of devices. While Wi-Fi might have been the model for TVWS, the fact that it could communicate 20 times further also meant if could only be deployed at a density that was $\frac{1}{400}$ of the Wi-Fi density, on a channel basis. That precluded the high volume consumer market that has made Wi-Fi so affordable.
3. The channels available were generally only 6 MHz wide. That was far below the typical 802.11g channel, and consumers saw TVWS as offering much less bandwidth than the advertised values on Wi-Fi labeling, even if these values were rarely achieved in practice by Wi-Fi. This placed TVWS at a fundamental disadvantage compared to its low cost alternatives.
4. The imminent incentive auction in the United States of America (USA) would likely severely reduce the already low amount of spectrum available through auctioning of

TV channels for cellular use. This created too much risk regarding future white space availability to support reasonable investment in TVWS products.

5. Even with a large investment in low volume equipment, there was no possibility of obtaining protection for this investment. While it might be acceptable to have a 100 USD investment at risk from interference, it was less so when the cost of equipment was many thousands of dollars.
6. The lack of volume resulted in the availability of only high-cost products, which further reduced the demand for devices, and so established an unattractive environment for the non-recurring investments required to bring low cost devices to the market. The unique aspects of the TV channels minimized the ability of industry to leverage existing, high volume, low cost components in the development of TVWS devices.

Many of these reasons were directly addressed in the PCAST recommendations, and follow-on FCC proposals.

1. A large enough extent of spectrum had to be available to provide reasonable density of use and a high likelihood of availability.
2. Higher bands with short propagation distances could support dense, high volume ecosystems that are viable for suppliers and users.
3. Provide channel bandwidths that are effective in achieving broadband levels of performance.
4. Provide an assured access framework that is not subject to reallocation in the near and medium term.
5. Provide the ability to acquire protection rights in order to assure the value of investment in the band.
6. Minimize the regulatory impacts on existing, high volume ecosystems that can rapidly exploit the band, and do so at the cost levels of the existing bands. The 3.55 GHz regulations permitted the unaltered use of Third-Generation Partnership (3GPP) Long-Term Evolution (LTE) Band 42 and 43 handsets with minimal modifications.

By contrast, in the less-developed world, a number of experimental deployments are underway using TVWS to provide wireless backhaul to under-served communities. Technical and operational success has been achieved in many of them. In many of these locations, the TV band is actually very unoccupied, or even vacant, so it is really not "white space."

Whatever the market success of TVWS turns out to be, it was instrumental in providing arguments for acceptance by several of the features of the eventual PCAST model, including:

1. Demonstrating the viability and use of an automated, centralized admission control process.
2. Ability of developers to implement the admission control systems, and of regulators to approve them.

3. Establishment of specific interference protection criteria, and regulations to enforce them, in the sharing regime.

## 2.6 European Two-Tier Model (ASA/LSA)

Europe was developing a two-tier, all protected, spectrum access model at the same time that the USA was developing its three-tier proposal in PCAST. This framework was originally known as Authorized Shared Access (ASA), and was later titled Licensed Shared Access (LSA) to reflect its licensed pedigree. From the perspective of carrier operation, the two concepts have some similarity. US carriers actually supported the use of ASA/LSA in the USA in preference to the PCAST approach in their initial positions after the release of the PCAST report.

ASA/LSA provides for secondary sharing of an allocation. In Europe, it is targeted for the 2.3 GHz band, which has many localized services in Europe. It is therefore a good candidate for sharing with International Mobile Telecommunications (IMT) applications. ASA/LSA would provide spatial reuse of the 2.3 GHz spectrum between the fixed incumbent operations and a dynamic IMT application. As such, it offers extensive reuse of the otherwise sparsely utilized spectrum, and in a manner that had minimal impact on the architecture of the cellular services.

The overall system architecture and high level procedures [5] of LSA is described in various European Telecommunications Standards Institute (ETSI) documents. LSA implements a sharing arrangement between two parties, a primary licensed entity operating a mobile or fixed network, which holds a licensed right to use spectrum, and the operator of a network that will share that spectrum. The characteristics of the spectrum which can be shared are provided by the primary licensee to the LSA Licensee through a spectrum resource description. Individual national administrations define the framework for this interaction, so the process is still under control of the spectrum administration. This description includes the time/space and frequency limits of the usage rights that are conveyed.

A comparison of the ASA/LSA regime and the three-tier structure is provided in Table 2.5.

The ASA/LSA approach is an important alternative to the three-tier architecture, so an in-depth discussion of it is useful at this point. The control architecture of the ASA/LSA structure is shown in Figure 2.2.

Unlike other secondary sharing arrangements, LSA has a specific system implementation architecture prescribed. The *National Regulatory Authority (NRA)* block denotes the national regulator, and the *Incumbent* block denotes the primary licensee of the spectrum. This provides an explicit role for the NRA in implementing and operating the LSA regime. Both incumbent and NRA access the Licensed Shared Access Repository (LR). This repository provides the terms of spectrum sharing that is offered to the LSA licensee, and receives acknowledgment and status from the LSA licensee. The LSA licensee operates a Licensed Shared Access Controller (LC) within their own network

**Table 2.5** Comparison of three-tier spectrum regime with the ASA/LSA approach.

| | Consideration | Consequences |
|---|---|---|
| 1 | Multiple protected tiers. | Two tiers have protection: the primary incumbent, and the geographically exclusive secondary user. |
| 2 | Lower tier that can share unused spectrum. | Only the secondary user use of the sharing opportunity, so there is no mechanism to share unused secondary sharing opportunities. It is an extension of exclusive use spectrum into a lower tier, but is not shared beyond that user. |
| 3 | Lowest tier with no protection. | No. |
| 4 | Small units of allocation. | The geographic areas that are available for sharing are quite large compared to the interference radius of a single node. |
| 5 | Automated transactions. | The link with the primary is automated, but interference management within the secondary sharing is handled by the system that is deployed there. |

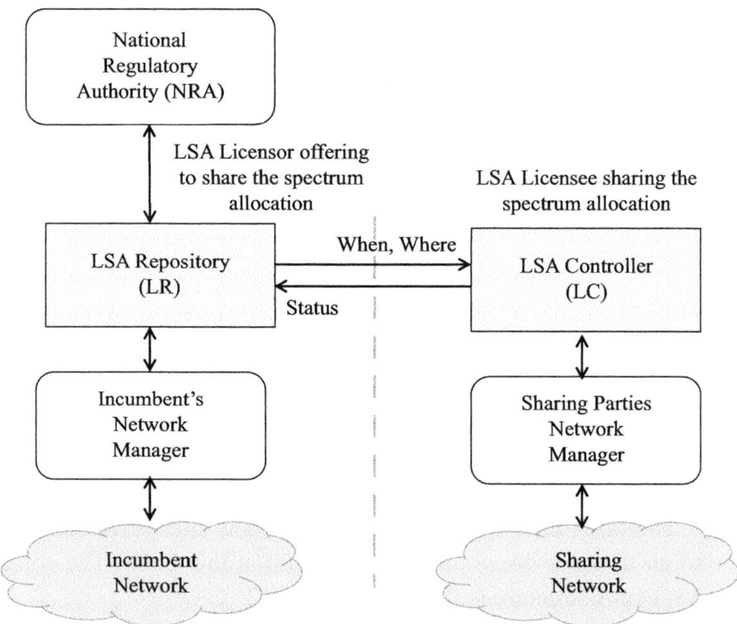


**Figure 2.2** ASA/LSA control structure.

domain. The LC interrogates the LR to obtain information on spectrum availability, and then updates the LR with its status of use by the LSA licensee.

The ASA/LSA framework was developed for a specific sharing scenario and band. Trials are underway in Europe in 2016.

One distinct difference between the intent of the ASA/LSA effort and the US PCAST is in the nature of the secondary user. In ASA/LSA, the management of the shared spectrum's secondary usage is provided by a system internal to the sharing carrier; depicted as the "LSA Controller" in Figure 2.2. If this carrier does not use the spectrum, it is essentially wasted. In contrast, the US PCAST also wanted to assure that there was no limit on the nature of the users of secondary spectrum; the democratization that had made Wi-Fi so successful. Therefore, while ASA/LSA allocates large areas of usage to an LSA controller (and thus a single user); the PCAST approach essentially allocates spectrum to any user that does not interfere with a higher-tier user.

Another difference between the three-tier and LSA frameworks is in the structure of the admission control system. In LSA, the responsibility of admission control is split between the public LR element, which presumably is transparently implemented, and the LC, which is within the LSA licensee's own domain, and presumably is a private implementation specific to the individual licensee, rather than a common facility representing all users of the spectrum, primary or secondary.

The contrast in approach is interesting. In the three-tier framework, all spectrum is automatically shared and usable, unless it would cause interference to a protected tier user. In LSA, spectrum is sharable if, and only if, it is released for sharing by the incumbent. In three-tier, usage by upper tiers constitute the exception case; in LSA spectrum sharing, the right to share is the exception. In LSA, the incumbent controls the availability and sharing terms of the sharable spectrum; in three-tier sharing, the government regulator mandates the availability and terms of sharable spectrum. None of these differences constitute correct or incorrect positions, but they do provide significant contrasts in both approach and objective. Simply put, three-tier shares all spectrum by default, while LSA exclusively protects all spectrum by default.

The differences between the ASA/LSA and the PCAST approaches have profound impacts on the nature of the deployments that occur in each regime. The ASA/LSA approach provides existing operators with the opportunity to make use of spectrum that would otherwise not be available for use. This is a meaningful and useful goal. The PCAST regime enables any entity to utilize the spectrum and deploy any technology they desire to. This feature is enabling of new network deployment concepts, such as neutral host operation and shared infrastructure, which will be developed later in this book.

## 2.7     Secondary Markets

Secondary markets may not appear on many lists of spectrum sharing techniques, but in principle, they offer many opportunities to dynamically manage and re-allocate the use of spectrum, to let it find its *"highest and best use,"* and support the entry of new, and even disruptive, entrants into the market place. Unfortunately, it has led to little of these benefits, to any significant extent. In developing the three-tier spectrum model, the issues faced by secondary markets had to be addressed structurally, and addressed to the maximum extent possible.

Secondary markets were envisioned by their advocates as the equivalent of a commodity exchange, except trading the rights to Hertz of spectrum, rather than pork bellies in the Chicago Board of Trade. This failure of secondary spectrum markets to emerge as a viable and significant mechanism is due to a number of causes including:

1. Spectrum is not really a fungible commodity. Different applications require different spectrum characteristics, and limitations in equipment availability hinders use of spectrum that has not been used elsewhere for the intended purpose.
2. High transaction costs and unpredictable delays.
3. The nature of traditional spectrum deployments.

Understanding these reasons for its limited impact, and addressing them in any future regime, is important. Some works in the field went so far as to describe spectrum portfolio creation, as if spectrum access was a set of equities and contracts, optimal pricing exercises, or open markets [6, 7, 8]. In these discussions, there is little treatment of the impact of limited system frequency coverage, harmonized usage, and the impacts to, and from, adjacent band users, among other very significant considerations that are often not present in the discussion.

In the 2000–2005 period, the US FCC did streamline regulations to make secondary markets more effective and to lower the transaction cost and delay. It enabled immediate processing of "certain qualifying" transactions, but these were generally limited to the direct transfer of licenses, with no changes in the usage conditions, and did not support the "repurposing" of spectrum that was envisioned by secondary market advocates. The FCC also established rules for "private commons," which would be a very open use of spectrum under a secondary license. However, in the end, the lack of well-understood rights, agreement on analytic methodology, and no-recourse decision mechanisms let the issues that had risen in the past largely remain intact. More recently (2016), Mexico's regulator, the Federal Telecommunications Institute (IFT) has approved a set of guidelines that would make secondary markets more effective. Mexico has had a history of spectrum secondary market transactions.

The first issue, the fact that spectrum is difficult to treat as a commodity, is one of the central issues in the lack of secondary market emergence. In some of the work cited in the suggested reading section, the advocates of secondary markets envisioned the application of the full range of financial instruments, such as hedges, options, and other derivative transactions that would be transacted in a transparent, open, and liquid marketplace. This, of course, was dependent on the fungibility of spectrum, and other factors that have, at the least, impeded the formation of these markets:

- Although the advocates of markets often used real estate as a model, spectrum also inherits the *"location, location, location"* principle that drives market-based valuation of real estate. Spectrum neighbors, like physical ones, have a significant amount of influence on the use of a band of frequencies. High power neighbors force any use of the band to have higher cost, higher power consumption receivers. Similarly, if the neighboring systems were designed on the assumption of low power occupants, then an attempt to place a high power application into the band will meet resistance, litigation, and may fail, just as attempting to site a dump in a residential neighborhood would. An example of this is the dispute in the USA between LightSquared and the Global Positioning System (GPS) community. LightSquared made a secondary market transaction with INMARSAT that would have placed high powered LTE systems into the band where only weak satellite signals were present before. The GPS community responded [9] by asserting GPS systems were not designed to tolerate that much energy in an adjacent band, and that it would be highly disruptive to the installed base of equipment. The high power deployment plan failed due to this very strong opposition, but only after the expenditure of billions of US dollars.

- When the primary demand for spectrum was to field networks of wide area coverage base stations, there was a belief that lower (around 1 GHz) spectrum was the "beachfront," and spectrum value was reduced as the frequency was raised, similar to a $\frac{1}{f}$ relationship. This assumption was reasonable for coverage networks, where longer range propagation reduced the need for cellular towers. However, the use of spectrum will become more heterogeneous. Interest in dense, capacity-focused networks intended to address the high, and increasing, demand of nomadic users is likely to increase. The value of spectrum is very much more contextual. Simple models of wireless spectrum value break down when they do not reflect the coverage versus capacity needs in each individual locale. This factor is likely to become increasingly more significant as mobile operators shift their investments between capacity (lower frequencies) and coverage (higher frequencies) networks.

- For consumer-facing applications, the availability of high volume, low cost devices is central to the utility of a band. Without an existing market for the band, there will be extensive delays and significantly increased costs for development of devices that are unique to the band. This is a significant disincentive to use of the spectrum that is otherwise equivalent to bands populated with consumer equipment.

The second impediment is the high transaction cost, delay, and uncertainty in the process of implementing secondary market transactions. This issue arises from a number of causes:

- For all of the reasons raised in the previous paragraph, there is incentive and opportunity for other users of the band, or adjacent bands, to raise interference protection issues to block secondary transactions. This opportunity is particularly likely due to the fact that existing band users and license holders have no clear definition of what the parameters of their protection rights are, or are not. In some cases, there are criteria for co-channel interference, but, with the exception of the OFCOM Spectrum Usage Rights (SUR) [10] effort, no criteria for adjacent band power levels is present in most licenses. These issues have been poorly addressed in the regulatory process, resulting in uncertainty, delay, and cost. Any number of positions can be argued to block secondary transactions.

- Often, secondary use is for a purpose other than that which the original license was issued for. Therefore, changes in license terms inherently are a matter for regulator decision-making and discretion, and it is likely to draw attempts to block the license changes needed.

**Table 2.6** Comparison of three-tier spectrum regime alternative approaches.

| | Consideration | Unlicensed | DFS | TVWS | ASA/LSA | Secondary Markets |
|---|---|---|---|---|---|---|
| 1 | Multiple protected tiers | No | Just one | Just one | Yes | Possible |
| 2 | Lower tier that can share unused spectrum | Yes | Yes | Yes | No | No |
| 3 | Lowest tier with no protection | Yes | Yes | Yes | No | No |
| 4 | Small units of allocation | N/A | N/A | N/A | No | Possible |
| 5 | Automated transactions | N/A | N/A | Yes | Partial | No |

- The process of resolving the interference possibilities, and any license terms, creates an opportunity for competitive considerations to be interjected into the process, with alliances formed around the issues raised by parties blocking or supporting the secondary market transaction.

Lastly, the nature of spectrum usage may be changing. The spectrum market has been considered as dominated by carrier needs for spectrum for wide area coverage networks, typified by the cellular base station and tower. Yet, the rate of growth in wireless demand is largely indoors, driven by the need to support users who are nomadic (rather than mobile). Additionally, the predicted massive uptake in machine to machine, or IoT communications will add significantly to the indoor bandwidth need. Meeting this demand may require radical changes in how this bandwidth is deployed.

The traditional Distributed Antenna System (DAS), tied to carrier spectrum, has had only a limited success in solving the indoor problem. Deployments by building owners, enterprises, and tenants are the logical answer to this need, yet traditional secondary market transactions are certainly not a practical model to enable alternative models for this deployment on such a micro-scale.

## 2.8    Summary of Spectrum Sharing Concepts

This section summarizes some of the material in the previous sections. Table 2.6 depicts how the collective set of spectrum management concepts addresses the five core principles from the three-tier spectrum model developed in Chapter 1.

These spectrum sharing efforts, in aggregate, had all of the attributes of a complete three-tier spectrum management regime. Each of them targeted specific use cases, and so only included the features essential for that one use case. But each of them also laid the groundwork for a more complete description of a sharing regime, which was band agnostic. That is the PCAST proposal that is discussed in the next chapter.

## 2.9 Suggested Reading

- Martin Cave and William Webb have an excellent work [11] on modern spectrum management that provides much of the detail that was not included in this book, and the reader is referred to this work for information on both exclusive and non-exclusive licensing.
- The Spectrum Policy Task Force (SPTF) report [12] was important in initiating a rethink of spectrum policy in the USA.
- The full text of the PCAST report [13] is a highly useful reading, as much of the detail of the proposal is beyond the scope of this book. Also, some topics, such as urban testbeds, that were developed by PCAST, are not addressed in this book at all.
- The Presidential Memorandum [14] that followed the PCAST report was important in providing the force of the administration support behind the sharing of Federal spectrum, which was critical to providing a spectrum segment in which the PCAST ideas could be implemented, as well as proving more spectrum sharing.
- The TVWS spectrum sharing was an important milestone. There are several services that provide graphic displays of TVWS availability in a given location, including one by Google that shows US availability at any specific area. Available at: www.google. com/get/spectrumdatabase/channel/.
- A large amount of work has been done to provide a detailed design for the LSA/ASA framework. These include the following ETSI documents:

  - TSI TR 103 113 (V1.1.1,07/2013) – "System Reference Document for LSA: Mobile broadband services in the 2,300 MHz–2,400 MHz frequency band under Licensed Shared Access regime."
  - ETSI TS 103 154 (V0.0.9, Draft 01/2014) – "System requirements for operation of Mobile Broadband Systems in the 2,300 MHz–2,400 MHz band under Licensed Shared Access."
  - ETSI TS 103 235 (V0.0.1, Draft 01/2014) – "System architecture and high level procedures for operation of Licensed Shared Access (LSA) in the 2,300 MHz–2,400 MHz band."

- The Wi-Fi Alliance Best Practices document [2], previously mentioned, is a very good discussion of the issues of implementing a "sense and avoid" regime with a number of different incumbent user emission characteristics. The problem is even more complex when a wide range of communications devices are included in the mix. These issues were a strong influence on the PCAST analysis.
- There is a large volume of literature on the subject of secondary markets. Some of the interesting work has been cited, and include Muthuswamy et al., describing the mix of primary and secondary spectrum that would form a portfolio sufficient to meet certain demands [6]. Mutlu offers a discussion of the optimal spot pricing strategy [7]. Jia develops concepts for auctions that optimize pricing and provide transparency

incentives [8]. All of these are interesting from the economics and system point of view, but generally fail to reflect the practical issues of radio design, limited frequency coverage of equipment and antennas, impact of adjacent users on the secondary users, and impact of the secondary users on the adjacent users.

## References


1 R. H. Coase, The Federal Communications Commission. *The Journal of Law & Economics*, **2** (1959), 1–40. www.jstor.org/stable/724927.

2 The Wi-Fi Alliance Spectrum and Regulatory Committee, Spectrum Sharing Task Group Regulatory Task Group, *Spectrum Sharing in the 5 GHz Band – DFS Best Practices*, (2007). www.ieee802.org/18/Meeting_documents/2007_Nov/WFA-DFS-Best-Practices.pdf.

3 FCC Office of Engineering and Technology, *Initial Evaluation of the Performance of Prototype TV-Band White Space Devices*, OET Report: FCC/OET 07-TR-1006 (2007).

4 ——, *Evaluation of the Performance of Prototype TV-Band White Space Devices Phase II*, OET Report: FCC/OET 08-TR-1005 (2008).

5 European Telecommunications Standards Institute, *Reconfigurable Radio Systems (RRS); System Architecture and High Level Procedures for Operation of Licensed Shared Access (LSA) in the 2,300 MHz–2,400 MHz Band*. Technical report (2015).

6 P. K. Muthuswamy, K. Kar, A. Gupta, S. Sarkar, and G. Kasbekar, Portfolio optimization in secondary spectrum markets. *2011 International Symposium on Modeling and Optimization in Mobile, Ad Hoc and Wireless Networks (WiOpt)* (2011), 249–256.

7 H. Mutlu, M. Alanyali, and D. Starobinski, Spot pricing of secondary spectrum access in wireless cellular networks. *IEEE/ACM Transactions on Networking*, **17**/6 (2009), 1794–1804. doi: 10.1109/TNET.2009.2019959.

8 J. Jia, Q. Zhang, Q. Zhang, and M. Liu, Revenue generation for truthful spectrum auction in dynamic spectrum access. *Proceedings of the Tenth ACM International Symposium on Mobile Ad Hoc Networking and Computing* (2009), 3–12. doi: 10.1145/1530748.1530751.

9 National Space-Based Positioning, Navigation, and Timing Systems Engineering Forum (NPEF), *Assessment of LightSquared Terrestrial Broadband System Effects on GPS Receivers and GPS-dependent Applications* (2011). www.gps.gov/spectrum/lightsquared/docs/2011-06-NPEF-lightsquared-report.pdf.

10 Office of Communications (Ofcom) (UK), *Spectrum Usage Rights; Technology and Usage Neutral Access to the Radio Spectrum*. Consultation (2006).

11 M. Cave and W. Webb, *Spectrum Management Using the Airwaves for Maximum Social and Economic Benefit* (Cambridge University Press, 2016).

12 Federal Communications Commission, *Spectrum Policy Task Force Report, ET Docket No. 02-135* (2002). http://apps.fcc.gov/edocs_public/attachmatch/DOC-228542A1.pdf.

13 President's Council of Advisors on Science and Technology, *Report to the President: Realizing the Full Potential of Government-Held Spectrum to Spur Economic Growth* (Executive Office of the President (EOP), Office of Science and Technology Policy (OSTP), 2012). http://obamawhitehouse.archives.gov/sites/default/files/microsites/ostp/pcast-stem-ed-final.pdf.

14 President of the United States, *Expanding America's Leadership in Wireless Innovation; Memorandum for the Heads of Executive Departments and Agencies*. Presidential memorandum (2013).

# Part II

# Three-Tier Dynamic Spectrum Models

# 3 Key Aspects of PCAST Recommendations

## 3.1 Movement Towards Flexible, Dynamic Spectrum Policies

The United States (US) President's Council of Advisors on Science and Technology (PCAST) is a US policy advisory board that directly advises the President and his Executive Office, with recommendations in science, technology and innovation. It comprises leading scientists and engineers, and performs a number of studies each year in a variety of topics selected by the committee. In 2011–2012, PCAST performed a study on Federal spectrum policy and innovation, which was released in July, 2012 [1]. This report provided a broad framework for managing shared spectrum, as well as specific recommendations for immediate policy initiatives to implement its recommendations in Federal-controlled spectrum. The broad framework was a fundamental departure from many of the approaches that had dominated spectrum policy.

The PCAST report did not spring up arbitrarily, or suddenly. It was the product of several prior evolutions of spectrum policy thinking in the United States of America (USA), and worldwide. Some of the prior efforts that set the stage for the development and acceptance of the PCAST recommendations include those shown in Table 3.1. There will likely be further extensions and enhancements to the PCAST three-tier model that will emerge, but it is an important starting point, and worth close examination.

This book utilizes much of the PCAST report as a baseline for the three-tier concept. However, the PCAST report is far from perfect. In part, the report is a political document, formed through compromise and consensus-building; reflecting not only technical realities, but political, and economic ones as well. Therefore, this concept of three-tier in this book is heavily influenced by PCAST, but is not identical. It reflects a further four years of industrial development, regulatory litigation, and the participation of innumerably more insights, perspectives, and points of view, than were reflected in the initial report.

## 3.2 PCAST Findings and Recommendations

A summary of significant findings of the PCAST report [1] are included in a single table in the report. This table outlines seven finding areas, and their associated

**Table 3.1** Significant prior spectrum management concept efforts.

| Effort | Significance |
| --- | --- |
| DARPA XG Program | The Defense Advanced Research Projects Agency (DARPA) Next Generation Communications Program (XG) program [2, 3, 4] produced significant research focus on Dynamic Spectrum Access (DSA), including performing some public demonstrations of interference-free spectrum sharing. The XG program also socialized spectrum sharing within the Department of Defense (DoD) and had outreach to the Federal Communications Commission (FCC). Much of the initial interest in the IEEE Dynamic Spectrum Access Networks Conference (DYSPAN) was from individuals involved in this program.[a] |
| FCC Spectrum Policy Task Force | The US FCC established the Spectrum Policy Task Force (SPTF) [5] after the FCC Chairman received a briefing from DARPA on the plans for the XG program, and hired the DARPA Program Manager to lead it. It proposed a number of changes to make the spectrum management process more flexible, and set the stage for consideration of ideas, such as those of PCAST. It made a fundamental argument against the "command and control" approach of current spectrum management, in which the Government regulators decide the best use of spectrum. |
| IEEE DYSPAN Conference | The IEEE DYSPAN conference was central to expanding the shared spectrum community within, and beyond, the initial DARPA and National Science Foundation (NSF) communities, and reached out to industry, economics, legal, and policy communities, and broadened to include the participation of European researchers, particularly those performing research under the European Union (EU) Framework program sponsorship. Unique among IEEE conferences, it included a policy track focused on the economic, policy, and legal issues and opportunities. By the time PCAST proposed its concepts, many of them had been extensively discussed and vetted in this interdisciplinary venue. There is an extensive set of writings on policy, economic and technical issues in the proceedings of this ongoing conference series. |
| UK OFCOM Usage Rights | The United Kingdom (UK) Office of Communications (OFCOM) developed the concept of Spectrum Usage Rights (SUR) [6] to describe the rights of emitters and receivers, and worse-case conditions that a receiver may be faced with. The intent of this was to provide guidance to receiver designers about the future environment that their systems would face, and provide the flexibility to place new applications in the spectrum without claims of harmful impact from receiver owners who had designed to the previous conditions only. Although it was difficult to place these principles into existing licenses, it was a framework that was reflected in the PCAST developed concept. Spectrum rights appeared to be a reasonable compromise to address receiver assumptions without regulatory imposition of receiver standards. |

[a] In full disclosure, the author was the Program Manager at DARPA for the XG program.

recommendations. Some are specific to the US spectrum management and policies, but most are inherently technical, and not nationally specific. These findings of the PCAST report are summarized (abridged) in Table 3.2. Although these findings were focused on US economic considerations, they are likely similar to many developed economies.

The writing is typical of a government report, and sometimes seems to hide the implication, or the true rationale of the point the recommendation is making. Some observations of the implications of the recommendations follow. The recommendations that are specific to US spectrum management policy for Federal and civil users, and will not be discussed further, as they are unique to the US regulation.

**1.1** Spectrum policy should not be focused exclusively on technical aspects, but should also reflect the necessity for nations to provide their industry with an opportunity to create the technologic leadership that is essential to create new industries. An overly narrow focus on incumbent interests may become a barrier to the introduction of new and disruptive technologies. Spectrum policy must be more broadly integrated with the aspirational aspects of the society and economy.

**1.2** The charter of this study was limited to Federal spectrum. However, the fundamental observation, that clearing spectrum for exclusive use was an unsustainable process, was equally applicable to non-Federal spectrum, just as to Federal.

**2.2** Lower power, indoor, low height capacity networks will be the major tool to address the rapidly expanding need for wireless capacity. The traditional approaches to spectrum management have had a focus on considering the worst case conditions when high power, high elevation base stations are a candidate for deployment. Policies that have been effective in managing spectrum for exclusively licensed Wide Area Network (WAN) applications may not be appropriate to the emerging need for local capacity networks.

**3.1** At the time of the execution of the PCAST study, the USA had been through significant conflict over receiver coexistence issues between cellular and public safety [7], and was in the midst of a conflict between a proposed cellular deployment and Global Positioning System (GPS) users [8]. These had reached the level of Congressional involvement, and had presented cost implications in the billions of US dollars. With low cost consumer devices dominating the ecosystem, receiver protection and performance considerations needed to be more prominent in the spectrum processes. An explicit understanding of the electromagnetic environmental conditions under which a device must be capable of operating in is key to flexibility in spectrum management and policy.

**3.3** The report praised the value of international harmonization of spectrum policies, but did not urge harmonization of bands due to the national differences and the flexibility of modern wireless technology. Common policies were viewed as more important than common frequencies.

**5.1** Lack of incentives for spectrum sharing by Federal users could apply to most civil incumbents, as well. Lack of monetization of spectrum efficiency and sharing has resulted in no incentives for government or civil users in many bands to make best use of allocated or assigned spectrum.

**Table 3.2** Summary of PCAST report policy findings (Paraphrased).

| Topic | | Finding |
| --- | --- | --- |
| From Scarcity to Abundance | 1.1 | Spectrum provides great economic opportunity for the nation, and is a foundation for American economic and technologic leadership. |
| | 1.2 | Clearing and reallocation of Federal spectrum is not a sustainable basis for spectrum policy. |
| | 1.3 | The fragmented partitioning of Federal spectrum responsibility has led to inefficiency, artificial scarcity, and the usage of the spectrum by both Federal and non-Federal spectrum users. |
| Policy Hurdles to Clear | 2.1 | Sharing of Federal spectrum provides an opportunity to deploy a wholly new approach to Federal spectrum architecture and policy. |
| | 2.2 | Wireless architectures have evolved from a single model of high-power, high-elevation base stations, to a mix of capacities with a mix of power, altitude and positioning. Different usage implies different spectrum policies. |
| Technology Advances and Challenges | 3.1 | Spectrum management and regulation is focused on the characteristics of transmission, whereas receiver characteristics increasingly constrain usage. |
| New Application Economy | 4.1 | Moving to a dynamic sharing model of Federal spectrum would unlock economic benefits by allowing the private sector to make intensive use of underutilized parts of the radio spectrum. |
| | 4.2 | Sharing of Federal spectrum provides an opportunity to deploy new spectrum management principles. The Federal Government is more likely to consider the "public good" than spectrum incumbents with no national mission. |
| Starting with Federal Spectrum | 5.1 | There is no incentive system today for agencies to be efficient in their use of spectrum, or to share spectrum. Incumbents have no "upside" from sharing spectrum. |
| | 5.2 | A Public Private Partnership (PPP) is the best mechanism to ensure optimal use of Federal spectrum and research investments. Such a framework could provide Federal agencies with incentives for more spectrum sharing. |
| | 5.3 | International harmonization of spectrum policies is essential to product innovation, interoperability and roaming, spectrum efficiency, and cross-border coordination. |
| Implementation | 6.1 | Insufficient opportunity is available to test the new architectures, policies, and the new systems proposed in this report. More incremental methods of providing spectrum are essential to innovation. |
| Immediate Selective GAA Sharing | 7.1 | Expansion of the white space system to include certain space to ground and radar-based Federal bands could allow immediate General Authorized Access (GAA) device usage while other recommendations are being enabled. |

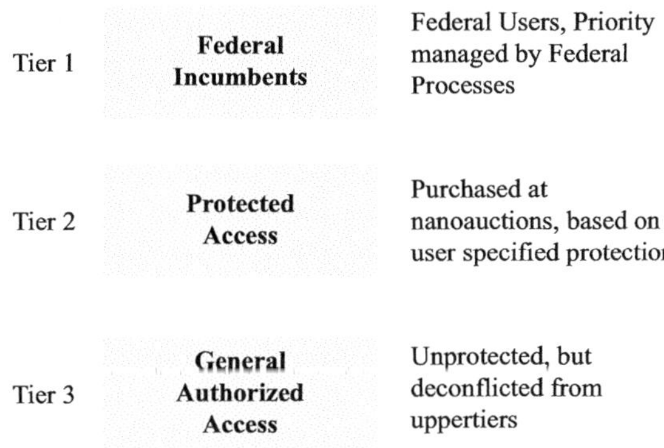


**Figure 3.1** PCAST defined general tiers.

**7.1** There are some very immediate opportunities to pursue spectrum sharing that require a minimum of regulatory action, or disruption to incumbents. These could be pursued prior to the full recommended regime being in place. In the USA, this led to FCC action to provide three-tier sharing of the 3.55–3.7 GHz band, and to auction sharing rights to the 1755 MHz Advanced Wireless Services (AWS) band.

Beneath these broad, high level, policy recommendations were a number of technical recommendations, implicit and explicit in the management regime it advocated. Some of the major technical recommendations are shown in Table 3.3. Several of these form the basis of the proposed three-tier spectrum sharing regime.

The auction concept of PCAST is almost diametrically opposed to the traditional values in spectrum auctions. The PCAST argument was that spectrum policy should attempt to make spectrum abundant and accessible for use by the wireless and sensing community. Successful policies should result in a low cost for spectrum rights, not a high one. Wireless innovation was of more economic advantage to a nation than a one time windfall of funds that tied up spectrum and limited innovation for decades to come. This is in contrast to the view that a "successful" auction was one which raised very high bids and payments. In reality, spectrum payments are just an indirect tax on the users of the services that pay the costs, and divert capital from investing in the infrastructure of wireless services.

## 3.3    Analysis of PCAST Recommendations

PCAST proposed a three-tier structure, as shown in Figure 3.1. The PCAST report was written to apply primarily to Federal bands so it did not address specific incumbents in the incumbent tier. The underpinnings of the PCAST approach was an automated admission control system, referred to as the SAS that would manage and isolate the multiple tiers proposed in the recommendations.

**Table 3.3** Summary of PCAST report engineering-level recommendations.

| Topic | Technical Recommendation |
| --- | --- |
| Protection Criteria at the Aggregate Interference Level | Typically, interference has been considered as a one-on-one issue. Therefore, the power of a potential interferer had to be reduced by the worst case density. Instead, PCAST recommended that the protection should be established based on the aggregate interference, and the admission control process should continuously increment the interference conditions at each protected node, and use admission control of new nodes to assure protection. This could enable more power to be used by sharing nodes, and still assure incumbent protection to any suitable criteria. |
| Spectrum Access System (SAS) Admission Control System | Previous spectrum management had generally established fixed criteria for entry into the spectrum. PCAST recommended a dynamic system that was stageful. Each time a device entered or left the spectrum it would implicitly adjust the entry criteria for a new node entering. This enables the Spectrum Access System (SAS) to offer aggregate interference protection, transition from fixed geographic protection to dynamic device protection, and to interact with devices to enable devices to adjust their operations to the specific location and spectrum that they desired to operate in. |
| SAS-Dynamic Market for Spectrum Rights | Traditionally, spectrum rights are obtained in a long term process, and are retained for a long period of time, as well. PCAST proposed that these rights be acquired dynamically, on demand, through "nano-auctions" (the author's term) that would compete against other users of the band when one device wished to acquire rights that would be competitive with other users at that location. This process would be triggered only by mutually exclusive demands for spectrum. |
| User Provided Protection Criteria | Traditional spectrum protection is provided to levels established in regulation. These generally stress controls on emissions, but provide little insight into what receiver protection will be provided in the future. PCAST proposed that requests for protection could cite arbitrary (although realistic) protection and emission requirements. These would include the out-of-band energy limits to be faced by the receiver, co-channel noise, and the classic emission rights. Requesting more than essential rights would raise the deconfliction distance, and likely raise cost. This was believed to have several effects: (1) clarify and explicitly state the environment that the receiver must tolerate; similar in concept to the OFCOM SUR proposal [6]. (2) It would reward technologies that were capable of increased coexistence, and (3) it would enable market forces to create the balance between increased device performance and the demands of specific services, in specific locations. |

The upper tier is occupied by the incumbent Federal users, whose priorities within the tier are established through the existing Federal processes. These systems would have no interaction with the admission control system used in the lower tier, with the exception of possibly reporting activity status to the system. To be acceptable to most incumbents, a sharing regime must have no, or negligible, procedural impact on the existing primary users, and no impact on the incumbent system design or performance.

Admission to the shared tiers was to be provided by a control system, referred to as the SAS. The middle tier, Priority Access (PA), was a protected, non-exclusive, access layer. These licensees would offer the right to be protected from interference from the lowest tier. The lowest tier is referred to as GAA. It is an authorized access, as compared to unlicensed or license-exempt, since every entry into the band had to be explicitly authorized by the SAS.

By contrast, an unlicensed device has the right to operate so long as it complies with the rules. In the framework proposed by PCAST, devices had no such right (regardless of whether they were second or third tier), and only had the right to emit if they were explicitly granted and maintained the approval from an admission control system. In terms of US regulation, it was considered "Licensed by Rule."

The proposed PCAST mechanism for acquiring PA status was not adopted by the FCC, but was interesting, and very flexible. Some of its attributes include:

1. Auctions were continuous, and were triggered based on mutually exclusive requests for rights. Devices or networks would request to acquire spectrum rights, and bid against any other users whose protection requests would conflict with theirs.[1] This framework resembled the structure of Internet advertising, where extremely rapid auctions determine the placement of advertisements in search results, or placement on web pages.
2. Protection criteria and emission characteristics were specified by the user, rather than by regulators. Requests for excessive power or receiver protection would cause more conflicts with other users, and likely raise the cost of acquiring protection. The level of the requested protection in these requests would be left to market forces to balance.
3. Receiver characteristics, such as adjacent band and channel protection requirements, were included in the request for protection. Rather than impose receiver standards, poor receivers would conflict with potential users over a greater area, and again, raise the likely cost of obtaining protection of poorly performing devices. This would continually incentivize improvements in receiver performance through market forces, rather than static, regulatory decision-making.

A prime objective of deferring protection specification to the device was to enable market forces to balance the many considerations of device cost, spectrum cost, service level, and other system and economic factors. PCAST rejected the concept that a single standard across large geographic areas of uneven demand, and in bands of varying interest, could be effective in best utilizing spectrum. Instead, it proposed that the use of the band should be dynamic, based on unique local conditions, which would drive cost, technical requirements and deployment. The band usage would tend to an equilibrium state; balancing the spectrum cost, driven by its scarcity, equipment

---

[1] A maximum percentage of the spectrum could be acquired with PA status, so a minimum amount of spectrum was maintained as freely available to ensure the ability of GAA users to enter the band.

cost, driven by its performance, and user demand for services. Instead of fixed criteria, these considerations would be adjusted as technology and demand dictated, and reach different solutions in different locations. Essentially, the band would be self-managing.

Another key principle was to avoid any of the negative incentives or behaviors in auctions that are perceived to be contrary to the interests of wireless users. The implicit incentives for these behaviors were viewed as a weakness of the auction process. The PCAST focus was to define policies that would have result in spectrum being made available for the maximum use, and thereby create wireless capacity. As an example of this PCAST concept, PA holders received no right to exclude other users and leave spectrum unused. They only have the right to be protected from interference from lower-tier users when they were actually deployed in the spectrum.

We can consider that a traditional exclusive use license provides two rights:

1. The right to be protected from interference.
2. The right to exclude other users from the band, regardless of whether it would cause interference to the license holder.

The PCAST approach essentially split these rights. A PA holder only had the right to be protected when a node was deployed. If they do not deploy in regions where they have PA rights, that spectrum is essentially treated as GAA. This effectively precluded spectrum warehousing or competitive preclusion. Anyone can purchase rights, but no one can block the use of the spectrum through purchases of licenses. In fact, this split was the intent of spectrum regulation from its onset, after the sinking of the ship RMS Titanic.[2] It was to ensure communications, not monopolies. The PCAST approach merely returned to these basic intentions, based on the use of modern technology to enable protection without exclusion.

Another issue in spectrum rights has been the duration of exclusive use licenses. The original theory of auction proposed by Ronald Coase [9] was that competitive auctions would result in spectrum being dedicated to the "highest and best use." Regardless of whether this premise (that winners of the auction had the highest and best use for the spectrum) is true initially following an auction, after decades of spectrum renewals,[3] it is likely that new applications, services, and providers would emerge if the rights had to continue to compete with newly emerging ones.

Yet, with the presumption of renewal, the spectrum assignments would be static after this initial auction. Instead, PCAST proposed that the nano-auctions it contemplated provide rights, or licenses for a very limited time duration, with no expectation of renewal. The spectrum environment would be continuously competitive. The spectrum rights granted would be long enough for capital recovery of reasonable investment, but not a long-term assurance of a line of business, or protection from new entrants.

A last point is that the objective of the report was to consider how spectrum policy could enable economic activity. In the context of this report, innovation was the key

---

[2] International spectrum management was initiated after radio frequency interference was blamed for the lack of rescue ships after the emergency radio alerts were sent from RMS Titanic.

[3] It should be noted that most auctions carry a fixed term, but have a "presumption of renewal," that effectively makes these licenses essentially perpetual.

to this economic growth. The mix of GAA and PA would provide the ability for new ventures to enter the spectrum incrementally, in a scalable growth path. For example, if exclusive use spectrum acquired at auction was the only way to experiment with, or deploy a new service, then no venture-funded startup could even consider it. Even if they bought a sliver of spectrum in one region, they would face an almost insurmountable task in expanding rapidly enough to justify the venture. These attributes might be considered to be the elements of a "venture-friendly" spectrum policy.

The PA and GAA structure enables systems to be deployed with no spectrum investment, or a much smaller investment to "rent" spectrum rights. If the venture works, the same strategy can expand to more spectrum and more locations incrementally. If the venture did not work, no capital had been tied up in spectrum rights that were not useful, and illiquid.

It is easy to envision entrepreneurs starting a pilot project in GAA in one city, perhaps expanding to PA after some success, and then making a decision to launch nationally, in advance of actually obtaining the spectrum. That is not imaginable, or even possible, in the exclusive auction market. Spectrum that is identified for clearing takes an average of 8.4 years before it is fully usable.[4] Add the time between auctions, and there is no reasonable way to think of the current process as providing a pipeline of spectrum for innovators.

Additionally, the only transaction possible is a perpetual purchase[5] (at a cost of billions of dollars) of a right that may or may not be needed in the future. It is the equivalent of asking a startup enterprise to first pay for a building, build the building, and wait for completion before a business could be started. With these timelines, costs, and uncertain future availability, anything requiring spectrum is essentially not able to avail itself of the venture ecosystem that was so influential in building the Internet, and its wide range of services.

There is a tendency by governments to consider spectrum auctions as a policy success, or not, by the revenue raised. The opposite is true. The fact that an auction raises massive capital is not indicative of a good spectrum policy. Instead, the policy has created a (likely artificial) scarcity of spectrum. The capital provided in the auction will not be available to deploy the infrastructure. The high cost of the spectrum will force the use of higher cost technology in the infrastructure, to minimize spectrum costs. High spectrum costs imply poor performance, or a perception of future performance, by the governments in meeting the spectrum needs of the industries that require it.

## 3.4  Ability of PCAST Proposals to Address Spectrum Sharing Issues

Chapter 1, Section 1.5 introduced a number of issues that spectrum sharing regimes must address in order to be effective in creating disruptive change in how spectrum is

---

[4] Based on a Table from the National Broadband Plan [10].
[5] Although not legally a perpetual right, the implicit guarantee is that there is a right of renewal. This is consistent with the view that spectrum is a property right, not a rental.

managed. The discussion below characterizes how the PCAST approach addresses these concerns raised in Chapter 1.

**Revenue** The auction of rights to operate in the protected second tier PA provides a source of ongoing revenue. This revenue stream would be a periodic one, rather than a one-time sale, but the present value of this, essentially infinite, revenue stream may be quite high. This present value may be comparable to the capital raised in the massive auctions of the past.

**Interference Assurance** The second tier assures that half of the spectrum is available on terms that are very similar to conventional exclusively licensed spectrum. There is the possibility of the primary tier reclaiming their spectrum rights, but the interference from other users is a condition that can be avoided, at a cost.

**Localization** Localization is certainly one of the virtues of the PCAST proposal. It enables each location to have a unique cost of spectrum, technical requirements (interference protection) and usage. In areas of low spectrum demand, technical solutions might balance in favor of lower cost, but less efficient spectrum utilization solutions, while in dense urban areas, the demand for spectrum would price it such that high cost, high efficiency technical solutions would be the most cost effective solution.

**Transaction Costs** Once the framework is established, and agreements on determining interference are specified, the system operates, allocates spectrum, executes auctions, and manages protection rights automatically through the automated admission control process. No human intervention or decision making is required to manage the spectrum. Also, the transaction delay and uncertainty are essentially eliminated. This requires that all stakeholders accept (or be compelled to accept, by regulation) a set of rules that determine if interference could, or could not, occur with a given situation.

**Incumbent Protection** Incumbents are protected in several ways. First, an explicit admission control assures that only authorized and "white listed" devices can enter the band. Second, if for any reason the incumbent does believe it is receiving interference, the secondary devices are known, and can have their operation immediately amended or terminated. Third, the protection criteria are specified at the aggregate interference level, not solely the impact of individual emitters. Therefore, there is no risk of higher than planned densities causing aggregate interference to a higher-tier user, as the admission control process would block additional users from using a frequency when aggregate total power at any protected node would exceed the protection criteria.

**Vague Rights** Rights and expectations in all three tiers are so explicit that they can be expressed in computer code, and certified. This is an extreme demonstration of a well understood, and complete, definition of rights. The emission characteristics, including the full range of emissions, are specified in the transmit rights provided to the admission control system. Protection of receiver operation is explicitly provided to protected tier users, providing a worse case co-channel noise level.

## 3.5  Incumbent Interests in Considering Spectrum Sharing

It is a fair question as to whether any incumbent user, or user group, would be interested in, or willing, to share its exclusive spectrum with lower-tier users. Three reasons stand out:

1. There are no benefits to the incumbent.
2. There is the risk of some failure of the protection mechanism, or of the devices in the band, that could result in interference to the incumbent's operation. Despite the sharing technology, some risk of interference will always remain, due to any number of possible conditions or failures.
3. If the technology or service that is sharing the spectrum became highly popular, or became societally or culturally essential or critical, there would be pressure to displace the original incumbent in order to free up spectrum for use by the more popular service, or, at the least, place limits on the ability of the incumbent to change characteristics or add new deployments that would displace the secondary user. At best, the incumbent would loose the flexibility they previously had to flexibly use the spectrum.

Spectrum managers in the USA refer to the third concern as the "Garage Door Openers" problem. When the DoD deployed Land Mobile Radio (LMR) trunking systems on military bases in Federal spectrum, it blocked, or directly interfered with, a number of nearby unlicensed devices, such as garage door openers [11]. Although those users legally had no rights, and were both required to accept interference from, and to not cause interference to, the primary users, the issues with garage door openers created significant public pressure to halt the DoD deployment of these systems.

In the end, DoD lost no spectrum or spectrum access. The DoD systems were deployed as planned, and the manufacturers and users of the unlicensed equipment were forced to upgrade or replace their equipment. Despite the policy success on the part of the DoD in protecting its rights, the issue left behind a concern that sharing spectrum would eventually result in the primary user being displaced from that spectrum, due to political pressure by popular secondary services, despite the formal rights of the Federal users.

It was therefore a surprise to many that the DoD supported the recommendation to share its spectrum. Certainly, some of this support could be attributed to the political leadership of the administration; which supported the concept of spectrum sharing to enhance wireless broadband availability. However, a strong DoD rejection of sharing would have likely overcome any political desire for a different position by DoD.

Instead, it is reasonable to assume that the DoD support of the PCAST position reflected a new and much more evolved spectrum incumbent strategy. In past spectrum sharing conflicts, incumbents had looked at sharing as the "nose of the camel under the tent," and to be avoided at all costs. In the PCAST case, DoD may well have viewed sharing spectrum as a method to acquire support for its continued control over the spectrum because popular secondary usage would deter any attempted reallocation.

This concept may appear counter-intuitive, but it is not contradictory. Spectrum holders also fear that the spectrum will be reallocated or reassigned from their control and use. Governments have learned that reallocating spectrum can create massive revenue inflows through auctions and leases. Other spectrum "needers" will try to get leverage to move out existing incumbents. Continued control of spectrum that is underused, or used in ways that have little public visibility and support, is therefore vulnerable to attack, despite the immediate regulatory protections.

Having other users share your spectrum creates support to the current status quo for several reasons:

1. The services sharing the spectrum have an incentive to support the status quo, and add their influence to that of the incumbents to resist any changes in allocation that would also cause their loss of access.
2. To some shared spectrum users, the right to use the spectrum under a shared arrangement, without the necessity to go to an exclusive perpetual-rights auction may be very attractive; more so than having it cleared for their use, but being auctioned with exclusive licensing.
3. It eliminates the argument that the spectrum is underutilized, and undermines the case that the band is providing minimal societal value.

Sharing spectrum may be unattractive to spectrum holders, but it is massively more attractive than losing the spectrum!

## 3.6    Criticism of PCAST Recommendations

It would be reasonable to assume that the PCAST concepts were immediately recognized as highly desirable new approaches to spectrum management. However, the PCAST report was met with (what seemed to be) almost universal criticism, primarily from the wireless operators and suppliers.

There were many objections to the concepts in the PCAST report raised. Some of the major objections included:

1. Exclusive licensed spectrum has been a highly useful tool for wireless, and has been successful in supporting rapid wireless deployments worldwide. The PCAST report presents an alternative to the spectrum clearing and relocation that was required for exclusive use auctions. This is both risky and unnecessary. It diverts attention from clearing spectrum for future auctions.
2. Without exclusive use of the spectrum, the control systems that have made modern wireless systems so effective could not operate with the same effectiveness.
3. Without assured exclusive access, no carrier or operator will invest in systems in shared spectrum. The PCAST approach would therefore not create any societal benefits through more wireless deployment, and the spectrum will remain underutilized. The spectrum should instead be cleared and provided for exclusive use.

4. Europe is moving to Authorized Shared Access (ASA)/Licensed Shared Access (LSA). This will be the standard for the wireless industry in Europe. Therefore, the USA should adopt this as its model, if it determines Federal spectrum can be shared with wireless services, but cannot be cleared. This would create a two-tier system, with no open access to unused spectrum.
5. Developing and operating the control system (SAS) would be too complex for realistic implementation and use. It will need to handle millions of devices in the band, the adjacent bands, and all of the interactions between them.

All of these concerns have a degree of validity, with the possible exception of the last one. Addressing them in the same order:

1. There has certainly been great success with exclusively-licensed spectrum. It is too early to know for certain whether three-tier creates more opportunity than exclusive spectrum. However, the PCAST report only addressed spectrum that was not a good candidate for clearing and auctioning. This argument is likely against an issue that had not yet arrived. It may have to be addressed when a situation arises where spectrum suitable for exclusive licensing becomes available, and it is proposed to instead manage it through a multi-tier framework. In effect, this is an argument between efficient, but homogeneous usage, and the more diverse, heterogeneous usage models envisioned by PCAST.[6]
2. Long-Term Evolution (LTE) operation in any congested spectrum is certainly an issue. Unlike Wireless Fidelity (Wi-Fi), LTE management assumes clear channels. However, the report provided for a protected tier that would operate with protection equivalent to exclusive spectrum, and thus should operate similarly to any other LTE system. For GAA users who choose not to acquire protection, the overall effectiveness would be optimized by creating coexistence between LTE systems in the unprotected tier. Yet, this condition could only occur after all of the channels in any given location were fully occupied, which would imply a significant level of success. To some extent, this argument is internally inconsistent, in that it both argues no one will use the band, and, at the same time, argues there will be so much use that it will have unacceptable interference. Additionally, many variants of LTE are being proposed for operation in the unlicensed bands, such as MulteFire, LTE-U, and LAA. The target spectrum for these technologies is more congested than the PCAST spectrum was likely to be.
3. While it is true that LTE investment is all in exclusive, assured spectrum, Mobile Network Operators (MNO), Mobile System Operator (MSO), enterprises and other users have invested billions in Wi-Fi deployments. Wi-Fi operates in the most congested spectrum that exists. Looking ahead in the story of the first three-tier band (3.5 GHz), there is no indication that there is any lack of investment interest, although this could not have been foretold when the PCAST report was released.[7]

---

[6] This situation does arise in Fifth-Generation Wireless Systems (5G), where there are few limitations from incumbent use, but the use of the three-tier framework will be proposed later in Chapter 17.

[7] In fact, many of the organizations that stated that no one would invest in the band have now expressed concern that the band would become too crowded, and that coexistence agreements need to be accelerated.

4. It is certainly correct that Europe is working to develop ASA/LSA. However, ASA/LSA does not address some of the other PCAST objectives, which include opening up bands for innovation, rather than simply supplying more spectrum to be used in the same manner as existing carrier spectrum. ASA/LSA is an agreement between a single primary, and a single secondary. Nor does it provide an opportunistic layer to assure full opportunity to use all of the available spectrum. It implies one, or at most, several, secondary users over a large geographic area. For spectrum that is suited to indoor use, a single controlling entity for the secondary spectrum is not a level of allocation that is appropriate in dense, three-dimensional urban environments, where many organizations could have control over placement of small devices, and operation of the many networks, and where clutter losses dominate interference considerations. The PCAST approach includes a solution to the issue that LSA addresses (sharing spectrum with a single secondary), LSA does not address the problem PCAST was addressing, which is to make all spectrum sharable and usable.

5. The argument that this regime is too complex to implement is hard to understand in this day of large-scale cloud computing. The computer analysis algorithms required[8] to implement the protection schemes are the same ones used for decades in addressing FCC and Federal, as well as spectrum issues; they are just implemented on a much larger scale, rather than one-on-one.

## 3.7    Consensus Formation Through PCAST

The issues raised by the wireless industry opposing the PCAST recommendations were perceived to be so significant that the House Energy and Commerce Subcommittee on Communications and Technology held a hearing *"Creating Opportunities through Improved Government Spectrum Efficiency"* on September 13, 2012 [12].

Figure 3.2 shows the author testifying at this hearing. Although the staff summary before the hearing was not supportive of the ideas in the PCAST report, the Congress took no action to block the FCC from proceeding with implementing any portion of the report.

The PCAST report came out only shortly before the FCC announcement that it intended to move ahead on implementing many of the PCAST report recommendations. Therefore, much of the public discussion regarding these principles was effectively moved to the more formal FCC proceeding, and the precise mechanisms the FCC proposed, rather than the general principles outlined by PCAST. One advantage of this process is that this created an informative, formal public record of the dialog on all sides of the issue.

Perhaps this shift in emphasis had an unfortunate aspect, as discussion moved to the specific proposals of the FCC, rather than the general principles in the PCAST report. Instead, consensus would be formed through the several-year-long process in resolving

---

[8] In fact, the models proposed by PCAST were drawn from past FCC proceedings, in which similar analysis was present in the record.

**Figure 3.2** The author at the house hearing.

the details of the FCC regulations for a single band, and in moving forward to develop industry methods to exploit the new rules. However, debate on the broad principles largely ended, and will likely be taken up outside of the USA, as other administrations consider the framework. Interestingly, within two years a very significant supporting consensus had formed, including even the most ardent critics of the PCAST report.

## 3.8 A Note on Capacity Versus Coverage Networks, and the Small Cell Model

One basis for the PCAST report was a fundamental belief that the necessary expansion in wireless capacity would come about through the implementation of small, localized, and very dense access points, rather than expansion of the traditional, wide coverage macro networks. Traditional macro cell deployments provided coverage, but their service was shared among a large number of users. Emerging localized capacity networks, however, were highly limited in their coverage, but provided almost identical bandwidth services to a much smaller area, therefore providing a much smaller number of users with greatly increased share of the bandwidth resource.

The PCAST report provided an analytic appendix, based on the author's capacity model included in a book that was then being released by the author [13]. This model argued that the measure of spectrum efficiency should consider the user bandwidth density that is created by a policy, not the peak bandwidth of a single cell. Small cells that could be spaced ten times closer than traditional macro cells, but provided only half the user bandwidth of the macro cell, still would provide fifty times the aggregate bandwidth per area (0.5 times $10^2$).

However, massive increases in small cell deployment would require a different ecosystem than traditional mobile operator owned and operated macro towers. Small cells would be located in buildings, enterprises, residences, and public venues that a single carrier could not dominate, or anticipate achieving access to, all of the necessary sites. A dense deployment would require the cooperation of the entities that controlled access and infrastructure at these facilities, and could address small cell deployment in a manner integrated with their other networking services. Therefore, more flexibility in spectrum rights was essential to enable the full range of deployment options to be achievable.

## 3.9  Interference Protection Priority

The willingness to provide such extensive spectrum sharing might appear as an abandonment of high quality protection of spectrum services on the part of the spectrum administration. Each sharing application adds risk of interference, and a policy that has the clear intent to maximize the utilization of the spectrum by many parties certainly carries more risk of interference than a policy that limits use of the band to a single party.

This move away from subjugating all other policy interests to maximize interference protection should not be considered unique to the policy community; and being thrust on an unwilling technical and operator community. At least in the wireless world, operators and engineers have recognized that high levels of assured interference protection lead to poor levels of spectrum utilization. Much of the system technology in LTE is included to enable high levels of spectrum reuse between LTE sites. Operators desire dense use of spectrum through their deployment, and have shown willingness to accept some risk of interference between the devices in their networks. Regulators should reflect this acceptance of interference-limited operations, and seek higher levels of utilization across different devices from different uses. An objective LTE design was to enable operation at "Spectrum Reuse One," where adjoining cells could share the same frequency assignments. Internally to cellular networks, the approach has become to manage co-site interference, not avoid it. But regulators have not adapted external interference protection levels to reflect the tolerable internal network interference levels.

This trend to compromise interference protection is immediately apparent when one considers the interference protection that is assumed in succeeding generations of cellular deployment technology. Succeeding generations have increased interference tolerance in order to maximize the spectrum sharing between adjacent LTE cellular sites. Engineers in the mobile industry have turned their attention to increasing density by applying a number of techniques within their managed systems:

- Reducing the sensitivity to the absolute level of interference in a channel. Dynamic management of the $S$ (Signal) power in relation to the level of $N$ (Noise) and $I$ (Interference) enables an assured $\dfrac{S}{I+N}$ level. This avoids the establishment of fixed levels of $I$, as in traditional regulation.
- Where Access Point (AP) sites are closely positioned, the systems manage time slots between adjacent usage regions, to avoid simultaneous transmissions by the adjoining systems.

- Where neither of the previous techniques are adequate alone, the systems can manage time sub-channel usage between adjacent usage regions to avoid simultaneous signals by the adjoining systems.

One way the PCAST recommendations might be viewed is that it creates two interacting laboratories for experiments by industry. The PA licensing provides a new spectrum sharing regime, which is highly dynamic, but provides fixed protection thresholds, similar to conventionally licensed bands. The GAA regime is an experiment in coexistence. To the extent that the users of this spectrum can address interference conditions dynamically and cooperatively, as described above, they avoid the cost and restrictions of the PA licensing. This difference in flexibility and cost in this approach is the effective monetization of interference management replacing interference avoidance. The end result could be a seismic shift in spectrum policy and spectrum utilization. The dense deployments of single operator networks show that this is technically possible; the question is whether this model can be extended across multiple deployers.

## 3.10 A Model for Spectrum Usage Through Opportunity Cost

The PCAST report included a quantitative analysis approach as an appendix to support its framework of spectrum sharing. This appendix was derived from a chapter in the author's then most recent book (*Scaling, Density, and Decision-Making in Cognitive Wireless Networks*) [13]. Because these principles are important to the three-tier regime, an abstraction of the author's original work is included as Appendix C for the reader that wishes to further explore the concept of spectrum opportunity cost as the metric of spectrum utilization.

The intent of this material was to argue that a preoccupation with only the impact of spectrum policy on incumbent technical operation had led to rejecting options that had significant societal and economic value. The proposal was to examine spectrum uses from a broader perspective:

- Receivers constrain the use of spectrum. The spectrum opportunity cost of a given application should include assessment of the loss of spectrum utilization that protecting that receiver required. For example, the poor filtering of GPS devices is a major consumer of spectrum, and essentially has reduced the opportunity for other uses in the vicinity of the GPS service, well beyond their actual GPS allocation.
- Density is key to determining the aggregate bandwidth provided by networks. Density is more important than assuring the lowest possible interference. Reducing the protection region around a node has the most significant impact in increasing aggregate bandwidth.

Measuring how one use of the spectrum impacts, and denies, spectrum to other uses is the most integrated manner in which to consider the relative impact of spectrum usage, and much more instructive than examining just the emission allocations.

## 3.11    Suggested Reading

- The PCAST report [1] itself is clearly an important document in three-tier spectrum. While the FCC has adopted rules that diverge from this vision, it still sets out ideas that will take further development, and several years to be integrated into the spectrum management regime.
- The SUR expression is very useful in developing an application- and technology-neutral mechanism for describing the rights and expectations of a device in the spectrum. Many of the concepts described here should be included in future three-tier frameworks, even though they were not included in the initial US version.
- There was a large amount of public dialog when the PCAST report was issued, most of it in the popular press, and accessible through the web. Some of the opposition pieces to the band were published by CNET, Cellular Telecommunications Industry Association (CTIA), and press releases from individual carriers. Information Technology and Innovation Foundation (ITIF) opposed the report, and had a number of press releases and events that criticized the recommendations [14]. In support of the report, there were academic responses, such as one from Linda Doyle of Trinity College, Dublin (TCD) [15], and the CSPAN interview with the committee co-chairs [16], among other supporting positions.
- The House Science and Technology Subcommittee of the Energy and Commerce Committee issued a pre-hearing statement that generally expressed opposition to the report. The hearing video is available on YouTube [12].
- One of the interesting aspects of the PCAST process was the support the PCAST proposals received from Jason Furman, Chairman of the Council of Economic Advisors (CEA), President Obama's Chief Economist, and a member of the Cabinet [17]. His comments examined the recommendations from the economic growth perspective, rather than the technical perspective most commentators applied.
- Nobel Prize winner Ronald Coase's work on market-driven spectrum allocation [9], which led to auctions, is an important work in spectrum allocation. Similarly, Tom Hazlett's commentary on the legacy of this work is important reading, even though it disagrees with several of the principles in this chapter [18].

## References


1 President's Council of Advisors on Science and Technology, *Report to the President: Realizing the Full Potential of Government-Held Spectrum to Spur Economic Growth* (Executive Office of the President (EOP), Office of Science and Technology Policy (OSTP), 2012). http://obamawhitehouse.archives.gov/sites/default/files/microsites/ostp/pcast-stem-ed-final.pdf.

2 M. McHenry, K. Steadman, A. Leu, and E. Melick, XG DSA radio system. *3rd IEEE International Symposium on New Frontiers in Dynamic Spectrum Access Networks* **3**/1 (2008), 497–507.

3 McHenry, E. Livsics, N. Thao, and N. Majumdar, XG dynamic spectrum access field test results. *IEEE Communications Magazine*, **45**/6 (2007), 51–57.

4 P. F. Marshall, Extending the reach of cognitive radio. *Proceedings of the IEEE*, **97**/4 (2009), 612–625.

5 Federal Communications Commission, *Spectrum Policy Task Force Report, ET Docket No. 02-135* (2002). http://apps.fcc.gov/edocs_public/attachmatch/DOC-228542A1.pdf.

6 Office of Communications (Ofcom) (UK), *Spectrum Usage Rights; Technology and Usage Neutral Access to the Radio Spectrum*. Consultation (2006).

7 L. Luna, NEXTEL interference debate rages on. *Mobile Radio Technology*, **21**/8 (2003), 26.

8 National Space-Based Positioning, Navigation, and Timing Systems Engineering Forum (NPEF), *Assessment of LightSquared Terrestrial Broadband System Effects on GPS Receivers and GPS-dependent Applications* (2011). www.gps.gov/spectrum/lightsquared/docs/2011-06-NPEF-lightsquared-report.pdf.

9 R. H. Coase, The Federal Communications Commission. *The Journal of Law & Economics*, **2** (1959), 1–40. www.jstor.org/stable/724927.

10 Federal Communications Commission, *Connecting America: The National Broadband Plan* (2010). http://download.broadband.gov/plan/national-broadband-plan.pdf.

11 ——, *Consumers May Experience Interference to Their Garage Door Opener Controls Near Military Bases*. Public Notice, 20 F.C.C.R. 3614 (2005).

12 House Energy and Commerce Committee, *Communications and Technology Subcommittee Hearing: Creating Opportunities Through Improved Government Spectrum Efficiency*. Video (2012). www.youtube.com/watch?v=ZvuOy_LtnUU.

13 P. F. Marshall, *Scaling, Density, and Decision-Making in Cognitive Wireless Networks* (Cambridge University Press, 2012).

14 Information Technology and Innovation Foundation, *ITIF's Bennett Says PCAST Spectrum Report Impractical*. Technical report (2012). https://itif.org/publications/2012/07/23/itifs-bennett-says-pcast-spectrum-report-impractical.

15 L. Doyle, *The Only Ones Who Will Drop Dead Are Those Who Stand Still*. Blog post (2012). ledoyle.wordpress.com/2012/07/31/the-only-ones-who-will-drop-dead-are-those-who-stand-still/.

16 C-SPAN, *Communicators with Craig Mundie and Mark Gorenberg*. Video (2012). www.c-span.org/video/?307379-1/communicators-craig-mundie-mark-gorenberg.

17 J. Furman, *Remarks on Public Sector Spectrum Policy*. Prepared remarks for presentation at Brookings Institution (2014).

18 T. W. Hazlett, D. Porter, and V. Smith, Radio spectrum and the disruptive clarity of Ronald Coase. *Markets, Firms, and Property Rights: A Celebration of the Research of Ronald Coase* (University of Chicago School of Law, 2009).

# 4   US FCC CBRS Regulations, and Other International Activity

## 4.1   Introduction

This chapter describes the rule-making process and formal requirements that were established by the United States (US) Federal Communications Commission (FCC) to implement the first three-tier band, the Citizens Broadband Radio Service (CBRS) band from 3.55 to 3.7 GHz [1, 2]. This chapter will describe the regulatory requirements only; methods for implementing them, and other three-tier regimes, are provided in detail in subsequent chapters.

The chapter closes with a discussion of movement towards three-tier spectrum management that has occurred outside of the USA. None of these is yet as mature as the USA regulations, but are different national approaches to creating a similar ecosystem.

This chapter is not intended to advocate for the specifics of the US three-tier framework. It compromised some principles of the PCAST vision of three-tier; these compromises may be undesirable, or may be improvements. They may or may not be relevant to the issues faced by other countries, and in other bands, when adopting the three-tier framework. Instead, the US three-tier framework is presented as a point of departure from which it is likely that consideration of other bands, and other countries, will depart.

Repeating a quotation cited in the first chapter, the intent of this band was stated by the FCC as [2]:

*"The Citizens Broadband Radio Service takes advantage of advances in technology and spectrum policy to dissolve age-old regulatory divisions between commercial and federal users, exclusive and non-exclusive authorizations, and private and carrier networks."*

The FCC actually issued its first interest in, and support of, the President's Council of Advisors on Science and Technology (PCAST) recommendations a few months after the formal release of the PCAST report [3]. There was an unprecedented amount of support in the administration, as shown in Figure 4.1. This figure shows the support for the PCAST release event, including the Chief Information Officer (CIO) of the Department of Defense (DoD), Assistant Secretary of the Department of Commerce (DoC) and Director of the National Telecommunications and Information Agency (NTIA), the Deputy Director of the White House Office of Science and Technology Policy (OSTP), the Chairman of the FCC, and the Chairman of the Council of Economic Advisors (CEA). This extent and level of participation demonstrated the broad agreement to both

**Figure 4.1** US Government support at PCAST report release.

share Federal spectrum and to implement the innovation-focused recommendations of the PCAST report, as well as a broad expression of support by the administration of President Obama. This was followed by the FCC press release stating support for the initiative recommendations, and stating that it would pursue regulation to implement them.

Following the PCAST report, the FCC moved ahead on two of the specific bands proposed by the report. This included expansion of the Advanced Wireless Services (AWS) band[1] to include 1 710–1 755 MHz through spectrum sharing, and to authorize three-tier operation in the band 3.55–3.65 GHz, with the option of extending the upper limit to 3.7 GHz. The FCC also decided not to make this an adjustment to an existing regulation, but instead, created a new FCC regulatory "Part" for the three-tier sharing regime.[2] Implicit in this was an expectation that more bands could be added to this regulatory framework with reduced process.

Although this band did not end up incorporating all of the features proposed by PCAST, it incorporates the spirit of the regime. The US regulatory process is very transparent, so the regulatory record includes an extensive set of arguments for, and against the three-tier regime, as well as features that were specific to the band and its incumbents. In this chapter, we will focus on the arguments that are generally applicable to other bands and nations, as these are most instructive to the analysis.

## 4.2 Summary of FCC Actions and Timeline

Table 4.1 shows the timeline of the culmination of the PCAST study period, and the FCC process to develop the rules for the first United States of America (USA) three-tier spectrum band. It provides the date of each major event, and a link to the document for readers desiring to research it in more depth than this treatment.[3]

---

[1] This concept had been developed prior to the PCAST study endorsing the concept of sharing this band under exclusive arrangements.

[2] FCC Parts are major classes of regulation, such as Part 15, for low power unlicensed devices, or Part 90 for WISP services in 3 650–3 700 MHz.

[3] Because of the future changes in the US Administration, several of these documents may not continue to be available at the links provided. The author will have many of the non-copyrighted documents cited throughout the book available on his web site at www.prestonfmarshall.com/ThreeTierReferences.

**Table 4.1** Timeline of major actions for implementation of the US CBRS band.

| Date | Event/Document/Location |
| --- | --- |
| Jul. 30, 2012 | White House Release of PCAST Report: *Realizing the Full Potential of Government-Held Spectrum to Spur Economic Growth* <br> Available at: www.whitehouse.gov/sites/default/files/microsites/ostp/pcast_spectrum_report_final_july_20_2012.pdf. |
| Sep. 12, 2012 | Press Release: *FCC Chairman Julius Genachowski Announces Plans to Initiate Formal Steps on Spectrum Recommendations from the PCAST* <br> Available at: apps.fcc.gov/edocs_public/attachmatch/DOC-316251A1.pdf. |
| Dec. 12, 2012 | FCC releases: *Notice of Proposed Rulemaking and Order: Amendment of the Commission's Rules with Regard to Commercial Operations in the 3550–3650 MHz Band* <br> Available at: apps.fcc.gov/ecfs/document/view?id=7022080889. |
| Apr. 23, 2014 | FCC releases: *Further Notice of Proposed Rulemaking: Amendment of the Commission's Rules with Regard to Commercial Operations in the 3550–3650 MHz Band* <br> Available at: apps.fcc.gov/ecfs/comment/view?id=6017612779. |
| Apr. 17, 2015 | FCC formally approves: *Report and Order and Second Further Notice of Proposed Rulemaking and Order: Amendment of the Commission's Rules with Regard to Commercial Operations in the 3550–3650 MHz Band* <br> Available at: apps.fcc.gov/ecfs/comment/view?id=60001029680. |
| Apr. 28, 2016 | FCC provides final CBRS rules in the *Order on Reconsideration and Second Report and Order: Amendment of the Commission's Rules with Regard to Commercial Operations in the 3550–3650 MHz Band* <br> Available at: apps.fcc.gov/ecfs/comment/view?id=60001696261. |

Compared to past changes in spectrum regulation, the process from the release of the PCAST report to the finalization of the two US Report and Orders was quite rapid. Some of this can be attributed to the support by the President of United States, as exemplified by the Presidential Memorandum [4] that framed the Federal spectrum sharing policy.

Another cause of the rapid action was the cultural changes in the spectrum management community. This included less support to incumbents attempting to block new spectrum initiatives, and an implicit understanding that innovation not only required spectrum, but it required "approachable," or "Venture Friendly," spectrum that could be made available for new applications. Spectrum access had to be available before these new concepts could generate the revenue streams that would make them competitive in exclusive licensing through auctions.

The previous Spectrum Policy Task Force (SPTF) efforts had introduced the concept of noise temperature[4] as a criteria for sharing spectrum, and a quantitative metric to provide to incumbents as guides for protection expectations. The FCC had also been aggressive in promulgating Television White Space (TVWS) regulations. Senior staff at the FCC had even been active in leadership roles in academic conferences, such as the

---

[4] Noise temperature was essentially an assured level of power/Hz, expressed as the equivalent black body radiation power.

**Table 4.2** Comparison of three-tier spectrum with CBRS.

|   | Consideration | Consequences |
|---|---|---|
| 1 | Multiple protected tiers | Yes. There is a primary protected tier, and a single protected tier available to secondary users. |
| 2 | Lower tier that can share unused spectrum | Yes. The one tier provides access for all users. |
| 3 | Lowest tier with no protection | Yes. The one tier is usable, regardless of interference to other same tier uses. |
| 4 | Small units of allocation | Somewhat. Non protected users have no allocations. Protected users acquire protection in units of census tracts and units of 10 MHz. Geographic units larger than proposed by PCAST, but similar principles apply. |
| 5 | Automated transactions | Yes. Completely automated admission control, consistent with PCAST principles. |

IEEE Dynamic Spectrum Access Networks Conference (DYSPAN) conferences, and the NTIA International Symposium on Advanced Radio Technology (ISART) events, which feature spectrum sharing. The FCC was set for a "perfect storm" to embrace technology-driven spectrum sharing.

Although the initiative was begun under FCC Chairman Julius Genachowski, the incoming FCC Chairman, Tom Wheeler, announced at the FCC event "3.5 GHz Spectrum Access System Workshop" that implementation of the PCAST report was the item that President Obama had told him was his priority.[5]

## 4.3    Comparison of FCC CBRS and Three-Tier Proposals

In the US governance system, the FCC is an independent regulatory agency, and so has no direct command and control relationship to the US President. Additionally, unlike PCAST members, the FCC had to be cognizant of very real political, economic, regulatory, and technical limitations on the ability to rapidly implement the full scope of the PCAST proposals.

A comparison of the basic three-tier principles and the final FCC CBRS regulations is provided in Table 4.2.

The FCC action invoked the PCAST study as its conceptual starting point, so it is worth a closer examination of how the CBRS regime compares to the specific PCAST study recommendations. Table 4.3 further described some of the areas where the CBRS differs significantly from the PCAST approach. Section 4.9 discusses the acquisition of protection rights in much more detail, as it is one of the most significant differences between CBRS regulations and PCAST, as in this area the FCC developed a unique approach to protection acquisition.

---

[5] Held on January 14, 2014, Materials and video are available at: fcc.gov/news-events/events/2014/01/35-ghz-spectrum-access-system-workshop.

**Table 4.3** Comparison of CBRS rules with PCAST.

|   | Consideration | Consequences |
|---|---|---|
| 1 | Multiple protected tiers | Essentially the same. |
| 2 | Lower tier that can share unused spectrum | Essentially the same. |
| 3 | Lowest tier with no protection | Essentially the same. |
| 4 | Small units of allocation | The CBRS geographic units are census tracts. These tend to be quite a bit larger than contemplated by the PCAST report, and their dimensions are not related to, or defined by, interference distances. Also, the FCC defined a fixed protection criteria that is provided to any tier 2 (PAL) user, rather than allow users to define the minimal protection they required. |
| 5 | Automated transactions | Admission control process is essentially the same, but there is no provision for real-time, or on-demand, auctions. Instead, auctions are run on an annual basis, and have fixed terms, with some rights of renewal. |

## 4.4    Specific Protection Provided to Incumbent Classes

Much of the attention paid to the 3.55 GHz band was focused on sharing with naval radars, but there is a very significant range and number of other users and system types in the 3.55–3.7 GHz band that had to be addressed in establishing the CBRS protection regime. In this section, we will not discuss the methods of achieving this protection; this is covered in detail in Part IV. Instead, this section will focus on the rationale for establishing these protections, and their implications for the band's utility.

We will consider the CBRS protection requirements from two perspectives. First, they do represent a very specific set of protections that were adjudicated by a major regulatory body, and therefore, form a baseline from which other administrations will consider in their own decision processes. Second, the concerns raised by the incumbents (and potential protected tier users) are instructive in the issues that concerned them. However adjudicated by the regulators, these concerns (those that were not proposed solely to obstruct action) are significant, at least from their perspectives, and thus are worthy of discussion in this work. Therefore, the different positions on some of the critical protection requirements will be discussed in more depth. More generalized discussion is provided in the later chapters.

Table 4.4 provides a summary of the protection that is afforded to all of the protected entities in the band. In addition to the primary user, naval radars, there are a small number of Fixed Satellite Service (FSS) sites that operate in the international extended C Band satellite frequencies (3 600 MHz–3 700 MHz). Additionally, there are tens of thousands of wireless Internet service provider (WISP) and utility licenses under FCC Part 90 in the upper 50 MHz (3 650–3 700 MHz) of the CBRS band.

**Table 4.4** Protection criteria for occupants of the US CBRS band.

| Occupant | Protection Criteria | Where Computed |
|---|---|---|
| Naval radars | Unspecified | Based on detection area |
| In-band FSS | −60 dBm blocking power | After RF filter |
| | −129 dBm/MHz | In channel |
| Adjacent band, TT&C FSS | −60 dBm blocking power | After RF filter |
| | −129 dBm/MHz | After RF filter |
| Adjacent-band, Non-TT&C FSS[a] | Can request interference resolution from SAS operator | N/A |
| WISP and Utilities | Grandfathered protection zones | Base stations only, no protection to specific subscriber units |
| PAL Client Areas | −96 dBm/10 MHz | Boundary of client service area |

[a]Devices not performing FSS TT&C, as defined in Part 96.

**Table 4.5** Methods for determining protected receiver location.

| Occupant | Location and Frequency Determination Methodology |
|---|---|
| Naval radars | No specific location is to be determined. Instead, large areas of CBRS operations are to be excluded upon detection of naval operation on a given frequency. In discussion with industry, the DoD has insisted that the sensor network not have any ability to even roughly locate naval vessels. |
| Adjacent-band FSS | C-Band FSS sites are already required to register their positions in order to be protected. They are now required to also report the specific FSS slot they are receiving. |
| Extended C-Band FSS | Same as adjacent-band sites, with an added restriction on any new registrations for protected status. |
| WISP and Utilities | Part 90 base stations (location, azimuth, beam width, power) are required to be registered. Customer Premises Equipment (CPE) do not have to be registered, and are protected to a default range within the base station coverage. CPE can be registered to extend the range of the base station. |
| PAL Service Area | PAL users declare a Priority Access License Protection Area (PPA) that defines the area over which its client devices may be located, and receive protection. The general topic of service areas is discussed in Chapter 6. |

The baseline premise of Dynamic Spectrum Access (DSA) is to use sensing to detect the presence of protected users, and avoid interfering with their associated receivers. This would be accomplished without direct interaction with the protected transmitters or receivers, or a linkage to provide awareness of their specific location. However, the Spectrum Access System (SAS) database approach requires that the admission control system service be aware of the location of all receivers in use by protected tier users. The database model of three-tier spectrum is therefore dependent on a reliable mechanism to locate protected receiver locations.

The methods to determine the location of protected receiving devices in the FCC CBRS methodology for SAS use is provided in Table 4.5.

Detailed discussion of the specific protection established for each of the incumbent services is provided in the following sections. A more detailed discussion of the methodology to implement these incumbent and Priority Access License (PAL) protections is provided in Part IV.

## 4.5    Protection Requirements for Naval Operations

The original PCAST assumption had been that Federal agencies would inform the SAS operators of Federal protection needs as they occurred, and the SAS would make the necessary adjustments in the CBRS deployments. However, it became clear that however obvious a naval vessel might be, the policy issues associated with informing non-classified systems of even the general location of a naval vessel made this strategy impractical. Therefore, the sensing model was adopted. SAS operators would be required to deploy naval radar sensing networks along the coasts in order to operate within areas where the Federal government stated interference could be caused to naval operations.

The regulations define two regions for protection of naval radars and the operation of Environmental Sensing Capability (ESC) systems:

**Exclusion Zones** No operation in these areas is permitted unless the SAS operator has an ESC deployed and active. The exclusion zones were extremely limiting on Citizens Broadband Radio Service Devices (CBSD) operation, so the use of a SAS or CBSD without an ESC capacity was greatly restricted. A graphic depiction of the extent of these exclusion zones is shown in Figure 4.2.

**Protection Zone** Once a SAS operator activates ESC coverage, the detection of a naval radar activates a Protection Zone over a large enough region to ensure protection of the radar on the frequency range in use by the radar, and over a large enough area to obscure the ship's actual location.

The exclusion zones shown in Figure 4.2 are quite extensive, and essentially preclude meaningful SAS operation in most of the populated areas of the USA. Therefore, although the rules do not require an ESC be deployed, in practical application, an ESC is an essential adjunct of any SAS that would operate effectively for a majority of the US population.

The timeline for clearing of a channel used by a naval radar has three components.

**Pre-Detect** There is a period of time between when the radar signal is at a detectable threshold, and the CBSD clearing is initiated. This includes the time for (1) the ESC to detect the presence of a signal, (2) communicate it to the SAS, (3) the SAS to process or fuse the reporting, in order to declare that a radar is present, and then (5) to determine the CBSDs to be cleared. In the Public Notice (PN) [6] for the SAS certification, it is clear that the ESC is defined as more than just the spectrum sensor; its timeline includes the data fusion and processing to determine if a radar is present. The FCC established no regulatory criteria for this, and it was left to industry and the US Navy to determine the sensing thresholds.

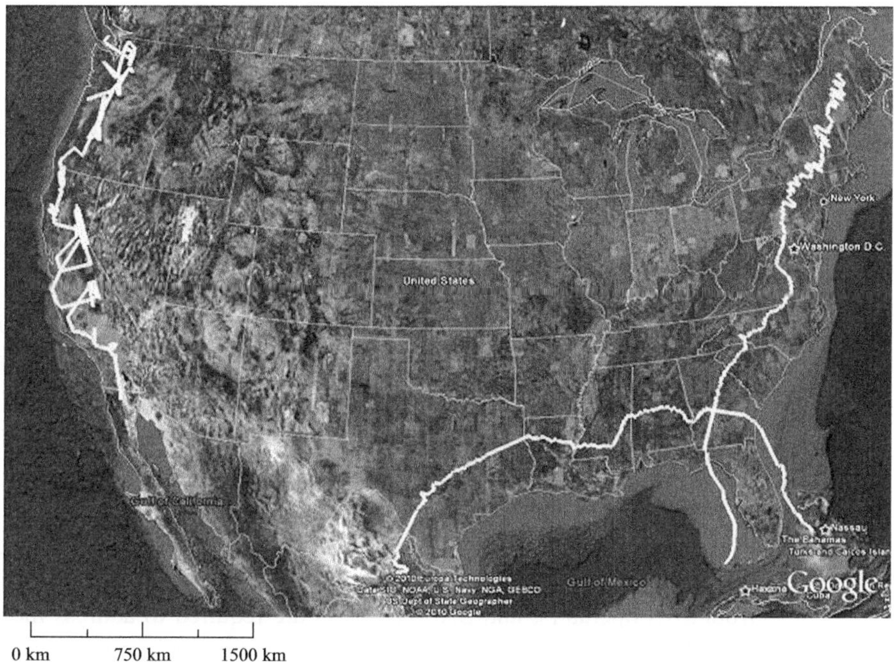



0 km          750 km          1500 km

**Figure 4.2** Extent of ESC exclusion zones limiting CBSD operation without an ESC active (from NTIA Quick Look Report [5]).

**Notify** The SAS has 300 seconds to notify the device that the radar detection has occurred, and to vacate the channel.

**Relocate** The device has a further 60 seconds after acknowledging the SAS's notification to vacate the channel. This one minute time period allows the network to perform a more graceful relocation of client devices by temporarily handing them off to other APs while relocating, or other technology specific measures.

Much of the specifics of the radar protection were deferred to the SAS developers' direct interaction with the DoD and DoN, and were not provided in the regulation.

## 4.6    Protection Requirements for Satellite Operations

There are several satellite sub-bands within the spectrum impacted by CBRS, and each has unique registration and protection standards. Although the intent of the CBRS regime was to not burden primary users with additional requirements to support spectrum sharing, the FCC did decide to increase the reporting burden, adding additional detail to the registration of FSS sites, and mandating annual updates, rather than once per decade as in the previous regulations. Chapter 9 will demonstrate the importance of this information in minimizing the exclusion footprint of an FSS site.

The newly required data included the antenna pattern, and the pointing elevation and azimuth. No protection is afforded until these requirements are met.

The FCC established unique requirements for stations in each sub-band, and for those performing Tracking, Telemetry and Control (TT and C). The registration protection and operating limits imposed by the CBRS regulations include:

**In Band, 3 600–3 700** No new registrations are permitted for equipment in this band for earth stations in the range 3 600–3 700 MHz dating from the publication of the original order. They are protected from co-channel interference in this range from CBSDs within 150 km to a level of $-129$ dBm/MHz, assuming a reference antenna system and RF filter with an insertion loss of 0.5 dB. They are protected from blocking energy from CBSDs within 40 km to a total power level of -60 dBm Root Mean Square (RMS), again using a reference filter and antenna, and an insertion loss of 0.5 dB.

**Non-TT & C 3 700–4 200** There are no explicit requirements for protection of non TT&C FSS devices. Operators of these systems can request interference resolution from SAS operators, and SAS operators are required to address these concerns in good faith. The SAS operators may assume that the earth stations are using the reference filter design: a high pass filter, with a sharp cutoff at 3.7 GHz. This filter is described in detail in Chapter 9.

**TTC 3 700–4 200** Sites performing TT&C are protected from blocking by CBSD deployments within 40 km of the TT&C FSS site. The SAS is required to protect the FSS site from adjacent band power to an RMS level of $-60$ dBm, assuming the industry standard (high pass) filter with a 0.5 dB insertion loss. They are protected from Out of Band Emissions (OOBE) from CBSDs within 40 km to an RNS level of $-129$ dBm/MHz, with the reference filter, and a 0.5 dB insertion loss in the FSS passband.

Although the FCC did not establish receiver standards in this band, the protection levels assume that there are minimum standards of filtering present in the receiver in order for the protection provided to be meaningful. In effect, this is similar to the United Kingdom (UK) Spectrum Usage Rights (SUR), in that it establishes an expectation of what level of adjacent and co-channel conditions a user of the spectrum can anticipate. Unlike passive regulation, it requires active intervention by the SAS to assure this criteria is met.

Although the intent of the entire Part 96 framework is to provide automated interference analysis and admission control, there is a provision for satellite operators in the 3 700–4 200 MHz band to report interference directly to the SAS operators. This is a "backstop" to the automated methods. The protection levels required for in-band protection are extremely low and will be difficult to validate in the field. So, it is reasonable to make resolution of interference claims a joint responsibility of the SAS and receiver operator.

## 4.7 Protection Requirements for Wireless Internet Service Providers (WISP) and Utility Operations

The USA had authorized operation of Part 90 devices in the upper 50 MHz of the extended C-Band. Naval radar operation in this 50 MHz was precluded when close to shore. The primary use of this band was for Wireless Internet Service (WISP) and electrical utilities for remote reading and control systems. These devices were to be provided protection and continued authorization for five years. No new licensing of devices was permitted after the first FCC Report and Order, and existing devices had to transition to operation under the newly established Part 96 rules.

The protection of the "grandfathered" devices was approached by incumbents and potential users of the band from very different perspectives. The potential users of the CBRS band proposed that the protection should be established specific to the registered endpoints of the networks (base stations and CPE) whereas the existing Part 90 users proposed that the protection be established as a protection zone around every registered base station, regardless of endpoint location.

The FCC provides the protection requirements in a public notice [7]. The FCC rejected explicitly protecting CPE, as the rules passed by the commission precluded this protection. Instead they established two protection mechanisms; one for base stations with registered CPE devices, and one for base stations without. Registration of CPE was optional.

The three protection rules were:

- For Part 90 base station sectors without registered CPE devices, the protection is for the full beam-width of the antenna, out to 5.3 km radius.
- For Part 90 sectors that have registered CPE devices are protected for their azimuth and beam-width, out to the location of the registered CPE, a (typically) maximum range of 18 km.
- Base station sectors that have no registered, or unregistered CPEs receive no protection.

The protection rules provided by the FCC are stated to be conservative; one might argue they were unrealistically conservative in providing protection.

- The FCC used line of sight path loss. At a distance of 5.4 km and at a frequency of 3.675 GHz, it is unlikely that realistic path loss will be at all close to the free-space loss. It will likely be 10–30 dB higher, implying protection radius values that have at least a ten-fold reduction in the protected area.
- The assumption of the lowest order modulation sensitivity values are also unrealistic in markets where wireless services have competition from any media, even cellular wireless.
- Even if one could argue or conceive that free-space loss was present at a distance of 5.3 km, the outer limit of a protection zone with registered CPW was set to be over 19 km. This requires that the path remain free-space for all of the 18 km range, and

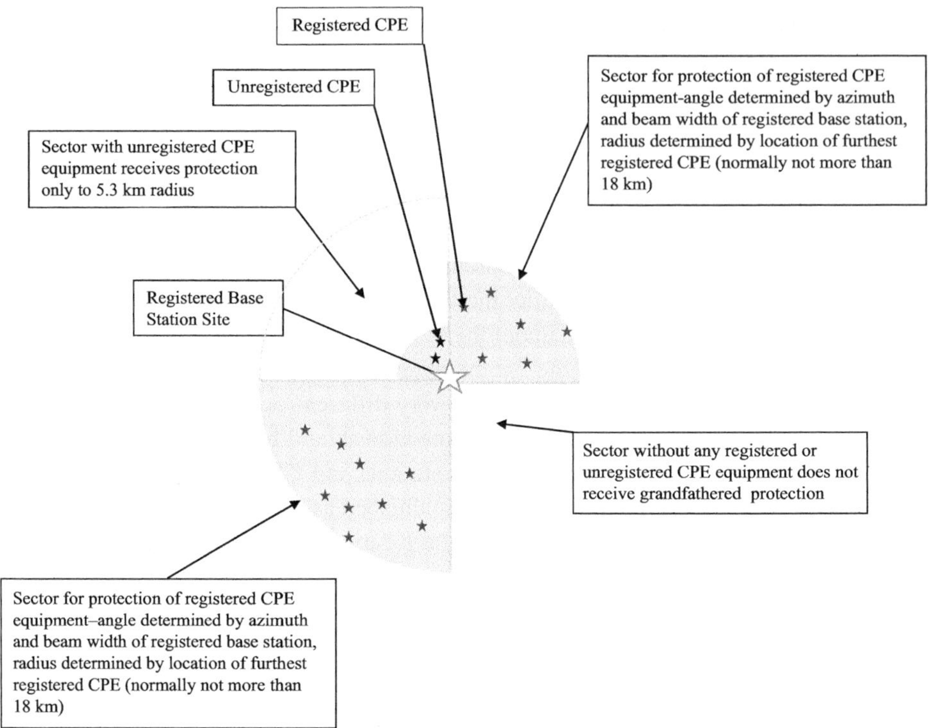


**Figure 4.3** The different Part 90 protection schemes for registered and unregistered CPE devices (from [7]).

have a further reduction in path loss from the already conservative free-space model of 5.8 dB!

This being said, the protection was only to be available for five years, so this incursion on usable frequencies should be short-lived. If PAL use becomes popular in areas with extensive Part 90 operations, the available GAA spectrum could be reduced to only 30 GHz (150 MHz - 50 MHz (Part 90) - 70 MHz (PAL)). The original assumption that seven PAL channels would leave more than 50% of the band available for GAA assumed that all 150 MHz was available for the General Authorized Access (GAA) user. In regions with extensive Part 90 operations, this will not be true.

A depiction of the different protection zones is provided by the PN [7], and shown in Figure 4.3.

## 4.8      Protection Requirements for Priority Access Operations

One of the primary features of the three-tier regime is that the middle tier (Priority Access (PA) or PAL) users only can exclude lower-tier (GAA) users, when their operation would cause interference to this protected use. These are designated as a PPA. Later, we will discuss the issue of client protection in more detail. The FCC agreed that PAL operations received protection of client devices throughout a service area surrounding the registered PAL device. The service area can be initially provided and updated by the PAL holder, or computed by the SAS based on an engineering estimate of the maximum service area. The SAS is required to assure that the operator provided PPA does not overly restrict the sharing of spectrum by GAA by extending beyond the geographic region in which the one Access Point (AP) could serve clients.

In keeping with the principle that protection only be provided to actual spectrum usage, the FCC also required that PAL operators inform the SAS if the device with the grant terminates its operation for more than seven days. The SAS can then let GAA share the spectrum until such time as the device is reactivated. This effectively blocks one potential strategy to "warehouse spectrum."

It was an objective of the regulatory regime to minimize the impact of the unique aspects of CBRS operations on client devices in order to make adoption more rapid and cost-effective through leveraging existing semiconductor and device ecosystems. To this end, client devices were not required to register with the SAS, as their operation was dependent to the decisions made by the AP, and therefore the SAS. Considerable discussion in the record pertained to how to ensure that these non-registered devices were not interfered with, nor cause interference, since the SAS itself was not aware of their location.

The FCC introduced the concept of a PPA, which provides the geographic contour for clients operating under a registered PAL CBSD. This area can be self reported by the PAL when the CBSD is registered, or, as default, computed by the SAS. The SAS uses whichever region is smaller. The PPA concept enables client devices to be outside of the scope of the registration process, as they were implicit in the registration of the AP. This mode of protection is discussed with more generality, and in more detail, in Chapter 6.

## 4.9  Second-Tier Protection Acquisition Method (Auction)

One of the fundamental differences between the PCAST recommendations and the CBRS regulations that were finally adopted is in the method that devices and users utilize to obtain protected status. Whereas the PCAST report had advocated protection be acquired on a device-by-device basis, the FCC elected to define protection through political geographic boundaries, rather than propagation contours.

PA protection rights are issued in individual census tracts. Census tracts comprise a geographic region that is developed by the US Census Bureau and are "designed to be relatively homogeneous units with respect to population characteristics, economic status, and living conditions." Census tracts average about 4 000 inhabitants [8]. In most cases, this region is much larger than the implicit recommendation by PCAST.[6] Some

---

[6] The primary reason a census tract could be smaller than a PPA is when the building has extensive three-dimensional development. Conventional spectrum practice does not reflect the vertical stacking of

are very small in area, as in a dense city, and others are much larger, such as one in the state of Alaska, which comprises one third of the area of the state.

In non-rural areas, a PAL may only be assigned if there is more than one applicant, so a single buyer cannot buy protection without bidding against at least one other entity. An exception is made in rural areas, where the necessity for two or more bidders is waived due to the likelihood that demand would be much lower.

The FCC also chooses to limit the aggregation of PAL spectrum control. A given license is permitted to control up to four PALs in any given census tracts. These limits carry over into secondary market transactions, as well.

## 4.10    SAS Competition and Information Sharing

One of the central recommendations of the PCAST report was that all aspects of three-tier spectrum should have strong competitive forces reducing costs, increasing capacity, encouraging and rewarding innovation, and ultimately, creating transparency. So, it is worthwhile to see how the FCC rules provide for competition in this critical ecosystem role: the SAS operator.

The FCC was quiet on the cost of spectrum access, which is certainly pro-competition. The FCC allowed SAS providers to make decisions regarding the scope of the rules in which they would participate, and this is certainly competition and free entry favorable.

Propagation models are one of the fundamental weaknesses in any spectrum sharing regime. Every participant in spectrum is aware that these models are not reflective of real world conditions. The range of loss that occurs in the real world is often tens of decibels from that predicted due to the complex relationship of the environment, the propagation paths, local building methods, and other factors that were beyond the ability of the developers of these models (typically in the 1960s) to incorporate into a general framework. Proving the flexibility to develop, validate, and employ new and more detailed models would have been an opportunity that the SAS could have provided, and monetized the benefits of improved models. However, the FCC in the second order asked industry to specify a consensus model that all SAS implementations must follow, at least initially. Whether competitive models will be permitted later remains to be seen, and the structure may permit anti-competitive self-interest to block competitive evolution in this critical aspect of spectrum sharing and SAS operation.

Unfortunately, we know that the requirement to form a consensus model also tends to be a "race to the bottom," with the lowest common denominator, with the eventual consensus choice being the option most readily available to all participants. It is understandable that the regulators would find that having each SAS produce the same results was attractive. It might not be correct, but it would be both irrefutable by other SASs, and it would preclude competition based on analytic models.

What remains unclear is how the FCC will respond to requests for future enhancements to the consensus model. These models might be non-standard, and utilize

devices, with the isolation of floor construction. Units 50 floors apart might show minimal path loss if analyzed by conventional models, such as Longley-Rice.

proprietary data or methods to reach different (presumably higher minimum path loss) answers than the consensus model. If the FCC is receptive to these proposals, then the SAS ecosystem can evolve to reflect ever increasing geospatial knowledge, the self-sensing of the SAS-controlled devices, and new experimental data that becomes available.

## 4.11 Secondary Markets

Section 2.7 discussed the issues that have caused secondary markets to be less effective in many situations where they would appear to have viability. Since a dynamic environment was one of the objectives of three-tier management, and also one of the claimed benefits of secondary markets, it is worth examining how secondary markets are enabled in the CBRS regulations.

There were those that believe that secondary markets encourage speculation and spectrum warehousing. It was an important question as to whether secondary markets in the CBRS band would encourage, or discourage, these less desirable practices. The short license period and assured availability of GAA spectrum make warehousing much less attractive than in longer term licenses.

Because the intent of the band was to foster innovation and dynamic uses of the spectrum, the FCC also decided to permit the sub-licensing process to proceed without prior commission approval of the transfer. The only requirements for secondary transfers are:

- The lessee informs the FCC that it is qualified to hold a PAL license, and provides the supporting material.
- The original PAL license holder notifies the SAS operator(s) of the arrangement.
- The SAS confirms that the lessee has provided notice to the FCC, that aggregation limits are not exceeded, and that the lease area is within the PAL, and does not encroach any existing PAL protection area.
- The SAS, rather than either of the parties to the transaction, reports the transaction to the commission daily.

This is an extremely lightweight or "Light-Touch" regime. The National Regulatory Authority (NRA) is monitoring the situation, but is not in-line to the operation, and has no prior item-by-item review, although it reserved the right to reverse such agreements within 40 days. This is consistent with the concept of automated admission control.

The FCC also decided to neither require nor prohibit spectrum exchanges. This leaves it open to either SAS operators or exchange operators to establish markets that could offer price discovery, transparency, and facilitate the broader objectives of dynamics in spectrum usage. The secondary markets that could emerge may more closely resemble the dynamic nano-transactions originally envisioned by the PCAST report.

Although the FCC is very prescriptive of the exact nature of SAS processing and decision-making, it has provided no direction on the operation of the spectrum marketplaces that could emerge, either within or external to the SAS services. Its regulations may therefore be considered to be enabling of a wide range of innovations in creating secondary markets. Presumably, market forces, rather than regulators, will determine the shape of these markets.

## 4.12    Spectrum Access System Operational Considerations

The entire regulatory framework is dependent on a new entity in spectrum management: a SAS, or an ecosystem of SAS operators. The SAS ecosystem is essentially delegated with the control of admission into the Federal spectrum. This is a unique aspect of this regime, and is a very significant extension of a classical band manager. Non-interference related considerations that are imposed on the SAS system and the SAS operator included:

- The SAS is responsible for authenticating that devices interacting with the SAS be devices that were actually certified with the FCC under Part 96 CBRS regulations. Industry made this requirement symmetric by also requiring that devices authenticate that they were interacting with an FCC-approved SAS. It might be argued that this (device authentication) requirement in some cases is counterproductive, in that if devices were illegal in the band, then it might be best if they used the SAS, rather than blindly transmit. However, if the SAS concept becomes more international, then NRA-by-NRA controls will be important.
- SAS administrators have a requirement to provide the public with information regarding the devices that are authorized in the band, similar to public licensing. This is a controversial requirement, in that cellular operators using exclusive spectrum are generally private with their specific deployments.
- There is a general obligation to maximize the opportunity for spectrum use by GAA and to ensure that the GAA operation has a maximum degree of coexistence. Specific powers of the SAS to implement these principles are not prescribed.
- Much of the specific design of SAS-to-SAS exchanges, PAL frequency assignments, authentication mechanisms and other implementing processes are implicitly and explicitly delegated, or assumed to be performed by industry organizations.

## 4.13    Device Classes

It was clear that the CBRS model had several potential deployment models. At one end, there were traditionally and professionally deployed outdoor small cells, with high Equivalent Isotropic Radiated Power (EIRP) and conducted power devices, and at the other end, consumer and enterprise devices that were like Wireless Fidelity (Wi-Fi) boxes, and deployed by consumers. The FCC established a set of device classes in order to differentiate between simple, low power and non-directive applications, and the high power, outdoor and likely directional deployments. The following Table (Table 4.6) is adapted from the second order §96.41.

**Table 4.6** Categories.

| Category | Maximum EIRP (dBm/10 MHz) | Maximum PSD (dBm/MHz) | Notes |
| --- | --- | --- | --- |
| End User Device | 23 | n/a | Must be under control of a CBSD. Does not require SAS registration. |
| Category A | 30 | 20 | Requires functioning self-location capability. Must be deployed indoors. Otherwise the device must be registered as Category B, and be installed by a professional installer. |
| Category B | 47 | 37 | Only permitted after an ESC is approved and deployed. Requires professional installer installation. |

## 4.14  Proposals for Three-Tier Outside of the USA

On April 14, 2016, the UK OFCOM issued a call for input titled *"3.8 GHz to 4.2 GHz band: Opportunities for Innovation"* [9]. Paragraph 1.10 of the inquiry clearly states a three-tier framework, with assured quality and access along with opportunistic access in the same band.

*"Given the characteristics of use of incumbent Fixed and Fixed Satellite Services, as set out in Section 2, we believe the 3.8 GHz to 4.2 GHz band has the potential to be accessed on a more intense shared basis by innovative applications with different requirements. In particular, some innovative applications might be able to use the spectrum on an opportunistic basis while other applications might, for example, require a defined quality of service but within a discrete area."*

OFCOM provides a graphic of the current deployment of fixed wireless and satellite sites in the 3.8–4.2 GHz band, as shown in Figure 4.4.

The density of FSS use is similar to that in the USA, on a site/km$^2$ basis. The structure that OFCOM proposed to manage the band closely resembles the US CBRS tiers. The document (para 3.7) describes that as:

**Tier 1 / Existing License Classes** would comprise the bands' incumbent services. Tier 1 licenses holders' spectrum access rights would fall within a given range of frequencies and geographic areas not accessed by the other tiers. Fixed and Fixed Satellite licenses would continue to be available on a first-come-first-served basis, subject to OFCOM's established coordination procedures, and UK Broadband would continue to access spectrum according to the coordination mechanism in which it avoids causing undue interference to the Fixed and Fixed Satellite services.

**Tier 2 / New Geographic License Layer** would comprise geographic licenses, as described in Section 3.3.1. Once a geographic license has been acquired, future Tier 1 licenses would need to be coordinated around these licenses.

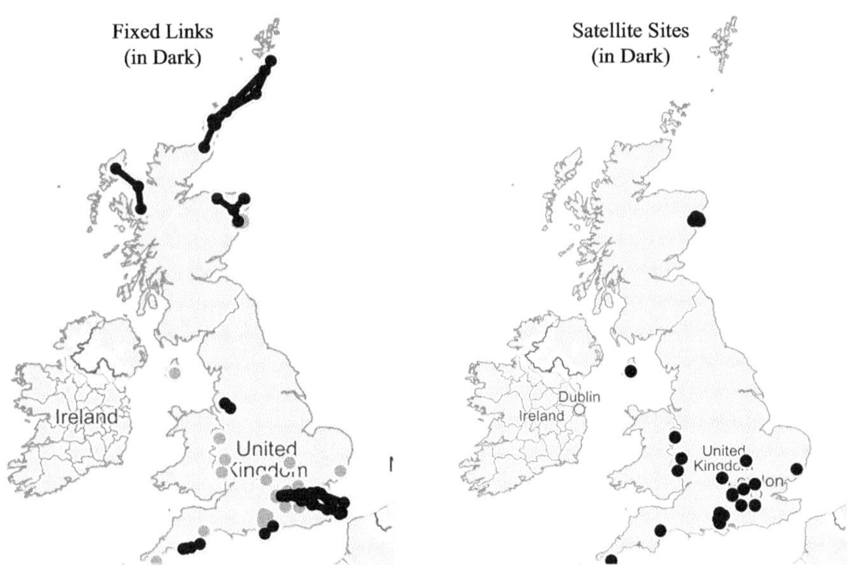


**Figure 4.4** OFCOM provided UK fixed and satellite sites in 3.8–4.2 GHz.

**Tier 3 / New Opportunistic Access Layer** would comprise a rules-based opportunistic access mechanism to the spectrum in a given geographic area. Stakeholders operating in this tier would need to not cause undue interference to both Tier 1 and Tier 2 licenses on a dynamic basis, possibly through a geolocation database or other technology enabler, and would not be able to claim protection from any other tier, or from each other.

Beyond the obvious similarities in the US regime and the possible framework of an OFCOM regime, there are some significant differences in the relative rights of the tiers.

- Tier 1 is not completely superior to Tier 2. In the OFCOM proposals, future Tier 1 licenses would have to coordinate with Tier 2. In the USA regime, Tier 1 has complete freedom to deploy, and Tier 2 is responsible for not interfering with the new Tier 1 usage and must adapt its usage if possible. Tier 2 has a first-come-first-served right over Tier 1 apparently.
- The US regime requires a database solution for protection of Tier 3, whereas the Office of Communications (OFCOM) framework appears open to other solutions.
- The primacy of Tier 1 users results in a US requirement for Tier 2 to use the database service. Since OFCOM requires new Tier 1 entrants to coordinate with existing Tier 2 users, there is no apparent requirement for Tier 2 to utilize the database system.

The public process to establish the UK rules is ongoing so we cannot judge what changes, if any, might occur as the concept becomes defined in regulation. The first difference (new Tier 1 use must be coordinated with existing Tier 2 users) significantly changes the Tier 1 rights. This was explicitly rejected in the US framework, due to likely opposition from these users.[7]

## 4.15 Summary

The US implementation of regulations for the CBRS band provides validation that the PCAST three-tier framework is capable of implementation in terms of discrete policy requirements, and can achieve general acceptance by both the potential user community and the band incumbents. There are certainly aspects of this regulation that can be challenged, and individual points of view would probably disagree with one or more of the regulatory requirements. However, none of this takes away from the accomplishment of creating the first viable regulatory structure for three-tier spectrum.

There are a number of references (such as §197, which discuss congestion management of GAA users. The FCC rules require that a SAS provide coexistence services to minimize density impacts on GAA users. The technology to do this, without discouraging innovation, creating implicit rights for interference free operation, and undoing the objectives of the entire regime is a challenge that three-tier spectrum sharing will have to address as it succeeds.

## 4.16 Suggested Reading

- The FCC documents, and the community comments on them, provide fundamental readings on the issues involved in three-tier spectrum sharing from the perspective of non-Federal incumbents, traditional spectrum exclusive rights users, and the potential users of the three-tier band framework. The two FCC Report and Orders [1, 2] define the regulations in this band. They not only provide the regulatory text, but also a very detailed summary of the positions of the responding community on the issues, and the rationale for the FCC's ultimate decision in each issue.

  Many comments on the first order opposed the entire concept, so subtle issues were not developed to the depths that were addressed between the first and second orders. Therefore the second Order discussion has another level of detail and nuance beyond the discussion provided in the first Order, but less of the fundamental opposition arguments.

- One of the topics that emerged in the community leading up to the second Report and Order was the subject of secondary markets in protected licenses. This discussion extends to forty paragraphs in the final order, but still only captures summaries of the arguments made by various parties. Positions filed by Cellular Telecommunications Industry Association (CTIA), AT&T, Cantor Telecom, Federated Wireless, and

---

[7] The FCC did block new deployments of extended C-Band sites, and additional Part 90 base station licenses, but these were secondary in the band already. So the USA did limit incumbent rights, but not the primary users.

Google, among others, provide different views regarding how secondary markets might emerge in dynamically allocated spectrum, and what the consequences of this emergence would be. These source documents are available at the FCC Electronic Comment Filing System (ECFS) (available at: apps.fcc.gov/ecfs/comment_search_solr/search) under proceeding 12-354.

- The NTIA Quick Look Report [5] provides analysis of the opportunities for clearing or sharing a number of bands, including 3 550–3 560 MHz. It includes an extensive discussion of the primary spectrum holders, perception of spectrum sharing and the issues associated with reducing the risk of such sharing. This report was instrumental in both the PCAST recommendations and the subsequent FCC regulatory actions.

- The process for certifying a SAS is laid out in a PN [6] that includes the scope of the approval process. Also, each potential SAS operator applicant had to respond to this PN with a description of how they designed their SAS, and how they intended to operate it. Both the PN and industry responses contain interesting material on SAS requirements and design.

- The Wireless Innovation Forum (WinnForum) is continuing to derive and evolve requirements for the implementation of the CBRS. This includes the SAS, the devices (CBSDs) and operational procedures. These documents are constantly evolving, so no specific document will be cited, and instead the entire directory of approved documents is suggested to the reader. These are generally accessible at: groups. winnforum.org/Reports. Some topics of interest include:

  - SAS-to-SAS Coordination [10].
  - SAS-to-CBSD Protocol [11].
  - SAS Architecture [12].
  - CBRS Threat Model [13].

- The record created by the UK OFCOM proposal to share the 3.6–4.2 GHz band is important reading to generalize the otherwise USA-dominated discussion of the regulatory concept. The original inquiry posting [9] provides documents summarizing the stakeholder positions.

## References


1 Federal Communications Commission, *Amendment of the Commission's Rules with Regard to Commercial Operation in the 3550–3650 MHz Band, GN Docket 12-354*, Report and Order and Second Further Notice of Proposed Rulemaking and Order (2015). http://apps.fcc.gov/edocs_public/attachmatch/FCC-15-47A1.pdf.

2 ——, *Amendment of the Commission's Rules with Regard to Commercial Operation in the 3550–3650 MHz Band, GN Docket 12-354*, Order on Reconsideration and Second Report and Order (2016). https://apps.fcc.gov/edocs_public/attachmatch/FCC-16-55A1.pdf.

3 ——, *FCC Chairman Julius Genachowski Announces Plans to Initiate Formal Steps on Spectrum Recommendations from the PCAST*. Press release (2012).

4 President of the United States, *Expanding America's Leadership in Wireless Innovation; Memorandum for the Heads of Executive Departments and Agencies*. Presidential memorandum (2013).

5 US Department of Commerce, National Telecommunications and Information Agency (NTIA), *An Assessment of the Near Term Viability of Accommodating Wireless Broadband Systems in the 1675–1710 MHz, 1755–1780 MHz, 3500–3650 MHz, and 4200–4220 MHz, 4380–4400 MHz bands*. Technical report (2010).

6 Federal Communications Commission, *Wireless Telecommunications Bureau and Office of Engineering and Technology Establish Procedure and Deadline for Filing Spectrum Access System (SAS) Administrator(s) and Environmental Sensing Capability (ESC) Operator(s) Applications* (2015). http://apps.fcc.gov/edocs_public/attachmatch/DA-15-1426A1.pdf.

7 ——, *Wireless Telecommunications Bureau and Office of Engineering and Technology announce Methodology for Determining the Protected Contours for Grandfathered 3650–3700 MHz Band Licenses GN Docket No. 12-354*. Public notice (2016).

8 United States Census Bureau, *Census Tracts and Block Numbering Areas*. Technical Report (2000).

9 Office of Communications (Ofcom) (UK), *3.8 GHz to 4.2 GHz band: Opportunities for innovation* (2016). http://stakeholders.ofcom.org.uk/consultations/opportunities-for-spectrum-sharing-innovation.

10 Spectrum Sharing Committee (SSC), *SAS–SAS Protocol Technical Specification*, WINNF-16-S-0096-V1.0.0 (Wireless Innovation Forum, 2016).

11 ——, *SAS to CBSD Technical Specification*, WINNF-16-S-0016-V1.0.0 (Wireless Innovation Forum, 2016).

12 ——, *SAS Functional Architecture*. WINNF-15-P-0047-V1.0.0 (Wireless Innovation Forum, 2015).

13 ——, *CBRS Threat Model*, WINNF-15-P-0089-V1.0.0 (Wireless Innovation Forum, 2016).

# Part III

# Components of a Three-Tier Architecture

# 5   Three-Tier Spectrum Admission Control Systems

## 5.1   Database-driven Spectrum Management Framework

This chapter will develop the basic building blocks from which various forms of three-tier spectrum management systems can be constructed. The technical building blocks are based on physics and fact, and therefore are not sensitive to nations or policies. The decision-making is nation- and frequency-dependent, and therefore is addressed separately from the technical tools that provide it with the basis for its decisions. This chapter is not intended to delve into specific algorithms or protection rules in detail. These will be addressed in later chapters specific to the issues involved in each.

It is easy to view each possible spectrum management regime as having unique requirements for the database system that implements its rule structure. This chapter instead will develop a set of constructs from which various spectrum management regimes can be constructed. In this way, we can readily focus on the common elements of all systems, and lead to common components. A common model of the technical infrastructure can make the construction of new regimes more practical to implement, effective, and verifiable.

In this chapter, we will use the general term of "admission control system." This is the general nomenclature for the functionality that is provided by the United States (US) Federal Communications Commission (FCC) Citizens Broadband Radio Service (CBRS) Spectrum Access System (SAS). Throughout this book, the term SAS will only be used for requirements that are specific to this US implementation.

Before going into other details, we can first broadly classify database systems into two classes:

**Stateful** Admission of nodes into the ecosystem changes the entry conditions for future entrants. The database system must be updated to reflect each admission decision or status change, and the sequencing of processing must ensure that decisions do not interact. Microwave coordination is an example of such regimes.

**Stateless** Admission of nodes into the ecosystem has no impact on the admission of future entrants, once the regulatory criteria is established. The system can make future decisions without regard to its own previous decisions. An example of this is determination of fixed contour based admission, such as fixed exclusion zones, or the Television White Space (TVWS) regime.

Next we consider the fundamental functionality of any database-driven admission control system. There are likely many possible organizations of these functions, and not all systems may require all of them. The partitioning used in the rest of book is shown below:

**Awareness** The database system will require knowledge of the characteristics of protected nodes, rights information, geographic restrictions, and other data that is fixed, at least in relation to entrants into lower tiers. The information can be declarative (facts through other sources), or organically sensed by dedicated or collaterally assigned devices. Additionally, if the system is stateful, it will have to update this awareness to reflect previous admission decisions it has made.

**Analysis** The database system will require analytic methods to analyze the impact of emissions. This will include tools such as propagation models, along with any supporting data, interference models, usage models and whatever considerations the database system is intended to implement.

**Decision Making** We separate the decision-making from the models that create the input to the decisions in order to isolate the purely technical considerations from those driven by policy.

**Interaction** The database system will require some method to interact with devices, collect information on their requests, and communicate results of analysis and any subsequent changes in their authorizations after their initial issuance. This has implications for the nature of the networks that can operate in these tiers.

**Coexistence** In addition to protection functions, the same database system may be tasked to also coordinate usage characteristics within tiers to minimize interference among tier users with peer rights.

There are many ways in which these components could be configured. Figure 5.1 illustrates a notional database access control system.

A summary of the principal transaction flow for providing spectrum to a three-tier user or device is described below. This is a very simplified model, but is the general

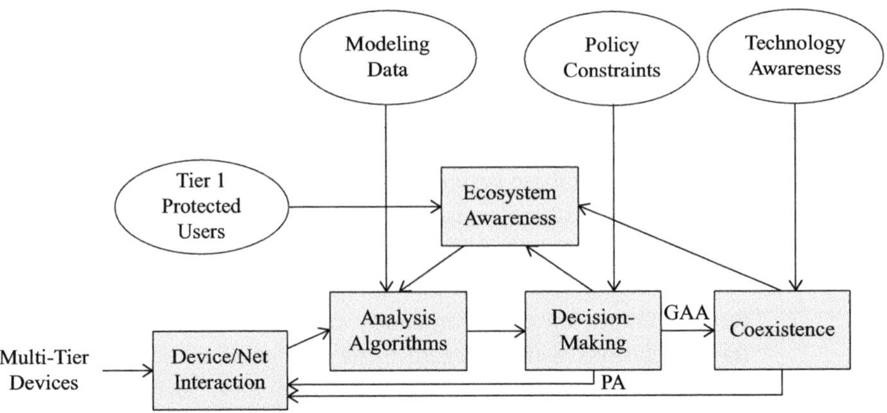


**Figure 5.1** Notional arrangement of database system components.

schema that will be examined in more detail later. Also, some steps may not be applicable to all regimes. For example, providing a coexistence service is not mandatory in a three-tier regime.

- The device or network requests a frequency assignment, or grant. This is either an explicit request for a specific frequency range, or a description of the bandwidth required. This grant request also includes the emission characteristics, and optionally the receiver environment constraints, antenna characteristics and other information needed to assess the interference potential of the grant, and if it is in a protected tier, the requirements for protection.
- The request is analyzed to determine the impact at any protected node within its area of influence. Depending on the policy, this could be a one-on-one analysis, or could aggregate multiple sources of interference at protected tier nodes.
- The policy constraints specific to the location and frequency are applied to determine if the grant request can be issued. This decision making function assures policy compliance. This includes (1) assuring that the entry will not violate any protection rights to higher or peer tier devices, (2) that the device, and the grant it requests, is compliant with the regulations in the band, and (3) that the device is authorized to make the grant request it has issued. An example of this last verification is whether the node that requests Priority Access (PA) protection is authorized to request it. In a complex environment, this process might have a ripple effect on other devices, if they must be relocated to avoid interference to an entering PA node, for example.
- If the request is for protected status, presumably the node will not receive interference, and will request a specific frequency, and follow the path to communicate the grant to the device.
- If the request was for a usable frequency, but it was not specified, such as in the lowest (General Authorized Access (GAA)) tier, some mechanism to maximize coexistence could be involved. This could be technology agnostic, such as de-conflicting channel assignments, or technology specific in aligning time slots, or sub-channel usage.

The basis for decision-making by an admission control system cannot be completely self-contained, and is dependent on a number of external sources, generally described below, and elaborated in later sections and chapters.

**Protected Users** One of the principles of spectrum sharing is that the sharing process should be minimally invasive on incumbent systems. Therefore, it is assumed that they are not required to develop new functionality to report status to an admission control system, but either statically declare their positions and characteristics or are dynamically detected. This data source reflects any information regarding their presence, either provided by the incumbent or detected dynamically.

**Modeling Data** Modeling data is used by the analysis algorithms, and may include terrain, structures, material, and other physics and geographic information. Data specific to regulatory limitations on operation are included in the policy considerations data source.

**Policy Constraints** Policy constraints define the specific rules used to make grant decisions in a given location and band. In this framework, they are isolated from the fact-based analysis process which provided the input values required by the policy regime, which does not reflect the regime criteria. Separation of policy constraints from analysis algorithms also enables the maximal reuse of components across regulatory regimes; reducing cost for each unique implementation.

**Technology Awareness** Technology awareness information extends generic channel deconfliction algorithms to reflect the specific capabilities of individual technologies. For example, Wireless Fidelity (Wi-Fi) spectrum sharing effectiveness is driven by the duty cycle of the usage, while Long-Term Evolution (LTE) is not. Time Division Long Term Evolution (TD-LTE) is capable of time alignment among access points, so long as they have high-quality time-of-day information available.

Each of these mechanisms is discussed in greater detail in the following sections.

## 5.2      Ecosystem Awareness

We have hypothesized that in our initial baseline concept of spectrum sharing, the sharing operation and mechanism will be invisible to the current users of the spectrum being shared. This is accomplished in two ways: assuring no interference technically, and minimal or no impacts operationally. Therefore, there are provisions to inform the admission control system and therefore, spectrum sharing devices of the existence of these primary users. The two default methods are:

- The protected tier user informs the admission control system of their location, transmitter, receiver, and antenna characteristics in advance. This is often information that exists in the licensing records currently managed by spectrum administrations.
- Sensing of the protected node operation by either nodes or dedicated sensors that inform the admission control system of the detections, and thereby invoke protection of the operation. This is appropriate when the usage is not fixed, or is intermittent. Additional sensing may also be used as an adjunct to intra-tier coexistence to estimate actual path losses, duty cycles and other aspects of tier incumbents.

Hybrid operation is also possible. For example, in the US CBRS rules (to be discussed in more detail in later chapters) the protected users include naval radars, which are to be sensed by sensors associated with the SAS, fixed C-Band satellite downlinks, whose location is already in databases, and Part 90 operations (such as wireless Internet service provider (WISP) base stations), which are also in regulatory databases. The Dynamic Frequency Selection (DFS) regime is another example. Although it does not comply with our criteria for a three-tier regime, it has an ecosystem awareness function for sensing of the many types of radars in the band, and alerting a local decision element. The TVWS system is all declarative by incumbent users, through registration of the broadcast station with regulators, or per-use registration of wireless microphones that use Television (TV) frequencies.

Another source of ecosystem awareness is the activity of the admission control system itself. The stateful admission control system is the competent authority for the location of all nodes that were permitted entry under the three-tier regime. Within the admission control system, a protected node is essentially the same regardless of its tier, so the awareness function can be abstracted independently from the method by which this awareness is granted. Each spectrum grant that is approved, expires, or is relinquished changes the state of the ecosystem.

Another practical consideration in the awareness function is the degree of deployment information that must be disclosed to have a viable sharing regime. In the US CDRS proceedings, this has been a subject of considerable controversy. Government users are very sensitive to having to disclose details of their operation. Public safety users fear disclosing investigations. Militaries are protective of their deployments and usage patterns. Commercial operators desire to keep deployment specifics confidential as proprietary business information.

All of these concerns make the technical process of organizing the ecosystem awareness more complex due to the competing requirements for accurate analysis of interference, and the maintenance of user confidentiality. In the case of the US CBRS band, the military users were given very wide protection areas to reduce the need for accurate admission control system location data,[1] while civil users of the shared spectrum were required to disclose Global Positioning System (GPS) accurate location information.

For now, we will maintain a very broad definition of ecosystem awareness. It is likely that these concepts will have to be expanded to include use cases that are specific to nations and bands that have not yet been examined for inclusion in a three-tier framework.

## 5.3    Analysis Algorithms

The analysis tasks of an admission control system are driven by the physics, not the regulations, so this element of the admission control system can be quite generalized in application to bands and locations. Specific algorithms will be discussed in a later chapter, but the components of the analysis component include models for the following:

**Device Characteristics** Detailed device characteristics are essential to the accurate estimation of interference conditions. For example, filter characteristics are critical to analyzing adjacent band and channel effects, out of band emission masks are essential to extending beyond co-channel users.

**Antenna Patterns** At higher frequencies, directional antennas become more common, and significantly impact the interference analysis. Antennas may reduce or enhance

---

[1]  In fact, the sensing system is prohibited from computing the location of naval vessels, recording accurate detection times, recording spectral signatures, or retaining the information about usage for more than the minimum needed to accomplish the protection process.

sensitivity. Satellite antennas operate in three dimensions, so they may have a very different response on the horizon than they do in the direction of the orbits or deep-space locations.

**Propagation** Propagation is one of the most significant factors in interference analysis. At the frequencies typically available for sharing, purely distance based loss models are overly conservative in estimating path loss.[2] Actual losses are driven by the obstructions in the path, more than the distance of the path. These models will be discussed in more depth later in this book.

## 5.4    Device/Network Interaction

In this section we will lay out a very general set of functions that are inherent in the operation of a three-tier admission control system. Different systems may arrange these in various schemes, but the essential elements are likely to be common across designs.

**Spectrum Inquiry** A spectrum sharing regime offers devices a large number of options. These options are very situational, as well. Therefore, it is important that the admission control system provide devices with the ability to inquire about what spectrum resources are (potentially) available to the device. This will be particularly important as more nations adopt three-tier spectrum, and do so in more bands, in order that devices, or device designs, can roam freely around the world and obtain availability information, without dependence on nation-by-nation, and regulation, programming. Wherever possible, policy should be implemented centrally in the access control system, not the devices.

**Device Registration** Device registration provides the static information regarding the device's identity, location and other information that is not impacted by the strategy implicit in its later grant request. This information could be integrated with a grant request. But, since a device may have many grants in its lifetime yet need only register once, it makes sense to treat this as a separate transaction.

**Grant Request** Grant requests are issued by devices and networks to obtain the right to use spectrum. Grant requests can be for a specific frequency or frequencies, or an amount of bandwidth. Grant requests result in either a grant being issued or rejected.

**Grant Relinquishment** Devices may relinquish a grant when no longer required, such as when the device wants to move frequency or terminate operation, among other reasons.

**Directed Relinquishment/Relocation** Since higher-tier users may enter the band, or change frequencies, the admission control system must have the ability to direct lower-tier devices to change frequencies to avoid interfering with these users. Depending on national regulations, devices would have to accept these relocation directives, or relinquish the grant it held and obtain another.

---

[2] This is conservative from the perspective of interference calculation, in that they provide estimates of path loss that are typically much lower than the observed values. These models would be considered optimistic from the perspective of communication link analysis.

**Grant Termination** Grants may need to be terminated in their entirety. Higher-tier users could utilize all of the spectrum, individual devices could be determined to be operating out of tolerance, interference events could be detected, or other reasons could result in a device being directed to completely terminate operations.

## 5.5  Protection Decision-making

Protection decision-making is abstracted as operating on the output of the various analytic tools. This can be considered the single-node-on-single-node interaction, or the more complex and meaningful many-on-one interactions, where the consequences of the entire ecosystem on each protected node are determined, and used, as the basis of decisions.

We abstract the decision-making from the analysis because we imagine a common set of analytic tools that are independent of the specific band and national administration. The decision-making component carries the administration-specific protection criteria, geographic constraints, and other limitations. Of course, this is an over simplification, but it is a starting point to avoid looking at each nation and each band as a unique situation for three-tier spectrum sharing. The commonalities likely outweigh the differences.

Protection decisions may impact more than just the node that initiated the transaction. The entry of a protected node, whether as a primary, or a secondary, can force relocations or removal of nodes in lower tiers. For example, the *"Use it or Share it"* principle permits a GAA to occupy spectrum to which a PA holder has rights, if the PA is not actually utilizing the spectrum. However, when the PA elects to exercise its rights, the lower-tier GAA must be relocated to spectrum that would not cause interference. So this is not a simple transaction model; one entrant could cause any number of chain reactions that might disrupt many other nodes operation, particularly in dense environments.

In Chapter 7, we will develop more detailed models of how the protection criteria that drive these decisions are, or could be, expressed.

## 5.6  Coexistence Optimizations

The admission control system could have the mission of assuring compliance with the interference protection rules. A well-functioning ecosystem requires an additional layer of services to assist devices in making decisions that result in the least interference impact. Although these services might be integrated into an admission control system product, they are essentially different from other aspects of the admission control system in a number of ways:

- The operation of the coexistence layer is not deciding the legality of operation, as in incumbent protection. It considers a number of alternatives, all of which are compliant with the regulations.

- The coexistence layer rules are likely developed by the members of the ecosystem, rather than national administrators.
- The coexistence layer may reflect technology-specific features of devices. Most regulations are technology neutral.

The coexistence function is not essential to the basic spectrum management regime, as it does not function in the interference protection of higher-tier nodes. However, it is highly beneficial to the lowest tier in distributing the use of the spectrum resource as evenly as possible across these users. Although it cannot create spectrum where there is none, it can provide more confidence that a reasonable level of service can be assured within a given spectrum, and minimize any "gamesmanship" in the use of the spectrum.

This function is particularly useful with technologies that have coexistence features inherent in their operation. For example, LTE was designed to operate effectively in "Re-use One" conditions where spectrum is shared between adjacent access points. However, these mechanisms were designed to operate within monolithic carrier infrastructures, with a single, fully coordinated, and connected Extended Packet Core (EPC), tightly coordinating the nodes operation.

However, in the use case we envision in the shared bands, there is not a single monolithic management layer, but a number of independent networks operated independently. These are not just one LTE network on a given frequency, but instead many. To mutually benefit from the coexistence capabilities of LTE, it is necessary to provide some coordination across these networks. This principle, *"Federated LTE"* provides a degree of coordination between sovereign network operators. Examples include: neighboring enterprises in an office building, adjoining network operators, apartment units, stores in a shopping venue, and the like.

At first glance, coexistence might appear to be relevant only to the protected (GAA) tier. However, it has utility in any tier to reduce interference between peers. For example, unless spectrum and area are set aside for guard bands, spectrally and physically adjacent PA users will have interference due to adjacent channel and co-channel interference. The same techniques that would enable mixed GAA users to operate in close proximity to each other will also be highly useful to enable PA operations to coexist along PA boundaries, or in adjacent channels.

## 5.7    Example Implementations of Database Systems

This section will discuss the two instantiations of database controlled spectrum management regimes. The first (TVWS) is not a three-tier management regime, but was one of the important precursors to the three-tier approach. The second is the US CBRS implementation of the structure. Each of these represents a deployed spectrum management admission control system that represents successful specification of the regulations, development of the software and its operation, development of devices to utilize the services, and validation sufficient for regulators to approve them for use. The specifics of the representative CBRS implementation of three-tier spectrum will be discussed in later chapters.

### 5.7.1　Television White Space Database Service

The architecture of the TVWS admission control system is shown in Figure 5.2, in the context of the components of Figure 5.1. The drawing shows the reduced complexity of not addressing aggregate interference protection, and not providing a protected secondary user tier. Essentially, admission is pre-computable statically, and admission requests can be compared to these pre-computed values. The ecosystem also is simplified by having all of the awareness of protection from the declared existence of TV broadcasters and wireless microphones.

- The TVWS regime does not consider aggregate interference. Therefore, there is no requirement to update the ecosystem as nodes are admitted to the band. Instead, the rules can be predetermined once the protected nodes are identified.
- There is no coexistence management services, so this function, and the supporting technology-specific models, are not required. Coexistence is deferred to the devices and networks to be accomplished, if at all.
- There is only one shared tier, so there is no requirement for multiple grant paths.
- All of the ecosystem knowledge required is provided by the registration of broadcast stations and microphones.

This architecture might be considered to be the simplest form of an automated admission control system. It has all of the essential elements of moving from manual determination of sharing opportunities to an automated transaction.

### 5.7.2　CBRS Spectrum Access System

The architecture of the three-tier CBRS [1, 2] SAS admission control system is shown in Figure 5.3, in the context of the components of Figure 5.1. The CBRS management fully implements all of the functionality of the model framework, with minor departures in implementation.

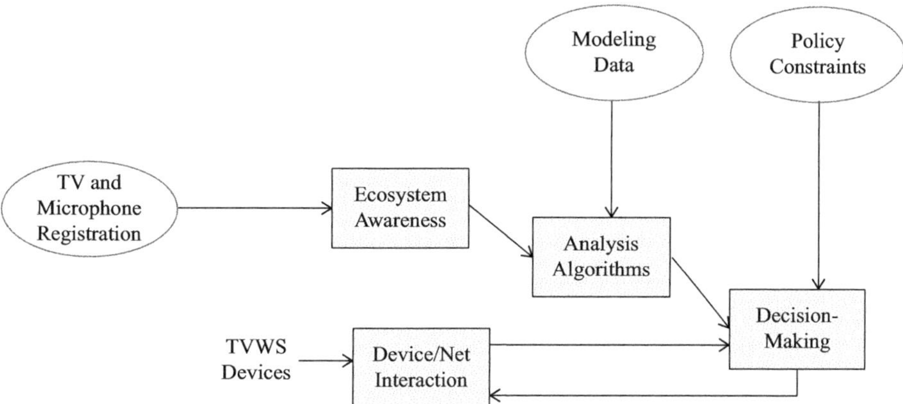


**Figure 5.2** Simplified depiction of the Television White Space admission control system architecture.

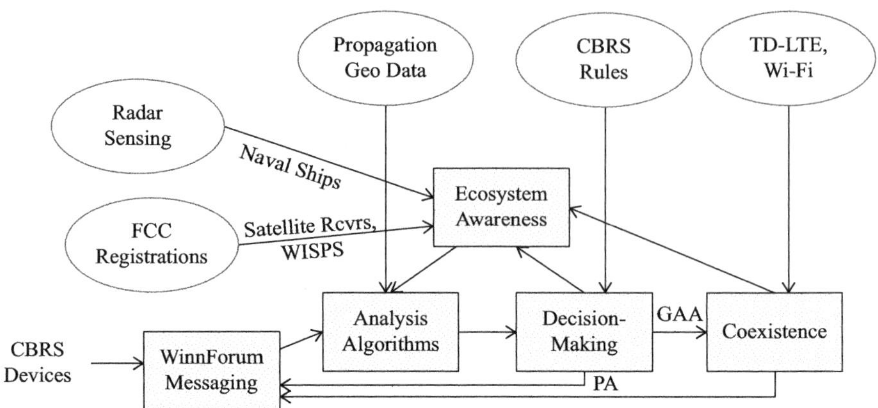


**Figure 5.3** CBRS admission control system architecture.

- The ecosystem has two sources of incumbent information. Radar sensing is used to create awareness of naval ship operations, and the existing regulatory regime provides records of Part 90 and Fixed Satellite Service (FSS) deployments.
- It is anticipated that there will be heavy band usage by TD-LTE technology, and it is possible that Wi-Fi and proprietary standards may be introduced, so the coexistence process is being tailored to support multiple technologies. This process is not a regulatory mandate, but is provided due to its utility and to reduce risk to users.
- The primary database supporting analysis is the geographic terrain data that is used in the propagation modeling, such as terrain, building structures, and attenuative materials.
- The specific device interactions are defined by the Wireless Innovation Forum (WinnForum) standards.
- The rule set is from the US FCC Part 96. These are complex rules due to the infancy of the technology, lack of operational experience with operating these systems, and protection of the incumbent users. The CBRS rules include a number of geographic considerations, so this regulatory data also includes a number of geographic products, such as census tract boundaries, urban/rural designations, etc.

## 5.8      Suggested Reading

This chapter introduced many broad concepts, but did not develop them into detail, as this is provided in the specific chapters later in this book. There are interesting works in each of the five functional partitions from Section 5.1. More specific readings are provided in the individual book chapters.

**Awareness** There has been a significant effort in the academic and regulatory communities to understand the capabilities and limitations of sensing by Dynamic Spectrum Access (DSA) and TVWS systems.

- The US FCC has provided some analysis results of the testing of TV sensors [3, 4].
- There has been a significant effort in the academic community to understand the capabilities and limitations of sensing by DSA and Cognitive Radio (CR) systems [5, 6, 7].
- The sensing of radars was central to the DFS regime [8].

**Analysis** Analysis considerations largely revolve around propagation models and their application to specific situations. There is an extensive literature regarding these models, most of which demonstrates why no model can be depended on in more than a few cases. The application of these models thus must be either highly conservative, or highly situational [9, 10, 11, 12, 13, 14].

**Decision-making** There are several sources that begin the definition of the decision-making of an admission control system, including:

- The only regulator input on admission control system requirements is provided by the FCC regulations for Part 96. These include two orders from the FCC [1, 2].
- The President's Council of Advisors on Science and Technology (PCAST) report [15] provides a somewhat more expansive, if less specific vision of an admission control system.

**Interaction** The WinnForum has developed a set of requirements for all of the functions within the CBRS ecosystem. Specific requirement documents have been developed for the admission control system, or SAS at several levels of generality [16, 17, 18, 19, 20].

**Coexistence** Most of the available literature addresses the coexistence of uncoordinated Media Access Layer (MAC) layers using Listen Before Talk (LBT), such as Wi-Fi [21], or the coexistence of Wi-Fi and non-LBT technologies, such as LTE or Long-Term Evolution Unlicensed (LTE-U) [22, 23].

# References


1 Federal Communications Commission, *Amendment of the Commission's Rules with Regard to Commercial Operation in the 3550–3650 MHz Band, GN Docket 12–354*, Report and Order and Second Further Notice of Proposed Rulemaking and Order (2015). http://apps.fcc.gov/edocs_public/attachmatch/FCC-15-47A1.pdf.

2 ——, *Amendment of the Commission's Rules with Regard to Commercial Operation in the 3550–3650 MHz Band, GN Docket 12–354* (2016). https://apps.fcc.gov/edocs_public/attachmatch/FCC-16-55A1.pdf.

3 FCC Office of Engineering and Technology, *Initial Evaluation of the Performance of Prototype TV-Band White Space Devices*, OET Report: FCC/OET 07-TR-1006 (2007).

4 ——, *Evaluation of the Performance of Prototype TV-Band White Space Devices Phase II*, OET Report: FCC/OET 08-TR-1005 (2008).

5 K. Sithamparanathan and A. Giorgett, *Cognitive Radio Techniques: Spectrum Sensing, Interference Mitigation, and Localization* (Artech House, 2012).

6 T. S. Dhope, *Cognitive Radio Networks Optimization with Spectrum Sensing Algorithms* (River Publications, 2015).

7 T. Yucek and H. Arslan, A survey of spectrum sensing algorithms for cognitive radio applications. *IEEE Communications Surveys Tutorials*, **11**/1 (2009), 116–130.

8 The Wi-Fi Alliance Spectrum and Regulatory Committee, Spectrum Sharing Task Group Regulatory Task Group, *Spectrum Sharing in the 5 GHz Band - DFS Best Practices*, (2007). www.ieee802.org/18/Meeting_documents/2007_Nov/WFA-DFS-Best-Practices.pdf.

9 A. G. Longley and P. L. Rice, *Prediction of Tropospheric Radio Transmission Loss over Irregular Terrain: A Computer Method - 1968*. Technical report, ESSA ERL 79-ITS 67 (Washington, DC: US Government Printing Office, 1968).

10 A. Molisch, L. Greenstein, and M. Shafi, Propagation issues for cognitive radio. *Proceedings of the IEEE*, **97**/5 (2009), 787–804.

11 H. Zhao, R. Mayzus, S. Sun, M. Samimi, J. K. Schulz, Y. Azar, K.Wang, G. N. Wong, F. Gutierrez, and T. S. Rappaport, 28 GHz millimeter wave cellular communication measurements for reflection and penetration loss in and around buildings in New York city. *2013 IEEE International Conference on Communications (ICC)* (2013), 5163–5167.

12 J. Lu, D. Steinbach, P. Cabrol, P. Pietraski, and R. V. Pragada, Propagation characterization of an office building in the 60 GHz band. *The 8th European Conference on Antennas and Propagation (EuCAP 2014)* (2014), 809–813.

13 C. Haslett, *Essentials of Radio Wave Propagation* (Cambridge University Press, 2008).

14 J. B. Andersen, T. S. Rappaport, and S. Yoshida, Propagation measurements and models for wireless communications channels. *IEEE Communications Magazine* **33**/1 (1995), 42–49.

15 President's Council of Advisors on Science and Technology, *Report to the President: Realizing the Full Potential of Government-Held Spectrum to Spur Economic Growth* (Executive Office of the President (EOP), Office of Science and Technology Policy (OSTP), 2012). http://obamawhitehouse.archives.gov/sites/default/files/microsites/ostp/pcast-stem-ed-final.pdf.

16 Spectrum Sharing Committee (SSC), *SAS Functional Architecture*. WINNF-15-P-0047-V1.0.0 (Wireless Innovation Forum, 2015).

17 ——, *CBRS Communications Security Technical Specification*, 1st edn (Wireless Innovation Forum, 2016).

18 ——, *CBRS Operational Security Technical Specification*, WINNF-15-S-0071-V1.0.0 (Wireless Innovation Forum, 2016).

19 ——, *CBRS Operational and Functional Requirements*, WINNF-15-S-0112 (Wireless Innovation Forum, 2016).

20 ——, *Spectrum Access System Requirements* (Wireless Innovation Forum, 2016).

21 N. Rupasinghe and İ Güvenç, Licensed-assisted access for WiFi-LTE coexistence in the unlicensed spectrum. *IEEE Globecom Workshops* (2014), 894–899.

22 C. Chen, R. Ratasuk, and A. Ghosh, Downlink performance analysis of LTE and WiFi coexistence in unlicensed bands with a simple listen-before-talk scheme. *IEEE 81st Vehicular Technology Conference* (2015), 1–5.

23 A. Mukherjee, J. F. Cheng, S. Falahati, L. Falconetti, A. Furuskär, B. Godana, D. H. Kang, H. Koorapaty, D. Larsson, and Y. Yang, System architecture and coexistence evaluation of licensed-assisted access LTE with IEEE 802.11. *2015 IEEE International Conference on Communication Workshop* (2015), 2350–2355.

# 6 Admission Control Management of Access Point Devices and Their Clients

## 6.1 Introduction

An important consideration in the spectrum management regime is the impact that the spectrum regime has on the flexibility of devices that would utilize it. If the constraints of accommodating the unique constraints of operating within the regime are too great, the benefits of spectrum flexibility will not be realized. In this chapter, we examine the impact of a three-tier regime on the range of devices that populate the secondary ecosystem in a client/Access Point (AP) network mode. In particular, we contrast standalone, independent endpoints, and those that are architecturally dependent on a "superior" node for operation and are therefore subservient to the spectrum decisions of the master (access point) node.

This hierarchal arrangement of access point and client is the fundamental design concept behind most of the current wireless Local Area Networks (LAN) and Wide Area Network (WAN) architectures, such as 2G, 3G, and 4G technologies, such as Long-Term Evolution (LTE), as well as Wireless Fidelity (Wi-Fi). In this chapter, we examine the implications of this architecture on three-tier regimes, and vice versa.

## 6.2 Management of Different Connectivity Models

It might appear that the secondary spectrum regime controls would have to be accessed by all of the devices that operate in a secondary status. In this section, we will examine at what level of the network to apply these requirements at, and therefore the ability to operate devices that are not three-tier aware in the three-tier ecosystem. The question is at what level of the network to apply these spectrum management techniques? The question of what is a controlled device, and what is not, is an important consideration and is not as obvious as it would appear.

We consider two treatments that can be applied:

**Endpoints** Link with fixed, independent endpoints of operation, either of which could transmit independently, or that have unique interference regions, such as with high gain antennas.

**Access Points** Links that may have more than one client device, when the clients cannot operate independently of the AP, and have overlapping interference regions, or whose locations would not be fixed, or knowable.

The endpoint case is relatively simple in the three-tier spectrum model. Each endpoint is an independent node in the admission control system. This matches the simple reference model. Unfortunately, AP networks are most complex.

A simplified representation of an AP deployment is shown in Figure 6.1. In this case, the access point is typically higher powered and better positioned in the center of an area. There is a service area in which mobile devices (clients) can roam and in which the AP can maintain control in terms of spectrum and mode decision-making. Since we are protecting this AP from interference, we are implicitly protecting each of its clients. Similarly, adjoining users are being protected from interference from the AP, and any of its clients.

For this reason, the admission control system must consider not only the location of the access point, but also all of the possible positions of clients, and the outgoing interference they could cause, and/or the incoming interference they would be subjected to. The logical extent of the AP protection is therefore not the Access Point Protection Area of Figure 6.1, but the Service Area Protection Area in the figure. This avoids the necessity of tracking each mobile client's location and making rapid changes in spectrum allocations, at the cost of likely protecting from conditions (device presence and positions) that do not exist at most points in time.

The maximum device density can only be reached with the exact protection provided by the end point protection model. This protection can be tailored to protect only the immediate region of each actual end point, and for certain directions of use in the path between the nodes. However, the AP model offers flexibility to accommodate mobile patterns of use with movement and variable usage density, using the AP service as a surrogate for the location of the mobiles.

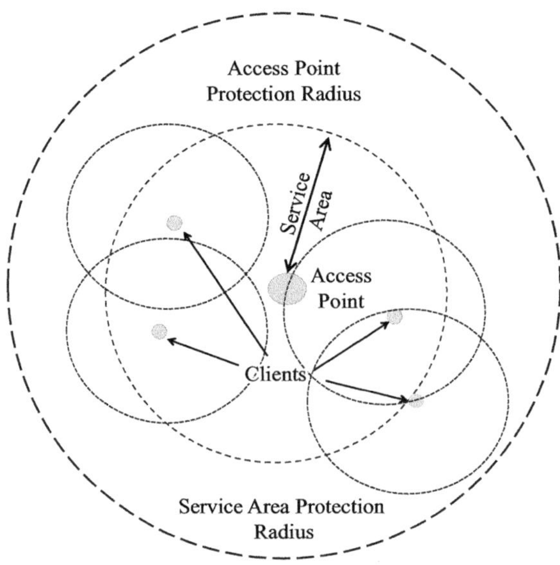


**Figure 6.1** Relationship of protection and service ares around access regions.

In the United States (US) Citizens Broadband Radio Service (CBRS) proceeding, the "definition of use" was a contentious issue. Some of those who believed they would likely be purchasing Priority Access License (PAL) rights, argued that the dimensions of use should match the geometry of their protection rights. Others, attempting to maximize the amount of sharable spectrum, argued for a minimal definition. In the end, the Federal Communications Commission (FCC) rules allowed the PAL owner to declare a dimension (a Priority Access License Protection Area (PPA)), but capped at the maximum radius that the node could serve, as determined by the admission control system. This is a plausible engineering definition of use, as compared to a policy-driven one.

Neither the AP nor point-to-point modes are explicit in the initial three-tier literature. In fact, there is implicit focus on AP operation in the US CBRS regulations, and it is clearly the reference model. Protection cases are shown in Figure 6.2.

Protection of AP operation is different from protection of client operation, and different yet again from protection of client devices from other client devices. Clients may be much closer to the potential protected victim, clients typically operate at lower transmit power when working with small cells, and they have less antenna gain, among other differences with the treatment of the AP.

In case 1, we protect the APs only; the clients are assumed to be close enough that the protection of the AP is sufficient to protect the client. In case 2, we protect the

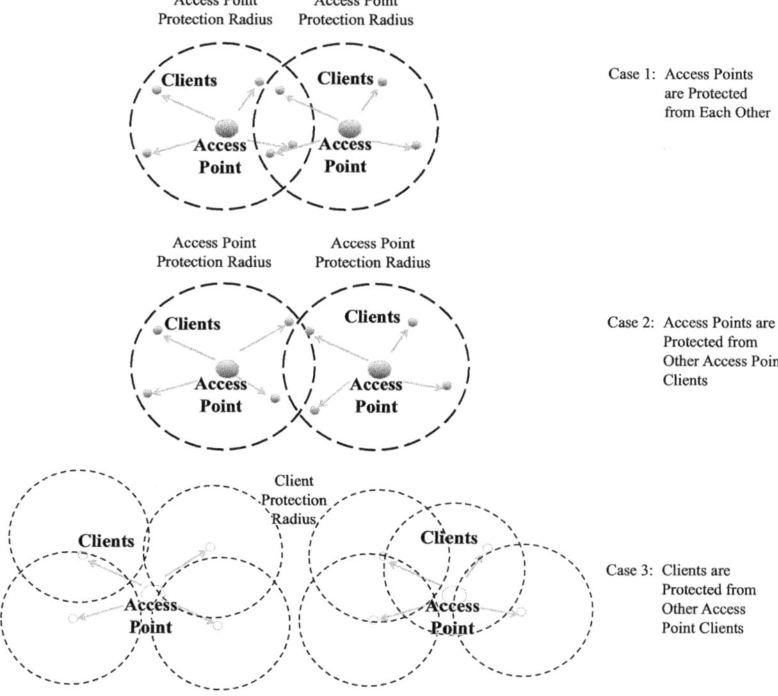


**Figure 6.2** Options for AP and client protection.

**Table 6.1** Client interference analysis variables.

| Variable | Meaning |
| --- | --- |
| $P_{AP}$ | Access Point EIRP (in dBm). |
| $P_{UE}$ | Client EIRP (in dBm). |
| $R_{AP\text{-}AP}$ | Range from AP to its closest permitted AP (in meters). |
| $R_{AP\text{-}UE}$ | Range from AP to its farthest client (in meters). This is the AP service area in this model. |
| $R_{UE\text{-}UE}$ | Range from a client on one AP to a client on an adjoining one (in meters). |
| $\alpha$ | Propagation exponent (as in $Lp \propto r^{\alpha}$). |
| $R_{AP\text{-}AP}$ | Distance between Access Points (in meters). |
| $P_{prot}$ | Maximum interference permitted at AP or client node (in dBm). |
| $f$ | Operating frequency (in MHz). |

client at the edge of a service area from interference by another AP, but the client is subject to interference from the clients associated with any other access point. This can be effective if a timing protocol can avoid these close proximity clients receiving and transmitting at the same interval. The most extreme protection is provided in case 3, where the admission control system enforces client separation sufficient to isolate client transmissions.

To analyze the consequences of these approaches, we will develop a closed-form model. Table 6.1 provides the variables that define the localized environment. For simplicity in this discussion, we will consider a constant propagation exponent, and symmetric AP nodes and clients, even though actual admission control system implementations would have to address heterogeneous device populations.

We know that for all of the protection cases, the path loss ($L_p$) must equal the difference between their EIRP ($P_{AP}$ or $P_{UE}$) and $P_{prot}$. A simple model of path loss over a distance $d$ in meters, frequency $f$ in MHz, and propagation exponent $\alpha$ is provided by Eq. (6.1).

$$L_p(d) = 10\alpha \log_{10}(d) + 20\log_{10}(f) - 27.55 \qquad \text{(in dB)} \qquad (6.1)$$

We can compute the required path loss between the node pairs to determine the path loss which must be achieved between these pairs. This also provides the AP density in this simplistic case. Eq. (6.2) provides the conditions that each case must meet.

$$\begin{cases} L_p(R_{AP\text{-}AP}) = P_{AP} - P_{prot} & \text{for Case 1} \\ L_p(R_{AP\text{-}UE} + R_{UE\text{-}UE}) = P_{AP} - P_{prot} & \text{for Case 2} \\ L_p(R_{AP\text{-}AP}) = P_{AP} - P_{prot} & \\ L_p(R_{UE\text{-}UE}) = P_{UE} - P_{prot} & \text{for Case 3} \\ L_p(R_{AP\text{-}UE} + R_{UE\text{-}UE}) = P_{AP} - P_{prot} & \\ L_p(R_{AP\text{-}AP}) = P_{AP} - P_{prot} & \end{cases} \qquad (6.2)$$

Note that case 2 and 3 have multiple conditions that must be met. Case 2 and 3 must meet the conditions of the prior cases, as well as the additional constraints, so only the first equation in each case is unique to the case and needs to be developed. These cases are inherently hierarchical.

The impact of the case 3 protection discipline varies greatly depending on the relationship of client power ($P_{\text{UE}}$) and AP power ($P_{\text{AP}}$). In case 1, the path loss for the AP power must occur over a path length of $R_{\text{AP-AP}}$. In case 2, this same loss must occur over the range of its own cloud, and the cloud to cloud spacing ($R_{\text{AP-UE}} + R_{\text{UE-UE}}$). Since these two distances must be the same (in the limiting case), we can see that the clouds cannot be allowed to overlap. This situation has a number of solutions, but the most obvious one is that $R_{\text{UE-UE}} = 0$, and that $R_{\text{AP-UE}} = \dfrac{D_{\text{AP-AP}}}{2}$. The actual dimensions of these distances are strongly influenced by the propagation exponent ($\alpha$), but the relationship among them is not $\alpha$ sensitive.

The case 3 conditions are obviously the most restrictive due to the condition that $R_{\text{UE-UE}} > 0$, since the UE power must be reduced to the $P_{\text{prot}}$ value across this range. Whereas case 2 allowed the maximum service area to be adjacent to another cell, case 3 requires that this loss be accomplished between the two service areas.

We can normalize the relationship of client and AP power by creating a variable $R_{\text{client-loss}}$ that is the distance for the client power to be reduced from $P_{\text{UE}}$ to the value of $P_{\text{prot}}$, and similarly $R_{\text{AP-loss}}$, which is the range for the AP power to be reduced from $P_{\text{AP}}$ to the value of $P_{\text{prot}}$.

We can express these two distance variables in terms of each other, and the propagation constant, as shown in Eq. (6.3).

$$\frac{R_{\text{client-loss}}}{R_{\text{AP-loss}}} = \left(\frac{P_{\text{UE}}}{P_{\text{AP}}}\right)^{-\alpha} \tag{6.3}$$

For example, if we had clients with a peak of 20 dBm, and APs with 30 dBm, in a close, free-space environment where $\alpha$ was around 2, the client-to-client spacing can be around 3 times less than the AP to AP spacing. We will normalize our results to the $R_{\text{AP-loss}}$ unit distance. The comparison is shown in Eq. (6.4).

$$\begin{cases} R_{\text{AP-AP}} = R_{\text{AP-loss}} & \text{for Case 1} \\ R_{\text{AP-AP}} = 2R_{\text{AP-loss}} & \text{for Case 2} \\ R_{\text{AP-AP}} = \left(2 + \left(\dfrac{P_{\text{UE}}}{P_{\text{AP}}}\right)^{-\alpha}\right) R_{\text{AP-loss}} & \text{for Case 3} \end{cases} \tag{6.4}$$

Case 3 is interesting in considering client protection strategies. If the ratio $\left(\dfrac{P_{\text{UE}}}{P_{\text{AP}}}\right)$ is small (clients have much lower EIRP than APs), client-to-client protection is not expensive, and almost reverts to case 2 in terms of spectrum density cost. However, when the EIRP is equivalent, then the spacing is three times that of case 1, and the density is reduced as the square of spacing. The density is less than half of what is achievable in the assumptions of case 2. There are very significant benefits in techniques, such as time alignment that avoids the necessity to protect clients from clients. A two times increase in spectrum effectiveness does not come this easily in other aspects of wireless design!

The protection of devices within secondary tiers has some flexibility. Table 6.2 illustrates the relationship of protection criteria, achievable density, and the methods

**Table 6.2** Density impacts of client protection.

| Case | Density | Impact | Mitigation |
| --- | --- | --- | --- |
| 1 | 9 | Clients are highly likely to be interfered with by adjoining APs. APs are protected. | Some timesharing of the band would be required to create the equivalent of a Media Access Layer (MAC) layer among adjoining devices. |
| 2 | 4 | Operation of clients will interfere with other clients far from the AP. | Time coordination to set agreed to client and AP transmit windows, in common across all of the devices. |
| 3 | 1 | Complete protection of clients and APs, with no constraints on their operation. | None needed. |

needed to ensure client reliability. In the analysis we assume clients and AP have same rights, they are in close proximity (free-space propagation) and the density is compared to case 3 protection.

Aggregate outbound interference is another issue that can introduce non-trivial considerations. The source of interference from the composite of the AP and its clients could be driven from the AP itself, or a closer client device, or it could come from a wide range of user devices, depending on the MAC layer.

- If the MAC layer is Time Division Duplex (TDD), then it is a reasonable assumption that the AP and the clients do not transmit simultaneously, so they do not need to be aggregated. If the power limits are identical, then the closest possible position to the victim is the location of the emitter, for purposes of outbound interference analysis. The AP can be ignored. This is typical of Wi-Fi.
- If the MAC layer is TDD (as above) and there is the possibility that multiple clients can transmit simultaneously, the problem is more complex. For example, LTE allows the AP to direct subcarriers (resources in LTE terminology), and therefore it could have multiple clients transmitting at the maximum power in the same channel.[1] Treatment of this has to be tailored to the operating concept of the network, and other than prescribing worst case, there is no single analytic treatment.
- If the MAC layer is Listen Before Talk (LBT), then it has the idealized behavior that is similar to TDD. If a node successfully senses that the channel is busy, it will not transmit on top of the existing transmission. There is a finite error rate in this detection.
- If the system uses Frequency Division Duplex (FDD), then each link has to be addressed differently. The downlink is the AP itself, and the uplink is presumably the same as the TDD case, with either a single or multiple devices utilizing a channel, per the two cases above.

---

[1] The assumption is that there is a power, rather than Power Spectral Density (PSD), limit. If there is a PSD limit, this is not a concern.

## 6.3 Device Considerations in Three-Tier Spectrum Management Regimes

The focus of most spectrum policy discussions are on the impacts of various policies on spectrum management structures. In three-tier spectrum management, there are significant impacts on the devices themselves. Some new software control functions are mandatory, such as the interaction with the admission control system. Others are desirable in order to provide high reliability operation in statistically available spectrum.

The device must include functionality that would not typically be included in a communications device, at least in a non-Cognitive Radio (CR). These include the following functional additions:

**Interaction** The admission control system imposes a new functionality onto devices. They must be capable of a new set of interactions to negotiate spectrum. Specific interaction requirements may vary by regime, but some sort of near-realtime interaction is required for admission control functionality.

**Sensing** Sensing provides the device (and through the device, the admission control process) with the ability to self-protect by being aware of the local environment. The admission control system is aware of all device positions, but it is not able to know the exact materials between devices, additionally, it does not even know if the nodes that might interfere are active. Local awareness is likely to be superior to global predictions, whether implemented locally or globally.

**Self-optimization** Spectrum decisions have previously been provided to devices. The CR field has researched how devices might make their own decisions, but there are few environments in which these technologies can be deployed. However, in an environment where a device might have multiple bands and channels available from an admission control system, the device must have the ability to match its mission to the spectrum that was available. This implies that the device be much more mission-aware than a "dumb box," so that it can make these decisions in an informed manner.

**Frequency coverage** Frequency coverage provides flexibility to the device to accept a wide range of grants, and ensure that it has the coverage to match the range spectrum the admission control system can provide access to. In the CBRS rules, devices are required to have coverage of the full CBRS band to avoid being unnecessarily restricted in spectrum access. As the amount of spectrum provided through an admission control system increases, device economics may restrict the range of coverage, and will be a strategic factor that is not present today.

**Recourse** The device will be faced with a unique number of situations. The admission control system may cancel a grant, and the device will have to decide if, and where, to ask for another grant. It should sense the environment, decide what channel is best in its local environment, and ask for a grant, rather than just use the frequency that was loaded by some external agent. The access control system may offer noncontiguous spectrum, none at all, or less than asked for. The more flexible the available frequency coverage and modes, the more recourse the device will have to provide high assurance operation in dense and constrained environments.

## 6.4    Database Interaction Capability

A requirement levied on a three-tier spectrum device is some mechanism to communicate with the admission control system in order to initiate operation. Fortunately, many of the applications for wireless involve backhaul to the Internet, so this capability is often inherent in both the device, and its deployment. However, there are other applications, such as public safety, Combat Net Radio (CNR), aviation air-to-air, and others where the use of backhaul Internet for authorization is not possible.

An example of an intermediate case is point-to-point microwave links. Typically, these links are set to a fixed frequency, and connect based on this installation parameter. In three-tier spectrum, this would not be possible. There are workarounds. One such workaround is for the Internet-connected endpoint to request its own channel, and then make a request as a proxy for the endpoint (which presumably has no Internet connectivity available). The first endpoint would have to obtain a compatible set of grants for each endpoint before it could initiate transmission and inform the other end of the link of its grant. If the link operates in TDD, the fact that the first node was present might be sufficient to indicate that it too had a grant for the same channel. If it operates in FDD, the far end would have to receive the grant over the first link, and then use that grant channel for its link transmit. In that way, it would never operate without a grant, and would have a mechanism to provide assured control over its operation. Proxy operation is required for this mode of access.

Another connectivity consideration is the control of a device after it receives a grant. There are any number of conditions that could cause the admission control system to alter or terminate the operation of a node during its grant period, unless it is in the highest tier. A higher-tier user could enter the band, and would be interfered with by the continuing operation of the node using its current assignment, an interference complaint may have been received, and changes may be either diagnostic or remedial.[2]

Higher-tier users are likely to insist that the framework provide for some form of positive control over the operation of lower-tier users that could cause interference to their operations. This could be provided through a number of mechanisms: push messages, pull authorizations, loss of duplex carrier, etc. The specification for this will consist of some specified time to terminate transmission, initiated from some triggering event, such as shown in Table 6.3.

In implementation, the cause of the abandonment requirement is probably not relevant. The design of the interaction between the device and the admission control system will be driven by the most stressing channel abandonment time requirement. Since the device is not aware of why it might be told to abandon the channel, a common mechanism can support all of the abandonment causes. Assuring the response times are imposed is a significant constraint on the device, and the topologies in which it can be deployed.

---

[2] Diagnostic changes would be to test and determine if changing this nodes operation changed the interference condition if there was some question regarding the source. Remedial changes would be for the purposes of resolving interference, when there was high confidence the node was responsible.

**Table 6.3** Example end-to-end channel abandonment time requirements.

| Event | Time Allowed | Abandonment Time Measured From |
|---|---|---|
| Detection of Fixed Location Primary Signal | 120 seconds | From transmit signal imitation. |
| Detection of Mobile Primary Signal | 120 seconds | From time signal is detected above abandonment threshold for 10 percent of any 5-second period. |
| Registration of PA in Same Channel | 30 minutes | Measured from admission control system grant issuance to PA node. |
| Interference Report | 300 seconds | Registered from admission control system Operator direction to terminate/change operation. |

The methods used in the US CBRS design are described in Chapter 13. The US design was driven by a very stringent abandonment time to protect naval radars, and may not be representative of other situations, of other applications.

## 6.5 Frequency Flexibility and Client Control

One key difference of a three-tier device from a conventionally managed one is that a three-tier device has no expectation of any specific frequency of operation. The wider the range of operating frequencies it has available, the more likely it is that it will be able to receive a grant. It is particularly useful if the available frequencies are not associated with correlated activity in higher-level tier spectrum users.

For example, if spectrum were shared with a forestry fire department, it might have a high availability percentage, but it would be completely unavailable at times and places with forest fires. It would have been better to share some frequencies with the forestry fire department and some with the flood control office, as these two demand events would be unlikely to occur simultaneously, and are probably decorrelated.

Implicit in the concept of operating in shared band is the requirement that devices have recourse to multiple options to connect. If the model is that shared spectrum is used as an offload network, then end user devices have recourse to some other method if a given channel is reclaimed. Devices intended to use shared spectrum have an obligation to be more flexible in band and/or service support.

This is not just an idealistic position. Spectrum incumbents have a well founded fear that users of the shared spectrum will become dependent on this access, and create political pressure anytime it is denied, as in the garage door scenario discussed previously [1, 2]. Even worse, there might be a willingness to utilize secondary spectrum for safety-of-life applications that create even more pressure on incumbents not to exercise their full rights. This issue was raised often in the discussion in the United States of America (USA) regarding the CBRS band.

The solution is that devices that utilize shared spectrum have an implicit requirement to support more connectivity options, implying more channels and bands, at the very least. Cellular offload is a very practical use of shared spectrum, as if spectrum is unavailable locally, the other cellular channels automatically are available and utilized, without any user involvement, or even awareness. Non-cellular applications need to have some equivalent methodology to assure that the devices can provide at least an adequate level of service in the absence of an available shared spectrum grant on any single band.

Band coverage is not enough. In dedicated spectrum, there are well understood limits on incumbent usages of adjoining bands and channels.[3] To use a US term, the shared spectrum might be the "*Wild, Wild West,*" not one of the regimented environments. Receiver analog performance, such as the linearity of the front end, filter width, tunablility, and selectability are much more important for shared spectrum devices than those intended for traditional, well-regulated spectrum.

## 6.6     Access Point Access Constraints

Previously in this chapter, there was a discussion of treating client devices of an access point as a *"cloud"* of devices, with no specific location, or even density. This is based on an assumption that client devices can find the AP channel and mimic it, and be compliant with the grant issued to the AP. Conditions for nodes to operate subservient to AP and under grant issues by the AP are:

- Make no emissions until able to validate that their controlling AP is emitting.
- Use the same frequency and less than the power limits permitted to the AP. If the AP receives a grant that is less than the power assumed by clients, it cannot utilize it, as the clients would violate the grant when they emitted.
- Have characteristics, such as EIRP and receiver protection, that are consistent with the grant requested by, and issued to, the AP. Typically, this would mean TDD operation.
- Immediately terminate operations if the AP terminates transmission.
- The reduction in maximum density caused by the lack of knowledge of the actual client node placement is acceptable to the network application. The alternative is 100% emitter registration.

There is a significant advantage to operating in the AP mode. The client need have no awareness that it is operating in the three-tier framework. Since it is likely that regulations differ in different countries, regions, and bands, having clients that do not implement each of these variants is highly effective. For example, in the US CBRS band, it is possible to use unmodified LTE band 42 and 43 User Equipment (UE) devices.[4]

---

[3] The emphasis here is on current issues, such as raised by NEXTEL Personal Communications Service (PCS) deployment and the proposed LightSquared LTE deployment. These demonstrate that there is inadequate design awareness and consideration of what might be in the adjoining spectrum segments in the future.

[4] With the possible exception of management by the AP to assure that the Out of Band Emissions (OOBE) are maintained below the regulatory limits.

The cost and complexity of implementing the unique aspects of the three-tier regime is therefore limited to the presumably less numerous AP devices. The speed of deployment of the US CBRS band will be greatly increased by the fact that this new band straddles Third-Generation Partnership (3GPP) band 42 and 43, and existing UE products could be planned for the band.

## 6.7 Point-to-Point and Multipoint Model

When the conditions described in the previous section (6.6) cannot be met, each network endpoint must register with the admission control system and receive grants individually. This mode has some advantages to the network:

- Each node can have characteristics that are unique from each other, and select unique grant conditions.
- Full registration can support more complex networking modes, such as FDD.
- Point-to-multipoint and mesh networks that have overlapping coverage can be supported, whereas overlapping mesh networks cannot be supported in access point mode, if deconfliction is required.
- Operation of each node is independent. If one node has its grant terminated, the other nodes are not impacted.
- The admission control system can be more exact in calculating protection implications, so a higher density can be achieved.

  The disadvantages that arise include:

- The network operator must assure that the two endpoints receive the same frequency assignment, even before they can perform two-way communication. This assurance can be provided by the admission control system or the device registration message contents.
- Twice as many admission control system transactions are involved.
- Endpoints must be capable of the admission control system interaction directly or through a proxy prior to entry into the band.[5]

## 6.8 Suggested Reading

- There are an incredible number of subtle issues that arise when integrating the non-symmetric usage, such as AP-client networked, into a three-tier structure that is intended to be technology neutral, and thus open to innovation. Individual technologies, such as Time Division Long Term Evolution (TD-LTE) have developed a large amount of standard material on sharing of spectrum across AP devices, yet this cannot be directly applied to three-tier without violating core principles. Methods of voluntary compliance with industry regimes appear to be a logical extension

---

[5] The FCC has language that requires authorization before beginning "service transmissions." The implication is that it may be permissible to make the registration and grant transaction over the air.

of the concept of an admission control system. Readings on LTE, and particularly TD-LTE provide considerable background on how such coexistence features could be implemented. Suggested readings are numerous, including many recent works [3, 4].

- The contradictory literature that has developed in the ongoing discussion over the proposals for Long-Term Evolution Unlicensed (LTE-U) and Licensed Assisted Access (LAA) highlight the issues involved in incompatible MAC layer concepts in a shared band. Whatever ones opinion is on the topic, the positions presented by all sides of this discussion are instructive in providing the maximal use of band with varying network concepts of operation [5, 6].

- Two previous books by the author addressed the issues of filter and Low Noise Amplifier (LNA) performance in the context of dynamic spectrum access systems [7, 8].

## References


1 Federal Communications Commission, *Consumers May Experience Interference to Their Garage Door Opener Controls Near Military Bases.* Public Notice, 20 F.C.C.R. 3614 (2005).

2 L. Luna, NEXTEL interference debate rages on. *Mobile Radio Technology,* **21**/8 (2003), 26.

3 A. Ghosh, J. Zhang, J. G. Andrews, and R. Muhamed, *Fundamentals of LTE* (NJ: Prentice Hall, 2011).

4 A. Ghosh and R. Ratasuk, *Essentials of LTE and LTE-A* (Cambridge University Press, 2011).

5 N. Rupasinghe and İ Güvenç, Licensed-assisted access for WiFi-LTE coexistence in the unlicensed spectrum. *IEEE Globecom Workshops* (2014), 894–899.

6 C. Chen, R. Ratasuk, and A. Ghosh, Downlink performance analysis of LTE and WiFi coexistence in unlicensed bands with a simple listen-before-talk scheme. *IEEE 81st Vehicular Technology Conference* (2015), 1–5.

7 P. F. Marshall, *Quantitative Analysis of Cognitive Radio and Network Performance* (Norwood, MA: Artech House, 2010).

8 ——, *Scaling, Density, and Decision-Making in Cognitive Wireless Networks* (Cambridge University Press, 2012).

# 7 Taxonomy of Protection Methods to Be Provided Across and Within Tiers

## 7.1 Introduction to Three-Tier Analytic Components

This chapter focuses on the specific algorithmic components that may be included in protection modes that three-tier regimes may implement. It provides a set of basic methods that will later be integrated to address different types of incumbent or tiered user protection, and to provide coexistence between peer devices within a tier.

The topics of this chapter are dealt with in great detail by a number of works, and a very detailed description of any of these topics is beyond the scope of a single chapter. This chapter will instead focus on the unique aspects of the use of these approaches in three-tier spectrum management. The reader is referred to the section on further readings for more detailed background and more extensive and detailed science on the topics.

The use of an automated admission control system opens up opportunities for much more sophisticated methods to simultaneously improve interference avoidance and density of use. The decision system discussion is addressed separately from the protection discussion in order to isolate the purely technical aspects of the control system from the policy-driven protection criteria discussion. This chapter will focus on developing a basic set of analytic tools that are applicable to all spectrum management regimes, independent of their structure.

## 7.2 Challenges in Ecosystem Interference Protection

Our traditional approaches to interference protection have generally focused on extensive and detailed analysis of the worst case conditions between a small number of possibly interfering nodes, and a small number of victim nodes. The results of this analysis are then generalized to more complex deployments of both interfering and victim nodes. In a three-tier regime, it is desired to change this analytic process in many ways:

**Ecosystem** The interaction of all of the nodes in the ecosystem must be considered. The interactions of all nodes upon all other nodes must be considered, rather than extend one simple case.

**Automated Analysis** The analysis performed in the conventional process may be based on science and physics, but the selection of cases and interpretation of results is a human process, and less scientific. For a three-tier system to operate effectively and flexibly, these decisions must be made automatically, and not be challengeable individually.

**Scale** The solution to not selecting single cases, and then generalizing them, is to consider all of the ecosystem simultaneously. This results in a computational problem of significant scale. For example, in the US Citizens Broadband Radio Service (CBRS) deployment, if we assume 10 million devices, then the number of possible interactions is $10^{14}$,[1] and assuming an entry rate of 1 node per 10 seconds, the admission control process must consider $10^{13}$ interactions (of varying complexity, as we will see) every second.

This chapter will discuss the analytic challenges in constructing a three-tier admission control system, but will not address the computational ones, such as algorithm selection, processing architecture or other internal features of such a system.

## 7.3    Propagation Modeling

The science of a three-tier spectrum regime is based on propagation estimates of the interfering paths. Most spectrum management approaches, including three-tier, focus on prediction, and avoidance, of possible interference, rather than detecting and mitigating the conditions. Spectrum management use of propagation tools is thus conservative by nature, since the incentive structure provides regret from being overly cautious, but significant regret from being overly aggressive.

Propagation loss has been studied extensively, and a large number of models have been developed of varying complexities, targeted environments, and frequency applicability, among other differences. These have been applied for interference analysis. However, it should be noted that these models were typically developed to perform communications link analysis.

There are fundamental differences between the modeling characteristics used for link closure and interference. In communications link analysis, the propagation estimate is intended to provide the maximum path loss that could occur to a given confidence probability. This is the maximum loss that could be anticipated. In contrast, determining if one node will interfere with another requires estimating the minimum loss that might occur between these nodes. These are not symmetric considerations.

Figure 7.1 illustrates the differences in communications and interference estimation in terms of the propagation loss curve. Communications analysis focuses on cumulative probabilities approaching unity, while interference avoidance focuses on extremely low probability regions of the curve. Whereas the possible loss is essentially infinite, the minimum loss is bound by the free-space path loss.[2] Assuming that the losses are

---

[1]   Some of these node pairs may be too distant from each other to require extensive analysis, but must be at least explicitly excluded from analysis.

[2]   This discussion does ignore any ducting of the high frequencies, which can occasionally result in path losses that are less than free space.

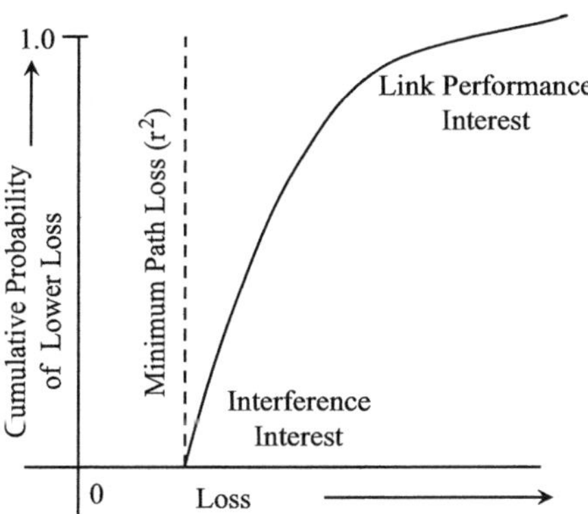


**Figure 7.1** The propagation loss curve.

distributed in any statistical manner around the mean loss is not effective, and, typically, underestimates the actual path loss significantly.

To build a classification system for propagation models, we can consider three fundamental levels of analysis

**Free Space** Most conservative approach, applicable to direct paths with no blockage or reflective surfaces along the path. Applicable to most ground to space, and elevated links.

**Terrain** Terrain models reflect the impact of terrain obstructions, curved earth, diffracted paths, and often troposcatter. The Longley-Rice model is one of the principle models of this category. This is the model most commonly used in spectrum management.

**Site Specific** Site specific models reflect the local scattering environment around the nodes in the path. The example in the next section would be an example of how site-specific data could be developed. The ITU-R P452 model allows for extension through site-specific scattering augmentation of the terrain-based analysis.

**Path Specific** Path specific models estimate the loss due to all of the items along the path, such as trees, buildings and structures, and any other sources of blockage.

**Ray Tracing** Ray tracing attempts to analyze all of the reflective and diffractive elements in the vicinity of the interference path. Ray tracing is particularly useful for estimating the time delays of the reflective paths, which is generally not an issue in interference analysis. For accurate estimates of path loss, ray tracing requires accurate models of objects at the level of resolution in the range of a wavelength, so large scale application in spectrum management has significant practical limitations.

The lack of realism of terrain-based modeling in urban environments is best exemplified by a metaphor proposed by Dr. Andrew Clegg (Alphabet)[3] who points out that the terrain-based models view Manhattan Island as unchanged from the 1400s

---

[3] Google Inc. at the time he made this suggestion.

**As the Longley-Rice Model Considers Manhattan Island**

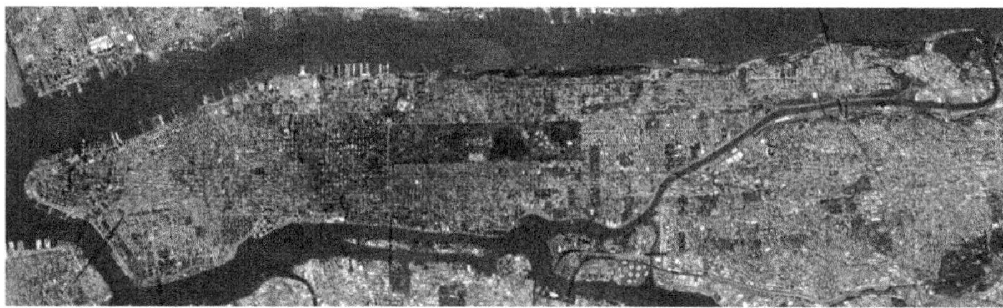

From Library of Congress

**As Manhattan Island Really is Today**

Courtesy of Google Inc.

**Figure 7.2** Terrain-based propagation model, and actual RF propagation view of Manhattan Island.

to today. Figure 7.2 illustrates the lack of realism in viewing terrain as the dominant factor in propagation. Propagation at the higher frequencies in use, in most settings, for wireless is dominated by the scatter loss along the path, not the terrain or even the length of the path!

It is clear that the propagation models available to us today are based on data points that were collected decades ago. The models were developed with relatively few data points and limited computation resources. They were typically developed by collecting data at much lower frequencies than are used and contemplated today. Yet these models still dictate spectrum availability, even though their results do not reflect the environment of many uses.

## 7.3.1    The Importance of Scatter Loss Modeling

An example of the impact of clutter or scattering on real propagation losses in the United States (US) CBRS band was provided in an ex-parte filing [1] by Google Inc. to the US Federal Communications Commission (FCC). Figure 7.3 illustrates the path loss measured over at least 1 000 000 unique paths in the Mountain View, CA area. Mountain View is a typically low-rise commercial area in a suburban setting. It is not

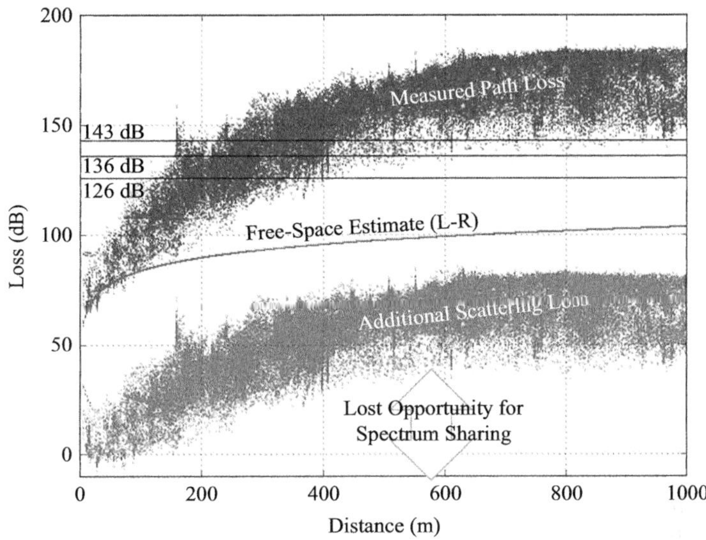


**Figure 7.3** Scatter loss measurements in Mountain View, CA (from Google ex parte filing on February 16, 2016 [1]).

dense or vertically developed as in a city center. It is an intermediate case in analyzing propagation, perhaps a bit more loss than a suburb, but much less than a dense city.

The upper distribution curve is the measured path loss over the distances on the horizontal axis. The intermediate line is the free-space ($r^2$) loss, which at distances less than one kilometer are similar to those of Longley-Rice in flat terrain. The lower distribution is the path loss that exceeds the Longley-Rice estimate.

For example, in Figure 7.3, the path loss at 400 meters does not have a single collection point with less than 30 dB of additional loss above the estimates provided by the Longley-Rice model, and typically is 50 dB more loss than the $r^2$ computation. In terms of range, use of this actual data would decrease required node spacing by a factor of 32.[4] In terms of density, this is approximately 1 000 times more devices! The region between the estimated path loss and the lowest value of measured path loss constitutes lost opportunities for spectrum sharing due simply to unrealistically conservative, and demonstrably incorrect, models.

There is no excuse for spectrum regulations to specify generalized, and high level terrain-based models in the day of high resolution models of most urban environments and accessible geographic databases, such as Google Earth. This is a legacy of developing specific rules for interference protection that had to be applicable in any environment. Therefore, any tool whose results were specific to an environment, such as clutter loss, could not be applied. However, when moving to an admission control system, such as a Spectrum Access System (SAS), every decision is situational, and

---

[4] In free-space propagation ($r^2$), each range doubling or halving accounts for 6 dB; −30 dB corresponds to a $2^5$ decrease in range.

therefore there is no reason not to include the situation specific aspects of the decision. We no longer need rely on "one size fits all" geographic protections.

## 7.4    Three-Dimensional Propagation

Traditional communications networks are two dimensional; they communicate over the land, and the bulk of their path is horizontal. In deployments that are enterprise, residential and indoor focused, many of their interference paths will be vertical, rather than horizontal. An admission control system that has almost identical *lat, long* pairs (and thus a distance of only meters between them) may in fact be hundreds of meters apart vertically, and have thirty 15 cm concrete floors between the nodes, as well.

Conventional models might show almost no path loss between these locations, but they will have so much loss between them that they may as well be on different planets. Vertical path loss is typically not driven by distance, but by the material loss along the paths. Two residential floors may have slight loss through wood-based floors, while the same distance in commercial spaces would have 50 dB of loss per intervening floor, due to the mineral-based material used.[5]

## 7.5    Density Modeling

One benefit of the use of an admission control is a much more precise determination of device density. Traditionally, if a given amount of interference power is acceptable at a given point, then the power permitted if one interfering node was present must be reduced by a factor reflecting the worse case density. In most cases, this density will not be present, and the use of the band will be reduced by this unnecessary level. With admission control, we can almost completely discard the necessity to estimate density. For nodes that are admitted as point-to-point, where every node is registered with the SAS, there is absolutely no need to estimate density.

The question of density estimation arises with the admission of nodes as Access Point (AP) networks. As discussed, the AP mode includes an assumed client area of operation and operating mode. Assuming that all of the nodes operating under an AP registration are on a common range of frequencies, it is reasonable to assume that the network is operating in Time Division Duplex (TDD) mode. Presumably the network has some mechanism to avoid collisions in time,[6] therefore only one node would be transmitting at a time. The issue of density therefore does not arise.[7]

---

[5] The representative losses at 3.5 GHz are discussed in detail in Chapter 12 and millimeter wave losses in Chapter 12 and Chapter 17.

[6] An example of this might be Media Access Layer (MAC) layer techniques such as fixed time slots, or use of Request to Send (RTS)/Clear to Send (CTS).

[7] The complication of sub-channel allocations, such as Long-Term Evolution (LTE) resource blocks is dealt with in later chapters.

## 7.6    Modeling Interference Aggregation

One of the unique aspects of the admission control system is that it can control admission to a channel based on the estimated aggregation of interference sources in order to assure a maximum interference level in protected receivers. Instead of "guessing" at the necessary restrictions on transmission rights in an attempt to limit interference, the admission control system estimates the impact of each individual emitter in the ecosystem and aggregates the total emissions of each of the emitters. This provides assurances to protected users that total interference will not exceed a specified threshold.

When we consider propagation, there is a probabilistic element. No analysis can estimate propagation exactly. Instead, they bound the possible range of path loss estimates to a given confidence. In a one-on-one analysis, assured protection requires that the path loss be determined with a very low probability that the actual interference path loss could be less than the estimate.

The logical way to perform aggregation would be to sum the interference power values that were determined in the one-on-one analysis. However, this runs the risk of being overly conservative. If we assume that there are $n$ independent and equal contributors to the aggregate emissions, then the probability that the aggregate emission level will be greater than the estimate will have a distribution $\delta$ of $\delta^n$.

We know from probability theory that if the power distribution $i$ were equal, normal, independent, and characterized by $N(\mu_x, \sigma^2)$, then the aggregate of $n$ equal distributions would be $N(n\mu_x, (\sqrt{n}\sigma)^2)$. Of course, none of these conditions are true, but even this simplistic case shows that addition of low probability points on the distribution is not a suitable methodology.

On the other end of the complexity scale, one could model the discrete distribution of each emitter path to the victim receiver. There is quite an extensive literature that addresses the distribution of path loss around the mean values form propagation models. A Normal (Gaussian) distribution is computationally convenient, but has probability tails at both end of its distribution. Barring the existence of a ducting or highly reinforcing multi-path condition (which is rare), the minimum path loss is the $r^2$ free-space loss. Using a Beta distribution above this minimum loss assures there is zero probability below this free-space loss level, and allows for the possibility of almost unlimited loss due to high clutter loss paths. This high-loss end of the distribution is of interest to the communications community, but not relevant to interference computation.

We will assume the loss is distributed between the losses associated with a propagation $\alpha$ that is in the range $2.0 \leq \alpha \leq 4.5$, for purposes of this example. The distribution parameters used are $\beta(\alpha, \beta)$. Using this model, Figure 7.4 illustrates the density distribution of a single path of 500 meters. The minimum path loss is 153.8 dB, the maximum loss is 220.3 dB, and the 3% bound (no more than 3 % chance of less path loss) is 159.8 dB.

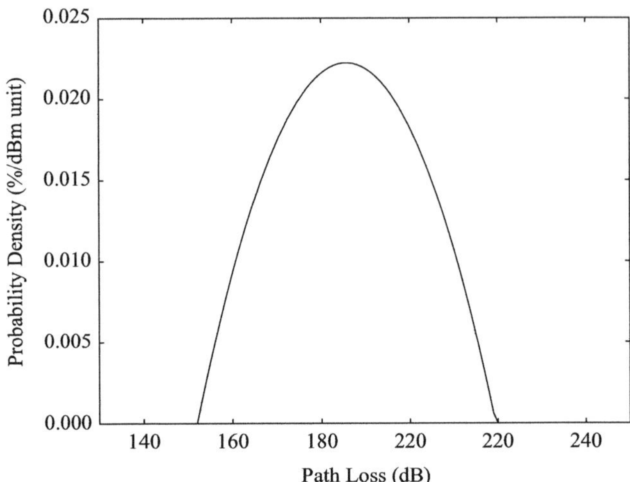

**Figure 7.4** Path loss probability density function distribution.

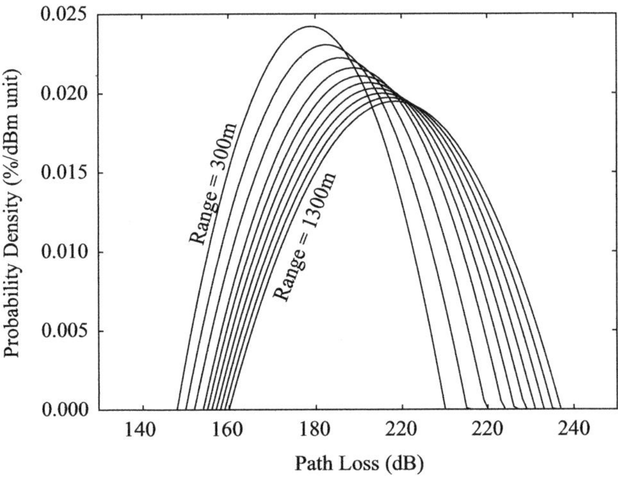


**Figure 7.5** Path loss distribution and corresponding power distribution.

If we consider a set of ten identical nodes uniformly distrusted at a range from 300 to 1 300 meters from a potential victim, we obtain a set of individual interference path loss (and thus interference power) distributions, as shown in Figure 7.5.

The power distribution can be determined from these path losses through aggregation. They create the probability distribution of the interference power that could be expected at the victim receiver. Figure 7.6 illustrates the Probability Density Function (PDF) of the aggregate interference. Figure 7.7 provides the corresponding Cumulative Distribution Function (CDF).

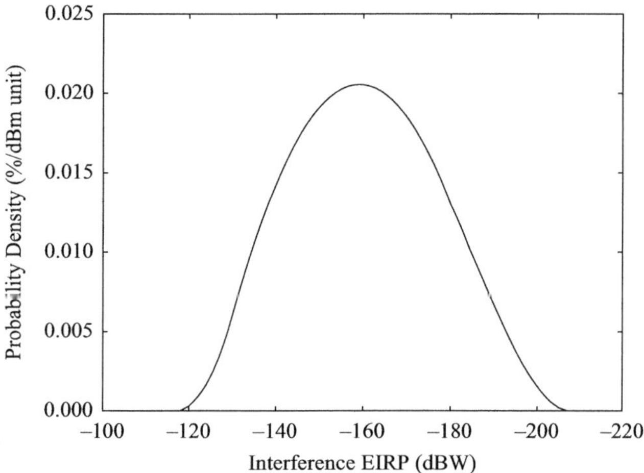

**Figure 7.6**  Interference power probability distribution of a distributed set of nodes at a victim.

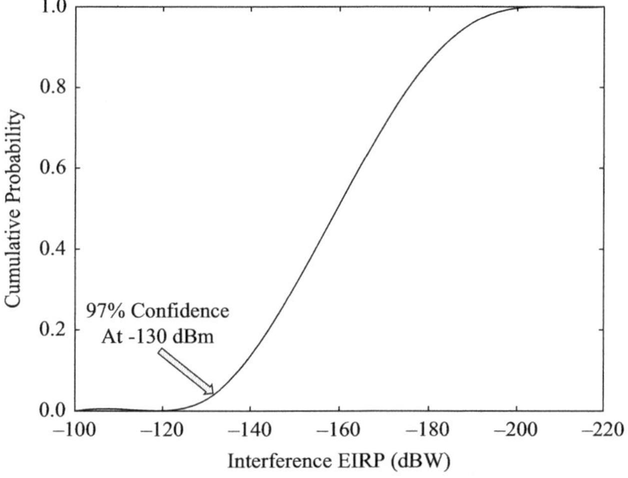


**Figure 7.7**  Cumulative distribution of the interference power distribution shown in Figure 7.6.

When we compare the potential interference power at a given confidence to the same confidence level across the aggregated distribution, the difference is approximately 10 dB, as shown below:

Sum of the 97% confidence values:                              −120 dBm

Enumerative determination of the 97% confidence:    −130 dBm

A 10 dB reduction in aggregated interference can provide significantly increased device density, and is a significant reduction in overly conservative interference practices. The increase in density is linear with the reduction in the interference estimate.

Therefore, ten times more devices could be admitted into the same region, using statistical aggregation rather than the standalone, worst case estimates.

The assumption of independence is particularly important. We can consider that propagation has a correlation distance, which is the distance out to where the land usage impact on propagation is independent. For example, if one path into a victim is over farm fields, it is likely that most paths will be over farmland. Similarly, if one path is urban, then in a capacity network, it is likely that the other paths are also urban, as well.

This enumerative approach is a computationally intense (although not complex) approach to aggregation. The use of closed form distributions, on the other hand, is not computationally intense, but makes assumptions about independence, distribution shape and other factors that are clearly not realistic.

However complex, the benefits of aggregation are significant in terms of opening up opportunities for sharing and for relaxing limitations on devices that share spectrum. The allowable increase in power is the difference between actual density to the worst-case, which was the basis for regulation typically more than an order of magnitude.

## 7.7    The Protection of Receiver Linearity

When the subject of interference is raised, the first reaction is to assume it means interference by two systems occupying the same frequency. If so, then a simple spreadsheet could just cross off frequencies as they are used. However, as the spectrum is used more densely, the interference is often not because of common frequency use, but from the impact of stronger than anticipated signals being processed by the front end Low Noise Amplifier (LNA).

The gain of an amplifier stage is given by Eq. (7.1) [2].[8]

$$V_{out} = G_{LNA} v_{in} + \frac{-2G_{LNA}^2}{IIP2_{volts}} v_{in}^2 + \frac{-4G_{LNA}^2}{IIP3_{volts}} v_{in}^3 \tag{7.1}$$

where $v_{out}$ is the output of the amplifier stage, $G_{LNA}$ is the gain of the amplifier stage in volts, $IIP2_{volts}$ is the Second-Order Input Intercept Point, $IIP3_{volts}$ is Third-Order Input Intercept Point, and $v_{in}$ is the input voltage to the amplifier.

The non-linear nature of this non-linear distortion response is obvious when examining the exponents of the voltage terms in the higher order intermodulation terms.

Figure 7.8 illustrates the effect of intermodulation distortion. This input is a set of signals with multiple frequency components, and with a total power equal to the Third-Order Input Intercept Point (IIP3) power. These effects are not subtle, and have had the effect of causing complete failures of otherwise effective communications systems. One example of this was in the United States of America (USA), where the effect of adjacent channel NEXTEL cellular systems caused localized failure of public safety radio systems, when they were in the proximity of these cellular systems [3].

---

[8] The first two distortion orders are shown here.

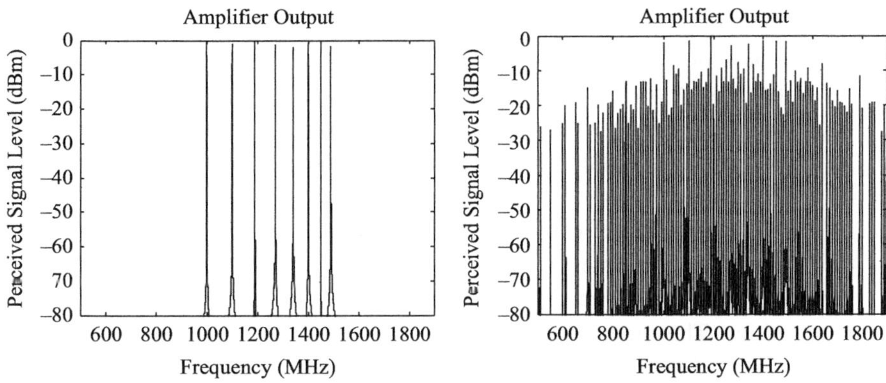


**Figure 7.8** Overloaded amplifier input and output frequency domain illustration (from [6]).

More recently a controversy arose over the use of spectrum adjacent to the Global Positioning System (GPS) spectrum, due to the effect on receiver operation [4, 5].

Intermodulation in the frequency domain is the mixing of signals, which creates the sum and difference of the signals. The second-order response is generally not a factor in receivers, since the sum of the input frequencies is twice the frequency of the signals. If $f_1$ and $f_2$ are the input frequencies, the intermodulation products are at frequencies $f_1 \pm f_2$. Assuming receive filters that are less than an octave, second-order distortion is not often an issue. However, the third order response of $f_1$ and $f_2$ are the same inputs mixed twice, or a series of responses including $2f_1 \pm f_2$ and $2f_2 \pm f_1$. If $f_1$ and $f_2$ are close in frequency (as when they would pass a bandpass filter), these intermodulation products appear on, or close to, the original input frequencies, and thus are a very significant cause of receiver failure.

The order of the non-linear effect is also significant. The third-order products increase at the third power of the input energy. A 10 dB increase in input energy creates a 30 dB increase in intermodulation products. So, if the receiver is close to compromise, even a small increase in input energy can make it completely fail. Accuracy in managing this interference is thus more of a challenge to spectrum management of blocking power than the traditional management of co-channel power.

Another consideration in non-linear response is the shift from narrowband to broadband waveforms. Two mixing narrowband waveforms create a relatively narrowband product. Two 6.5 kHz narrowband audio signals create a product that is 19.5 kHz. It is likely that this product may not impinge on the channel being used. On the other hand, two 20 MHz channels mixing create a 60 MHz wide mixing product; one much more likely to fall onto a channel in use.

## 7.8  Methods for Specifying Receiver Protection Thresholds

Typically, spectrum regulation is performed by establishing limits on the rights of transmitters, such as power, out of band emissions, etc. This is the only practical manner in which to establish rules that are common to all applications and locations. The intent was to protect receivers from interference, but the approach was indirect. However, with admission control, it is now possible to directly implement the management structure by starting with what the original intent was; to assure a given set of protections that receivers can assume to be present, regardless of how transmitters are used, deployed, or designed.

Interference threshold establishment is the basis for quantifying the protection methods. To some extent, it is inherently adversarial, as incumbents desire maximal protection from interference, and secondary users desire maximum spectrum opportunities. The methods differ in the degree that they require the knowledge of receiver installation characteristics.

In this section we will address three questions:

- What are the types of protections that should be provided? Typically, protection of the receiver could include the co-channel noise, and many variants of adjacent channel and band protection.
- Where at the receiving location, or within the receiving system, is the criteria to be met?
- How are the protections to be specified?

### 7.8.1    Types of Receiver Protections

We will investigate methods to specify two broad classes of receiver protection:

**Co-channel** Co-channel interference is other signals or unintended emissions raising the noise floor of the receiver in the range of frequencies that the receiver is operating on.

**Adjacent Channel/Band** Adjacent channel or band protection limits the amount of power in frequencies that are adjacent to the intended receiver operating range, or channel.

The criteria for co-channel interference is the impact on the ability of the demodulator to recover the transmitted information as if the interference was not present. All receivers have a natural noise floor which serves as a reference for additional interference noise. Typical terrestrial systems point at the horizon, and their antenna noise temperatures are generally driven by the equivalent black body radiation from the ground and objects in their field of view; a temperature that is typically in the range of room temperature, or about 290 K. The spectral noise density ($N_0$) that is created by any body is given by the thermal noise equation, shown in Eq. (7.2), where $T$ is the temperature in Kelvin, $B$ is the bandwidth in Hz, and $k$ is the Boltzmann constant, approximately $1.38 x 10^{-23} \frac{W}{Hz} K$. Eq. (7.2) computes the thermal noise level.

$$N_0 = kTB \tag{7.2}$$

The permitted elevation of noise can be expressed as a ratio of this natural noise level, or Interference-to-Noise ratio ($I/N$), which equates to a power spectral density. The computation of this density ($P_{\text{int}}$) for a given $I/N$ ratio is provided by Eq. (7.3).

$$P_{\text{int}} = \left(\frac{I}{N}\right) kTB \qquad (7.3)$$

Protection of receivers from adjacent band power is specified as the maximum power in a range of adjacent frequencies. Determining a specific value is difficult, as the signal characteristics have a significant impact on how the intermodulation noise is distributed across the receiver's spectrum.

As an example, in the US CBRS proceeding a technical analysis provided by Alion Science and Technology [7] proposed a blocking, large-signal limit of –80 dBm. This equated to an intermodulation noise floor elevation of 0.1 I/N for an assumed IIP3 level. This is a reasonable value, and will be used in the satellite receiver protection examples in this book. Assuming a satellite system noise temperature of 78 K, the thermal noise equivalent of a 0.1 $I/N$ ratio is a noise temperature elevation of 7.8 K.

### 7.8.2    Location of Protection Application

The location at which the receiver protection criteria must be met is itself a significant question.

**Flux in Space** Flux in space is the most natural expression of protection because it is independent of the receiver system design, and can be applied.

**LNA Input** The actual impact of noise is at the first active stage of the receiver. Specifying this location accounts for the impact of antenna pointing, front end filtering, as well as the flux in space values.

Flux in space is the basic building block for each of these options. Flux in space is simply the amount of power flowing through a given area of the sphere surrounding the transmitter, at the receiver location. Typically, it is specified in relation to an omnidirectional (zero gain) antenna across a range of frequencies.[9] This value is readily determined using transmitter characteristics and a path loss model.

The problem with the flux in space criterion is that it fails to recognize the impact of the receiving system on the occurrence of interference. An antenna may have 30 dB of gain in one narrow direction, and –10 dB in all others. Establishing a fixed flux in space criteria inherently reflects the worst-case situation, with the antenna pointed at the secondary transmitter. This would reduce the allowable power by 10 000 times, in the cases where the antenna was pointed away from the potentially interfering node, or the protected node antenna was pointed away from the interferer, which is the most likely case. Using knowledge of the antenna pointing and antenna gain curves, the admission control system can compute the energy in the LNA, and assure it is protected.

---

[9] As an example, the US CBRS regulations define the protection criteria for a Priority Access License (PAL) as –90 dBm per 10 MHz channel.

Another, and more complex, rationale for this criteria is when there is necessity to protect the receiver from overload or blocking, and the receiver has filtering in some of the band of concern. In this case, the admission control system needs to know the frequency domain characteristics of the filter, and can convolve these with the frequency domain emission, determine the total energy that will pass through the filter, and assure the performance of the total system of transmit and receive antenna, Radio Frequency (RF) filtering, and feed performance is all reflected in the protection regimes decisions to both maximize sharing and assure protection.

RF component specifications (or at least the mechanism to establish their provenance) must be included in the interference analysis methodology provided by the regulator. The US CBRS regulations specify:

*"The aggregate passband RF power spectral density at the output of a reference RF filter … calculate antenna gain using §25.209(a)(1) and §25.209(a)(4) and a reference RF filter between the feed-horn and LNA/LNB[10] with 0.5 dB insertion loss in the passband."*

This regulation requires the admission control system to calculate the combined response of the antenna (in spatial gain, and frequency), the filter response and loss, applying the industry standard product, characteristics, to determine the predicted Power Spectral Density (PSD) provided to the LNA/LNB first transistor. Rather than regulate cause (power), the regulation defines the results (received interference and blocking levels).

Traditionally, spectrum regulations have mandated some level of protection that was constant across a large range of spectrum. In reality, the receiver response to signals is quite different when the signals are adjacent to the received frequency or are further away. By using the actual response of the entire RF system, it is possible to both best protect the receiver, and to maximize the use of the spectrum.

The processing flow of this model is shown in Figure 7.9. This shows a frequency domain representation of the requested emissions, including all of its emissions, even unattended ones as the start of a process. This is then processed through the antenna response curve, which includes the actual gain in the specific direction of the emitter, and the frequency response of the antenna, and the response curve of the receive filter.

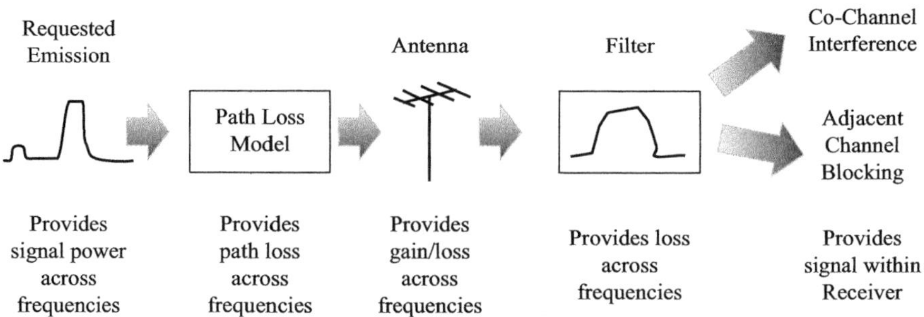


**Figure 7.9** End-to-end processing of a complete receiver model.

[10] A Low Noise Block Converter (LNB) is a device which combines an LNA stage and a down converter stage to an Intermediate Frequency (IF), such as the 900 MHz typically used in satellite systems.

At the completion of this process, the admission control system is able to predict the maximum power across all of the range of frequencies that were modeled. This then yields the in-channel power, which elevates the noise floor, and the total energy in adjacent frequencies, the sum of which might drive the receiver LNA into non-linear operation, and therefore also raise the perceived channel noise floor.

These approaches will be quantitatively applied in Chapter 9, when we examine the satellite receiver protection process in detail. This chapter will examine its impact on acceptable densities in a fixed transmit criteria and a receiver LNA admission control policy.

## 7.9    Summary

The adoption of three-tier spectrum management does not simplify the nature of interference protection. In fact, it makes it somewhat more complex, as the regulator is now shifting from expressing constraints on the use of the spectrum, to providing limits on the impacts of this use. This approach has been embedded in spectrum management analysis since its inception, but has not been the expression of the resulting policy, as there was no means to operationalize these limits, as is now possible with an admission control system.

The situational nature of admission control system processing provides a number of new methods to express regulatory requirements for protection within the spectrum tiers. These are new options for regulation and coexistence that can be invoked in a three-tier admission controlled regime. They include:

- Interference protection can now be based on path loss estimates that are specific to each individual propagation path considered by the admission control system, rather than generic, and worst-case paths, as in past practice.
- Density can be managed explicitly through admission control, permitting control over aggregation of interference.
- Probability of interference can be managed directly through understanding of the path loss probability distribution of each individual contributor to interference at protected nodes.
- The same techniques used to provide co-channel protection can also address non-linear receiver response, and the resulting issues with receiver blocking and reduction in sensitivity.
- Interference impacts can be expressed at any point in the receiver signal processing chain, enabling the impact of filters, antenna directionality, and front end response to be incorporated into a protection scheme.

## 7.10    Suggested Reading

There is an extensive body of scientific work that has been done to evolve the physics principles of spectrum management. A deep understanding of these principles is not essential to understanding and implementing the ecosystem of three-tier spectrum, but is useful to understand the weakness of the current regime.

- The starting point in any spectrum occupancy and interference analysis technology is an effective propagation estimation tool.

  - The Longley-Rice Model is used extensively in regulatory activities [8].
  - Several of the communities within the International Telecommunication Union (ITU) utilize the ITU P452 Propagation Model [9].
  - There are many closed-form models proposed, such as the Hatta, and Extended Hatta model for use in wireless planning. These models tend to be tailored to the deployment of macro-scale networks.

- The United Kingdom (UK) Office of Communications (OFCOM) Spectrum Usage Rights (SUR) concept was described previously. This developed a formal description of license rights in their proposal for explicitly including rights statements in licenses [10]. This is one of the first regulator proposals for a standard expression of rights, and particularly the adjacent signal environment which has created so much controversy in the past.
- The clutter loss example was derived from an ex parte filing by Google in the CBRS proceeding. A complete description of the experimental design and processing is provided in the Google filing [1], available from the FCC website.
- The details of the operation of receiver components are certainly an important consideration in the development of spectrum management regimes. There is an extensive set of books on circuits and design. The design material in this section is largely covered in Ulrik Rhode's circuit book description [2].
- An illustrative interference analysis for multi-tier spectrum interference to Fixed Satellite Service (FSS) sites is provided in a report by Alion Science and Technology [7] which was submitted to the FCC in support of a filing by a number of content companies essentially opposing the CBRS rulemaking. The report was submitted to argue there were almost insurmountable issues with strong-signal overload in the adjacent band. It also demonstrates that a high pass filter would be effective in protecting the FSS receivers. This was the solution the FCC eventually chose in the rules.[11]

## References

1 Google Inc., *Ex Parte Filing: Amendment to the Commission's rules with Regard to Commercial Operations in the 3550–3650 MHz Band, GN Docket No. 12–354* (2016). http://fcc.gov/ecfs/document/view?id=60001462642.

[11] The FCC did not mandate [11] that FSS operators install the filters, but it did define the protection provided to them was to assume the existence of a reference filter, Therefore the FSS operators would be at risk of interference if they did not install the filters, and had no recourse against the CBRS operation.

2 U. L. Rhode and D. P. Newkirk, *RF/Microwave Circuit Design for Wireless Applications* (New York: John Wiley & Sons, 2000).

3 L. Luna, NEXTEL interference debate rages on. *Mobile Radio Technology*, **21**/8 (2003), 26.

4 Federal Communications Commission, *Comment Deadlines Established Regarding the LightSquared Technical Working Group Report.* Technical report, DA 11–1133 (2011). http://apps.fcc.gov/edocs_public/attachmatch/DA-11-1133A1.txt.

5 National Space-Based Positioning, Navigation, and Timing Systems Engineering Forum (NPEF), *Assessment of LightSquared Terrestrial Broadband System Effects on GPS Receivers and GPS-dependent Applications* (2011). www.gps.gov/spectrum/lightsquared/docs/2011-06-NPEF-lightsquared-report.pdf.

6 P. F. Marshall, *Scaling, Density, and Decision-Making in Cognitive Wireless Networks* (Cambridge University Press, 2012).

7 Alion Science and Technology, *Effects of the Proposed Citizens Broadband Service to CBand DOMSAT Earth Stations.* Attachment to filing by the Content Industries to the FCC on Proceeding 12–354 (CBRS), Consulting Report ESO-13-011-v3 (2013).

8 A. G. Longley and P. L. Rice, *Prediction of Tropospheric Radio Transmission Loss over Irregular Terrain: A Computer Method – 1968.* Technical report, ESSA ERL 79-ITS 67 (Washington, DC: US Government Printing Office, 1968).

9 Radiocommunication Sector, *Prediction Procedure for the Evaluation of Interference Between Stations on the Surface of the Earth at Frequencies above about 0.1 GHz* (International Telecommunication Union, 2015).

10 Office of Communications (Ofcom) (UK), *Spectrum Usage Rights; Technology and Usage Neutral Access to the Radio Spectrum.* Consultation (2006).

11 Federal Communications Commission, *Amendment of the Commission's Rules with Regard to Commercial Operation in the 3550–3650 MHz Band, GN Docket 12–354* (2016). https://apps.fcc.gov/edocs_public/attachmatch/FCC-16-55A1.pdf.

# Part IV

# Protection Processes for Incumbents and Peers

# 8   General Process for Database System Operation

## 8.1   Introduction

This chapter will describe how the tools and models discussed in the last chapters are collected and organized to develop a comprehensive admission control process for three-tier spectrum bands, such as a Spectrum Access System (SAS). This will include an inventory of the models needed to construct an admission control system, and a high level description of the stateful information needed to perform the functions that are required to support the spectrum management regime. By necessity and for clarity, there will be some overlap with the discussion of methods in prior chapters.

The concept of the admission control system is a simple transaction processing system with a persistent state. The main transactions processed by the admission control system are described in more detail in other chapters. In this chapter, we will examine the fundamental functions of any admission control system that supports three-tier spectrum, and make this discussion as independent of the specific regime as possible. Also, we will abstract the complex algorithms that are the heart of the admission control system processing and only consider their functionality, not their internal design or level of implementation.

Later in this book (Chapter 13), the specifics of the admission control system for the US implementation of the Citizens Broadband Radio Service (CBRS) SAS and rules will be developed in detail, on a message by message, and object by object basis. This design is only one possible implementation of the fundamental interaction needs that are developed in this chapter.

## 8.2   Major Admission Control System and Device Transactions

The transactions that are described in this chapter generally mirror those that will be described in greater detail in Chapter 13. The functional grouping of the admission control system device transactions capabilities are:

**Inquiry** Provide device with information regarding spectrum environment, but make no changes to spectrum usage.

**Grant Request** Request a spectrum grant, and make any necessary adjustments to peer or lower-tier users.

**Authorization** Provide continued authorization to utilize a spectrum grant for a fixed duration of time.

**Relinquish** Relinquish a spectrum grant, and adjust the aggregate interference to reflect this removal of a node.

**Reclaim** A higher-tier user declares right to spectrum, or is detected, and that right is invoked.

In presenting these transactions, we will adopt a standard graphic representation. In some cases, the processing is quite simple, in others it interacts extensively with a large number of entities. The emphasis in the following sections is on the general strategy for the interaction, and the decision-making on both sides of the interface. The specifics of the interfaces themselves are discussed in more detail in Chapters 13 and 14.

## 8.3    Admission Control System Persistent State

The heart of the admission control system algorithms is the persistent store that maintains the state of the devices that have been authorized in the band, and the existence and interference conditions of each of the protected devices (regardless of their tier) that are registered and/or active in the band.

For purposes of this discussion, we consider that the admission control system has two categories of persistence. The first, static data, is provided externally to the admission control system, and includes awareness of devices that have achieved a protected status through some means other than the admission control system. For example, existing licenses, geographic boundaries, terrain features, and regulatory mandates. The second is the dynamic data that is created and maintained by the admission control system as a consequence of its device authorization/deauthorization and sensing processes. The dynamic data essentially constitutes an inventory of the devices under management by the admission control system, and their interference consequences. A summary of some of these information categories is shown in Table 8.1.

We can envision the admission control system transactions as having some interaction with each of these persistent data sets, and abstract away the specifics of the algorithms used. These will be addressed in later chapters, especially Chapters 13 and 14.

## 8.4    Admission Control System Initiation With Static Data Process

There are a number of different static data sets that an admission control system will require in order to perform its protection mission, such as those in Table 8.1. Although the dynamic processes are the most visible admission control system processes, the success of the admission control system is dependent on accurate awareness of many environmental aspects and detail regarding the entities to be protected. The process for

**Table 8.1** Categories of admission control system persistent data retention.

| Consideration | | Explanation |
| --- | --- | --- |
| Static | Geographic | Terrain data regarding terrain height, clutter, and other physical world representations for propagation loss estimation. |
| | | Defined protection or exclusion zones, regions, borders, and other geographic features. |
| | Protected Entities | Definition of all entities that entered other than through the admission control system, or regions that are provided protection. This is linked to the dynamic data associated with each of these entities. |
| Dynamic | Interference State | Emissions are aggregated for each protected node. The basis of these aggregations is also maintained, as it is necessary to remove interference from nodes that leave the channel. This can be a list of contributors to the aggregate, or sufficient information to recreate the same values identically as when they were incremented into the aggregate. |
| | Device State | Registration data that describes the node, its location, capabilities, owner, and fixed rights. |
| | | Grant data that describes the spectrum grant that was issued to the node (one or more). |
| | | Authorization data to maintain track of the expiration time on the current reauthorization (if any). |

acquiring this information is likely specific to the National Regulatory Authority (NRA) implementing the regime, and the existence and organization of the incumbent records, or the methods for obtaining them.

Different three-tier regimes may have very different methods of being informed of incumbent activity. The United States (US) CBRS band has passive sensing of naval radars, and static reporting of Fixed Satellite Service (FSS) sites and wireless Internet service provider (WISP) positions. Other regimes may have active reporting by protected primary users of their location and usage characteristics.

Although protected incumbents' data is considered to be static in this model, it does have dynamic implications. Lower-tier nodes may have to be dynamically cleared from the spectrum. This is addressed through the continuation authorization of devices, and is addressed in Section 8.6.

## 8.5 Logical Flow of an Admission Control Process

The admission control process has two considerations: one to determine the node that can be entered, consistent with the protection of peer and higher-tier users, and the second to determine any adjustments in lower-tier usage to both protect the entering node and to reconcile any aggregation interference issues. This consists of two phases:

**Interference Analysis** Before committing to the entry of a node, ensuring that the admission can be performed without causing interference to a higher or peer tier protected node.[1] This phase determines that the node has, or does not have, the right to enter the spectrum.

**Interference Elimination** After committing to the entry: (1) assure that the entry does not create aggregate interference to other protected nodes, which is resolved by terminating grants to other nodes or adjusting power limits or other aspects of existing grants; and (2) if the node is entitled to protection, ensure that any nodes that would cause interference to it have their grants terminated.[2] This function is also performed when the admission control system is notified of the presence of a protected incumbent by any of the external notification processes, such as through regulatory databases or sensing. This process, for both entry through, or external to, the admission control system is addressed in the next section, 8.6.

Both the inquiry, registration, and grant process require significant amounts of data on the device location, capability, and intent. This can be provided before an inquiry/grant request, as in the US CBRS implementation, or after the inquiry/grant request but before, or with, the grant request. Figure 8.1 shows how these functions are sequenced for a grant, with the phase two processing continued in Figure 8.2. The sequencing of grant and inquiry data provision is arbitrary.

The admission control system needs to determine if the aggregate emissions limits to a higher or peer incumbent is exceeded. If it is, the admission control system needs to determine if this is solely due to the new node, plus its tier peers. If the tier peers and the new node do not exceed the interference protection criteria, then the node can be admitted, although lower-tier nodes may need to be terminated.

Another way of viewing this transaction flow is to consider it as restricted or divided into two phases: pre-commit and post-commit. Pre-commit activities include all of the analysis performed on the assumption that a grant can be issued. The post-commit phase performs the accounting to reflect the issuance of the grant, and any other action that is taken to resolve interference conditions that the admission creates.

The grant process interference analysis is partitioned into two phases:

1. Before committing to the entry of a node, ensuring that the admission would not cause interference to a higher (or possibly peer) tier protected node. This phase determines that the node has the right to enter the spectrum. If this phase has no conflict, entry is assured. This analysis is a one-on-one analysis, and need not consider aggregation, unless the policy provides superior rights to nodes that had already entered the tier.

---

[1]  This entry may require removal of lower-tier nodes in order to meet aggregation limits.

[2]  An effective admission control system would also provide new, non-interfering grants in the event of relocation.

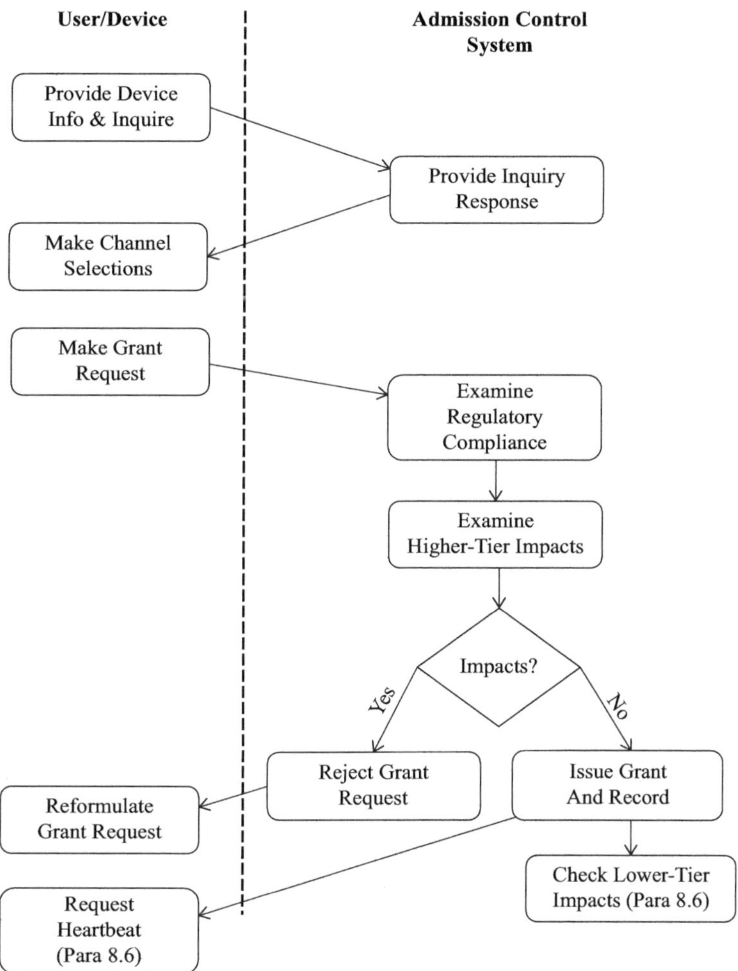


**Figure 8.1** Logical flow of spectrum grants three-tier admission control process.

2. After committing to the entry: (1) assure that the entry does not create aggregate interference to other nodes, which is resolved by terminating or adjusting grants to other nodes; and (2) if the node is entitled to protection, ensue that any nodes that would cause interference to it have their grants adjusted or terminated.

Grant status is maintained persistently, and now permits the device to request authorization to actually emit (next section). The persistent store is the linkage between these two processes. Grant issuance also must be synchronized with any other system that has a requirement to be aware of the ecosystem in this location and band. This could be a regulator, or another admission control system, and is discussed later.

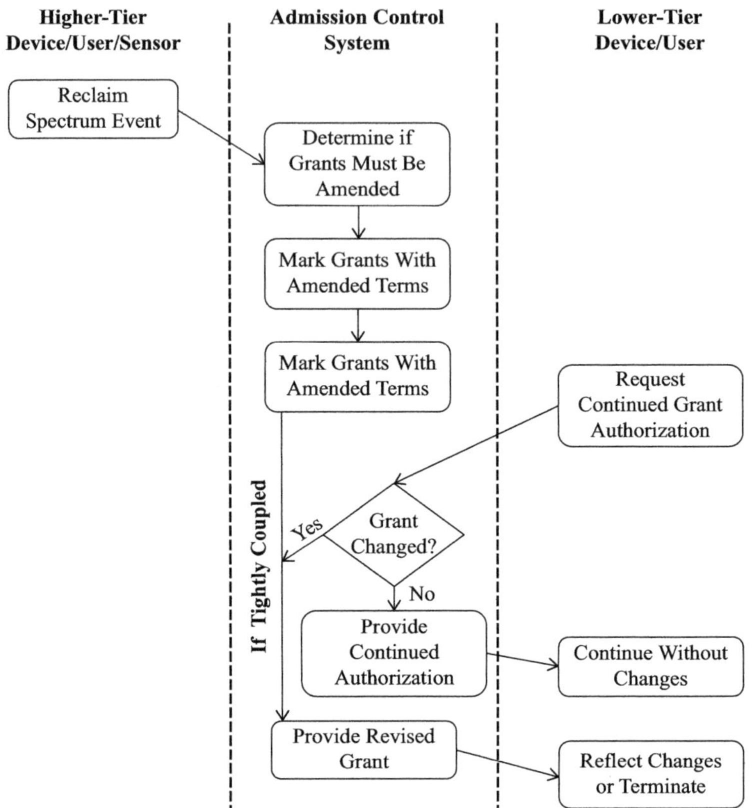


**Figure 8.2** Logical flow of authorization to continue to emit process.

## 8.6     Logical Flow of the Interference Resolution, Incumbent Notification, and Continued Authorization

The logical flow of the process of addressing protected node entry, or reports of incumbent activity, is shown in Figure 8.2. Different three-tier regimes may have very different methods of being informed of incumbent activity. The US CBRS band has passive sensing of naval radars, and static reporting of FSS sites and WISP positions. Other regimes may have active reporting by protected primary users of their location and usage characteristics.

Higher-tier devices and sensing informs the admission control system there has been some action to potentially reclaim spectrum by a higher-tier user. The system determines if this usage conflicts with any existing lower-tier grants. This is depicted by the block titled *"Determine If Grants Must be Amended."* We do not consider the case that the notification might impact a higher-tier user, since that condition would have been recognized prior to the notifying event grant being accepted. We only have to consider the "trickle down" impacts.

Regardless of whether the transaction was initiated external to the access control system (such as registration of a tier 1 incumbent user through a regulatory process), or through a protected tier registration, the process is similar. The activity is reported through a transaction that ensures interference protection to the newly entered protected node. The admission control system needs to determine if the aggregate emissions limits to the protected node is exceeded. If it is, the admission control system must decide which nodes have their grants amended or terminated. There is no technical basis for deciding which grants are to be amended or terminated, but the most logical sequence is to terminate the nodes that have the highest level of interference power into the incumbent. Whether there is a preference between the second and third tier in this process is a question for regulator policy.

It should be noted that if preference is given to any one tier in his removal process, it can essentially preclude operation of lower tiers over a wide area. If a node, or set of nodes, in the preferred tier uses most of the interference margin into the protected node, then the lower tier may be starved, and have very little interference margin to work with, and even insignificant amounts of interference may be the cause for removal or denial, as the admission control system seeks to bring the interference below the limits. The impact on these tiers can be quite substantial.

The logical flow has two options: a closely coupled mode, and a loosely coupled mode. In closely coupled, all devices have continuous and reliable connectivity with the admission control system so that the admission control process can reliably push a change in grant status to the device. In the case of the loosely coupled process, the contact is intermittent (such as polling or operation behind a firewall or Network Address Translation (NAT) proxy) and the device asks for authorization for continued operation.

If any grants are impacted, such as a requirement to terminate, lower power, give up some of the grant spectrum, or other changes in their terms, the admission control system must inform these devices and requires them to accept these revised terms, or terminate operation. If the device is closely coupled, the change can be immediately sent to the device. If the device is loosely coupled, the changes in the grant must be marked in the persistent store. When the device requests periodic reauthorization, it is provided with the revised grant terms. If no grant changes have occurred since the last reauthorization, the device is reauthorized under the existing terms of the grant.

When provided with a revised grant, the device has essentially two options. It can accept the terms of the grant, which may also include suspension. It can decide to not accept them, which means it must shutdown when the current authorization expires, and during that period, may attempt to obtain a new grant from the admission control process. If the process of providing reauthorization fails for any reason, such as loss of connectivity or admission control system failure, the device must consider that the authorization has expired, and must suspend use of the grant until it can obtain reauthorization

The continued authorization process can be adapted depending on the technology, and the need for rapid clearing due to incumbent activity. This is likely to be a local decision based on protected incumbents. However, the general problem is the same regardless of time limits for clearing.

## 8.7    Logical Flow of the Database Synchronization Process

One of the three-tier principles is that competition benefits all participants in an ecosystem by encouraging innovation and cost effectiveness, and it creates confidence in the viability of its use. The admission control system should be no exception. It is certainly more convenient to anoint a favored supplier with the right or duty to provide this service, but history has shown that few monopolies innovate and improve cost effectiveness, particularly with government-assured monopolies.

It is certainly inconvenient to distribute admission control functions among not only multiple computers or data centers, but among different operating entities. However, the advantages are readily apparent. In the USA, early in the President's Council of Advisors on Science and Technology (PCAST) consideration, there were those that suggested a *"Federal SAS"* to manage the spectrum. Fortunately for the spectrum community,[3] the perceived failure of a government-directed Information Technology (IT) staff to develop a nationwide health insurance registration website (www.healthcare. gov) compellingly argued for the private development and operation of the SAS service, and precluded a government monopoly on its operation.

The market-driven decision to permit multiple service operators creates a requirement to synchronize grant data among operators. The synchronization is driven by a number of considerations, but one of the most significant is that likely interference protections are based on aggregate interference. This implies that all system operators must accumulate interference to all protected nodes, regardless of whether they were the admission control system that admitted the node or not.

The exact design of multiple data center/multiple operator services is beyond the scope of this book, and not relevant to the design of spectrum management regimes, so we will not develop the specifics of the design, except to provide some principles that should be reflected in whatever design and synchronization structure emerge.

- The failure of any one database system should not block performance by the others. Each admission control system must be capable of performing basic grant and handshake processing autonomously.
- Provision should be made to enable new entrants to *"catch up"* with established services at any time. Established services should be able to resynchronize after catastrophic events.
- The distributed nature of the service should not diminish the accountability of service providers for the grants they issue and continue to authorize.

The implementation of the database synchronization is a matter of computer science, not wireless or spectrum engineering. Any number of methods can be utilized so long

---

[3] Perhaps not so fortunate for those who tried to use the healthcare.gov site.

**Table 8.2** Probability of synchronization-induced collisions in independent admission control system operation, using the US CBRS as an example.

| ASSUMPTIONS | | | |
|---|---|---|---|
| Factor | Value | Units | Notes |
| Area of USA | $10^7$ | $km^2$ | |
| Area with CBSDs | 10 | % | Areas with sufficient population density to justify deployments |
| Registrations | $3 \cdot 10^6$ | /year | Estimated third year deployments over 12 months |
| Active time of day | 12 | hour | Registrations at any one region will generally be within working hours |
| Impacted area | 0.09 | $km^2$ | Typical area impacted by a single CBSD in typical environments |
| Synch time | 2 | sec | Time delay to synch grants between admission control systems |

| COMPUTATIONS | | | |
|---|---|---|---|
| Factor | Value | Units | Notes |
| Arrival rate | 0.19 | /sec | Estimated based on 3 million devices in last year, concentrated in 12 hours. |
| Deployed area | $10^6$ | $km^2$ | Area with devices deployed |
| Probability of impact | $0.9 \cdot 10^{-8}$ | | Any two random grants being within same impact area |
| Prob unsynched entries | 0.38 | | New admission request occurring while synch of previous registrations pending |
| Prob synch impact | 3.4 | | Probability of unsynchronized grant in new registration impact area |
| Synch failure rate | 0.10 | /year | Synch failure rate per year |

as they assure a very low probability of assignments being made in the absence of awareness of actions by other admission control systems. Using the US CBRS band as an example, the probability that two admission control systems make uninformed decisions regarding potentially impacting assignments can be computed as shown in Table 8.2.[4]

In most cases, a synchronization failure will have no impact, as it is largely irrelevant for General Authorized Access (GAA) grants, and PAL owners are likely to utilize a single admission control service, which would inherently be up-to-date. But even in the extreme case, the likelihood of impact is very small compared to the variances

---

[4] The numbers used in this computation are examples, and have no empirical validation. Additionally, the assumption of even distribution within usage areas and over time is probably not valid, but serves as an example.

of propagation, location measurement, azimuth estimates and all of the other factors that impact the assignment process. For example, the propagation estimates used in computing interference are typically in the range of $10^{-2}$ to $10^{-3}$ likelihood, so the probability of not recognizing an interference condition may be another two or three orders of magnitude less than even the low estimate (one per ten years) provided in Table 8.2.

## 8.8 Computational Consistency and Admission Decision Correctness

In this section, we examine the question:

*"If two admission control systems reach different decisions about admitting a node, must one of them be wrong?"*

The answer to this question is one of policy, not of science. The admission control system is essentially answering the question: *"Can the system assure that no interference to protected nodes will be caused by this admission?"* If they have a different basis for making decisions, then it is not unreasonable that one could provide the assurance, while another could not. The admission control system is not determining if interference will, or will not, occur. The system does not know this. It only knows if it can assure non-interference.

One way to consider this decision is as a logical proposition. We let *admission* be the decision (true or false) to let a node enter, and *interference* be the resulting, real-world, measurable occurrence of interference. It might be obvious to consider the decision of the system to be equivalent to the existence of interference, as shown in the logical proposition in Eq. (8.1).

$$admission = interference \tag{8.1}$$

In this case, it is clear that any difference in admission control decisions would mean that one was in error, as the value of *interference* is a measurable fact; as if we can perform an experiment with exactly these conditions present.

The correct manner in which to view the assertion made by an admission control system admission decision is better expressed as the implications shown in Eq. (8.2).

$$admission \implies \overline{interference} \tag{8.2}$$

Since this relationship is an implication of not causing interference, but not equivalence to interference itself, the converse assertion need not be true. It is therefore also correct to make the assertion in Eq (8.3). This is due to the fact that implication is not a symmetric relationship, like equivalence.

$$\overline{admission} \;\not\!\!\implies interference \tag{8.3}$$

Not admitting a node does not imply that the admission would cause interference, it only implies that the admission control process could not ensure that it would not cause interference. It does not imply that it would definitely cause interference. This is a very weak assertion, and driven by knowledge of the environment, as much as the physics of the node placement.

It is worth considering why two different admission control processes would come to different conclusions about their ability to assure non-interference.

**Propagation Model** Propagation models vary in their realism. More complex models typically can create high loss estimates at equivalent confidence levels than lower complexity ones. Free-space models typically create the lowest estimates of loss, and site-specific clutter loss models generally result in much higher estimates.

**Propagation Data** The level of propagation data, such as terrain resolution can yield better, and less conservative, estimates for path loss, as an example.

**Clutter Data** Transitioning from general clutter models to data that is specific to each site, and even each path, can increase estimated path loss very significantly.

**Usage Data** Understanding the statistics of usage may enable the admission decision to be based on more exact statistics of usage, particularly when aggregating a number of emitters, and where statistical treatment of a large number can be utilized.

We can consider that in each category (such as propagation estimation) there is a hierarchy of awareness detail that can be provided to the algorithm. With little information, assumptions must be close to the worst-case values. As detail is provided, these assumptions can be either confirmed, or relaxed. When relaxed, they should increase the loss lower bound estimates, and therefore permit more entry into the band. It is an essential test that estimates are monotonic with detail awareness.[5] Assuming loss is monotonic with additional awareness, the additional information should never cause a node to be rejected. If an ignorant algorithm would admit a node that a more informed one would not permit, it is indicative that the ignorant algorithm was not appropriately conservative in its assumptions.

The idea that the admission control decisions could differ is potentially bothersome to many. It precludes cross-checking systems, and means each must be individually, and independently, validated. It runs the risk that the competition among systems becomes a *"race to the bottom"* sacrificing protection mechanisms for customer adoption due to higher admission rates. It provides (invalid) arguments for protected users to claim the admission decisions were clearly wrong, since another admission control system had said so.

## 8.9  Logical Flow of a Spectrum Inquiry Process

Figure 8.1 illustrated the role of the spectrum inquiry process in the overall strategy of a device or user to obtain a spectrum grant. The spectrum inquiry transaction is

---

[5] By this we mean that if the algorithm was provided with more information, its estimates of the lowest possible path loss or similar metrics also increase, or at least do not decrease.

**Table 8.3** Levels of complexity and utility in spectrum availability inquiries.

| Attribute | Level | Content |
| --- | --- | --- |
| Coverage | One Nation/ One Band | The entire admission control system can only address one band, and all responses are implicitly from one rule set, and from one country. |
| | One Nation/ Many Bands | The admission control system can address operation in more than one band, and informs the device of the rule set for each of the multiple bands. The device can then select the bands/rules in which it decides to operate using this regulatory information. |
| | International/ One Band | A single, normalized, band coverage, with the likelihood that the admission control system will inform the device of specific rules in the location it reports. |
| | International/ Many Band | The admission control system provides the device appropriate sets of band rules and options in any country supporting at least one three-tier band. |
| Depth | Status Only | The admission control system informs the device of the rules at the location and range of frequencies indicated, and the status of the spectrum in any of the included bands (Priority Access (PA), GAA, protected, excluded, etc.). |
| | Availability | The admission control system provides the previous entry response, but also analyzes the ability of the device to actually receive a grant, given the current usage of the band. |
| | Coexistence | The admission control system provides the previous two entries' responses, but also suggests the choices that it determines to be most likely to be usable by the device. |
| | Secondary Market | Offer of spectrum access, protection, or other increased rights by spectrum license holders as a secondary market for access. |

not essential to any of the interference protection missions of the admission control system, but is highly useful to devices, particularly when there are opportunities for three-tier spectrum worldwide, and/or in multiple bands. For purposes of discussion we can consider several levels of inquiry, which are a function of both the service, and the international context.

Table 8.3 illustrates levels of spectrum inquiry in terms of both coverage, and of depth of analysis.

It is unlikely that all of the worldwide NRA processes will, or should, agree on a common set of rules, and it is improbable that a single set of rules will work in many different bands of spectrum, with different incumbents. Therefore, the spectrum inquiry provides the bridge between a worldwide ecosystem of devices, and a nation and band-specific approach to spectrum management. It is the alternative to harmonizing spectrum sharing, which is likely a challenge well beyond the technical ones.

So long as there is general agreement on a naming system for spectrum rules, and a standard for authenticating the compliance of a document with a specific rule set, we can envision an ecosystem with a number of rules and bands, mixed in arbitrary ways.

One possible logical flow of the process for a device to perform an inquiry and use the information to select its method of grant acquisition is shown in Figure 8.3. The process depicted here essentially works through layers of the decision-making process. It is initiated at the highest level, with the admission control system providing a list of bands and required credentials (licensing approvals) for each band and regulatory regime. The device filters these by matching licensing credentials it is compliant with, prioritizes these by the utility of each of the viable options to meet its mission needs.

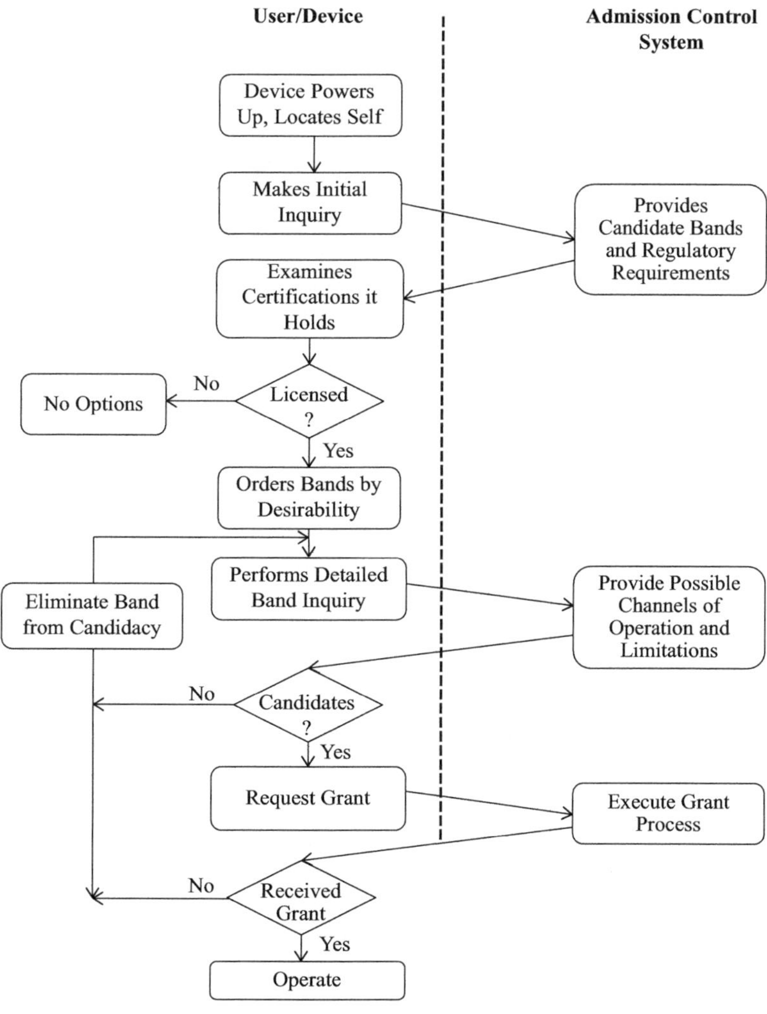


**Figure 8.3** Logical flow of an inquiry transaction process.

The device can then perform a detailed, band-specific inquiry. This might yield no channel options, or unacceptable ones, in which case the band is eliminated from candidacy, and the next one is considered. If the results are favorable in a band, the device performs the grant request process, and either receives a grant, or reiterates the process for a new band. Options certainly exist. The device could run through all of the bands and identify options in each before considering which to request, and other devices may only have one band or national certification, in which case it can enter the band specific process immediately.

Different vendors would select the bands they thought were most useful to cover, and the rule set(s) for which they would seek regulatory compliance/certification. Some might be NRA specific, such as the US CBRS Part 96, and others might be internationally recognized, such as an European Union (EU) rule set. Vendors could also differentiate their products by the sophistication of their algorithms to select candidates for consideration.

Since it is our goal to automate the spectrum process, and make it invisible to the user, we assume no human interaction with the device. The device would appear at some location and be turned on. It has geolocation ability, so it can determine its location. With that, and an available Internet connection, it contacts whatever admission control system it has authority to use. The device need not authenticate itself, but it does authenticate that it is communicating with an approved admission control system through the admission control system certificate. In Chapter 18, Figures 18.1 and 18.2 provide examples of how such information might be provided by a world-aware admission control system in support of internationally roaming three-tier devices, and is used in this example.

The inquiry function provides the device with a list of bands, and the operating authority required for operation in each band. Along with this information, it might also provide the actual availability of spectrum, and recommendations on coexistence, if its inquiry function had the depth of processing. The device would compare the specific bands available and the certifications needed to utilize the bands, and select the intersection of its capabilities and certifications and the available bands and required certificates. It would then enter the process described earlier to register and obtain a grant, presenting the certificate appropriate to its selection.

A reasonable extension of this model is to not limit the spectrum choices to Government-provided access. The admission control system could provide the same inquiry functionality to also access private secondary sharing markets, using the same models as for Government-provided rights. The authentication in these cases would be a pre-existing agreement regarding ordering terms and conditions which would provide assurance to the spectrum holder offering the sublease. In this case the inquiry functionality would include the contractual and financial terms and conditions, as well as the technical ones.

Harmonizing spectrum access has been a major goal of spectrum management, but it has severe challenges in a nationalistic world in which wireless technology is so key to national interests, and where spectrum usage often reflects local, rather than global needs. Automated admission control addresses many of the drivers for harmonization without impacting the sovereignty of national decisions. As band coverage is less of a technical obstacle, regulatory compliance complexities will become the barrier to device roaming the world. The use of automated admission control system becomes an attractive alternative to disruptive, and increasingly impractical international harmonization. Harmonizing admission control frameworks may be more attractive than harmonizing the band usage itself.

## 8.10  Absolute Versus Relative Protection Criteria

The traditional approach to interference management has been to focus on the level of interference relative to the noise floor. This is certainly a reasonable approach in macro networks, where the devices are operating at the range of the lower power uplink, and at the very range of the cell. Since such cell towers are extremely expensive, any degradation in range of coverage has significant impact on the network operator.

However, in capacity focused small cell operations, particularly when indoors, or with overlaid macros coverage, these conditions are not present. Since operation at the noise floor is not the objective of (or necessary to) many of these deployments, a better measure of tolerable interference is the worst case ratio of the intended signal ($S$) to the aggregate interference noise interference ($I$) plus receiver noise ($N$). Establishing a minimum assured threshold for $\frac{S}{I+N}$ assures the network operator of operation, and maximizes the ability to utilize the spectrum. This measure may be particularly relevant in establishing methods for coexistence among devices in peer tiers. In a dense network, with $\frac{S}{N}$ rations that are 40 dB or more, this metric could enable interference levels that were tens of dB higher than noise floor relative criteria would permit.

Another rationale for this metric is that the coverage of capacity networks tends to be constrained by clutter loss of buildings and other obstructions, rather than $r^2$ path loss. The gradual reduction in range at edges of a lower frequency macro network does not occur. Instead, the boundary of network coverage is more likely to be a wall! A 6 dB increase in the noise floor of a coverage network would reduce range by half, and coverage by 75%, whereas it might have no impact on the utility of a capacity network at all. Capacity networks tend to operate at high values of $\frac{S}{N}$, or not at all.

## 8.11  Statistical Considerations

The reliability of communications is inherently a statistical consideration. Links are not designed to certainty of operation, they are designed to achieve a reliability and a maximum error rate. Packet processing on the Internet is not designed for absolute reliability, but to achieve a certain probability of delivery of individual packets. Yet, the interference components of this reliability have generally been approached as if it was possible to achieve certainty. This has the effect of shifting unnecessary cost and other burdens onto the spectrum users.

**Table 8.4** Example statistical considerations in probabilistic interference determination.

| Consideration | Explanation |
| --- | --- |
| Power Level | Many modern wireless networking technologies effectively manage device power to minimize interference. This is particularly true for client devices of an Access Point (AP). |
| Network Usage Duty Cycle | The usage of network is not always 100 percent of capacity, and a mix of networks can be anticipated to have a range of demands, which is reflected in the duty cycle of their emissions. Note: design induced duty factor limits are in the last item. |
| Transmit Antenna Orientation/Gain | Transmit antennas are often poorly positioned to cause interference. This is particularly true for User Equipment (UE), which is handheld, and often not on its optimal axis. |
| Path Loss | This was addressed extensively in Chapter 7. |
| Receive Antenna Orientation/Gain | Same as the transmit antenna. Applicable to interference into mobile devices. |
| MAC Layer Correlations | The operation of most networks is not just random transmissions by devices. The Listen Before Talk (LBT) and Time Division Duplex (TDD) Media Access Layer (MAC) layers do not transmit 100 percent of the time. Across a large number of nodes, it is unlikely that all, or even more than one, transmit at once. Client devices do not overlap with AP transmissions in many protocols. Clients do not overlap other clients in LBT, as examples. |

Chapter 7 introduced the concept that interference aggregation of propagation estimates of a given cumulative probability of occurrence did not preserve that probability, but resulted in increasingly more conservative estimates. In this section, we will expand that concept to include a wide range of other statistical factors that can impact the likelihood of any given estimate of interference actually occurring. There are any number of factors that could be considered in an analysis.

Typically, interference analysis has considered only the worst case of each distribution, since in a one on one analysis, this is the only assumption that can provide any assurance of non-interference. But in a regime where many interference impacts are aggregated, a more sophisticated analysis is justified. While it is possible that any one possibly interfering node may be in its worst-case configuration, it is much less likely that all sources of interference will be in their worst-case conditions.

Some of the more probabilistic contributors to interference are shown in Table 8.4. These considerations are highly dynamic, and not correlated, across users, so they form a valid set of assumptions for a statistical treatment of likelihood based on their nature as independent events.

It should be noted that the US CBRS rules consider none of these factors, with the possible exception of MAC layer correlations, which is implicit in the rules permitting unregistered UE devices. The first implementation of the regime is likely too soon to also move to a more statistical treatment of interference. However, as confidence in the regime itself is gained, this is the next challenge.

Although acceptance of a statistical model may be a significant departure from more risk averse models, management of risks is a more appropriate metaphor for spectrum management. A cursory examination of many of the parameters indicates some may not be statistical in a one-on-one analysis, but in a many-on-one analysis, the statistical assumptions may be significant with an impact of 10 dB or more.

For example, if we make the following assumptions for the UE outbound side of the interference case:

**Power Level** -10 dB, typical power control conditions in average range networking.

**Network Usage** -3 dB, typical networks are efficient only if typical offered load is below 50%.

**Transmit Antenna Orientation** -3 dB, typical antenna response in most of the omnidirectional positions.

**Path Loss** Statistical considerations of propagation estimation were addressed previously.

A mean aggregate interference power reduction of 16 dB equates to a range reduction of $2^{\frac{16}{6}} = 2^{2.67}$, or a density increase of a factor of over 40. This is an important factor to be considered in planning density constrained deployments.

## 8.12　Site- and Path-Specific Considerations

The performance of an admission control system can be greatly augmented with localized knowledge of the deployment area. For example, how much attenuation is present between floors in an office building. Similarly, very detailed geo-information on individual building entrances, exits, interior space, and outdoor features (open field, deciduous forest, coniferous forest, etc.).

Figure 8.4 illustrates the analysis of propagation loss using a geo-aware data set that can decompose the path into its constituent path components.

The distribution of typical site-specific path loss was discussed in Chapter 7, and illustrated in the additional site-specific path loss shown in Figure 7.3. A one-on-one analysis might have to use the lowest sound of the path-specific loss model, but an aggregation analysis can utilize the distribution of losses shown in experimental collections across a wide range of interference contributors.

## 8.13　Apportionment of Interference Margins

The SAS process could be viewed as a distributed (multiple SAS service providers) system that allocates portions of the interference permitted to protected nodes. There are issues of ongoing protection, adaption, and fairness that can arise through this

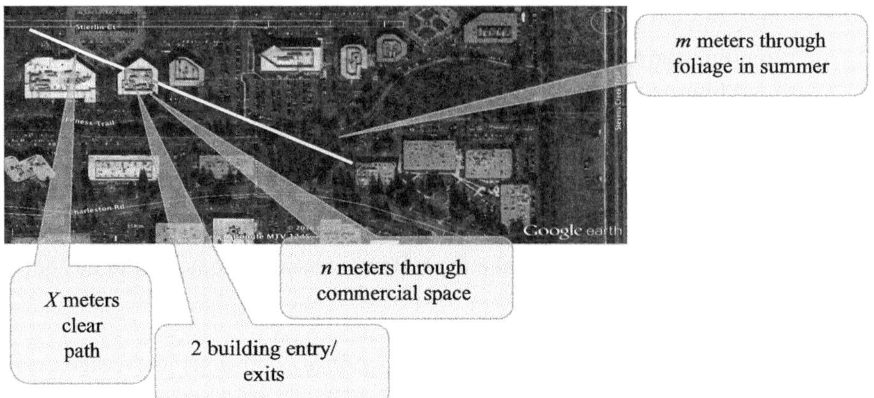


**Figure 8.4** Geo-informed, site-specific propagation estimation.

aggregated model of interference protection. Examples of conflicts that can arise include:

**Admission Sequence** We consider that there is a pool of interference margin that is "consumed" as three-tier devices enter the ecosystem. At some point this "pool" is consumed. The admission control system has essentially two choices: block any new entrants that would exceed the interference limits, or admit the new node, but reduce the power permitted by the existing grants to limit aggregation. If the admission control process does not permit entry, it has essentially established a new right: that of first entry. Even if the decision is made to reduce existing grants, it leads to the question of whether to reduce each by a fixed ratio, or lower the maximum power across all contributing grants.

**Changes in Protection** Incumbents protection requirements may change after nodes have been admitted to the sharing regime. A satellite antenna may change its pointing (elevation and azimuth) that would increase its susceptibility to interference from existing grants. Similar to the new entrant problem discussed above, this may require the admission control to either rescind existing grants and terminate nodes in the band, or to proportionately reduce power to all of the nodes that contribute to the interference.

**Interference Resolution** Even after nodes admissions have been analyzed through whatever algorithms are approved, there is a possibility that higher-tier users would report interference, and whether true or false, it is likely that the admission control system would have to take notice of these reports, and implement mitigation action. This again forces a choice between removing devices, or reducing the power to multiple devices.

The US CBRS rules are quiet on these issues. These situations may not arise in the specific structure of the rules in that band. However, a general regime for three-tier must provide explicit direction to the admission control systems on the treatment of these conflicts. Creation of "first-in" rights is a policy, and not technical concern.

It is useful to consider scenarios in which these conflicts may occur. We imagine an FSS site has an allowance of a certain interference power level, and that limit is reached by twenty distant nodes, each evenly contributing to the aggregate that is just short of this limit. A single, very close device requests entry. This device would almost reach the interference power limit by itself. When all 21 nodes are in the band, the interference limit would be exceeded. The admission control system would face a decision on how to address the admission of a complaint single node, that results in a non-compliant aggregation of nodes.

- If the 20 distant nodes were in the band first, it would be the most natural, and simplest, decision to not permit the close node to enter at all. There would be no impact on the existing node grants.
- If the grant was rejected, and the interference was co-channel, then the close node would have the option to attempt to request a grant on another channel. The success of this would be dependent on the specifics of the emission masks, other users, etc.
- In the same conditions, the policy could state that all nodes have a right to be admitted. In this case, the channel with the least aggregate total emissions would be selected, and would require that the power levels (or at least EIRP in the direction of the FSS site) be reduced by 3 dB, including the grant request of the close device.
- The most complex approach would be to hold the initial grant power limits intact, and treat reduction in the aggregate emission as a coexistence issue, and collectively manage the duty cycle of the devices to avoid instantaneous violations of the interference threshold. This would be a technology-specific approach.

In these examples, the same process can be used, even if the initial grant of the close node would exceed the threshold, it would just increase the level of power reduction required by all of the nodes that contributed to the aggregate interference.

None of these choices is particularly attractive. However, this is apparently the cost of sharing spectrum in a three-tier framework. There is apparently some inherent loss of certainty whatever mechanism is selected. This is a trade of certainty for capacity. It is the cost, of massively increased spectrum opportunities.

## 8.14    Suggested Reading

- There is an extensive discussion of potential admission control system algorithms applicable to the SAS algorithms in the US CBRS proceeding [1, 2]. In many cases, the discussion leads to the Federal Communications Commission (FCC) rejecting the complexity, but even the rejection has interesting background on the rationale for limiting SAS complexity in this first implementation.
- The Wireless Innovation Forum (WinnForum) developed a consensus document describing the requirements for the operation of both a SAS [3] and a Citizens Broadband Radio Service Devices (CBSD) [4] in the US CBRS band. Requirements

for operational security protection [5] of sensitive operations, and supporting communications security (including authentication mechanisms) [6] are also documented in WinnForum documents. These documents are publicly available.

# References


1 Federal Communications Commission, *Amendment of the Commission's Rules with Regard to Commercial Operation in the 3550–3650 MHz Band, GN Docket 12–354*, Report and Order and Second Further Notice of Proposed Rulemaking and Order (2015). http://apps.fcc.gov/edocs_public/attachmatch/FCC-15-47A1.pdf.
2 ——, *Amendment of the Commission's Rules with Regard to Commercial Operation in the 3550–3650 MHz Band, GN Docket 12–354* (2016). https://apps.fcc.gov/edocs_public/attachmatch/FCC-16-55A1.pdf.
3 Spectrum Sharing Committee (SSC), *Spectrum Access System Requirements* (Wireless Innovation Forum, 2016).
4 ——, *CBRS Operational and Functional Requirements*, WINNF-15-S-0112 (Wireless Innovation Forum, 2016).
5 ——, *CBRS Operational Security Technical Specification*, WINNF-15-S-0071-V1.0.0 (Wireless Innovation Forum, 2016).
6 ——, *CBRS Communications Security Technical Specification,* 1st edn (Wireless Innovation Forum, 2016).

# 9 Protection of Incumbent Satellite Operations

## 9.1 Introduction

This section describes the methodology to protect one of the most sensitive incumbents in the spectrum community, satellite up links and downlinks.[1] This chapter also serves as the computational exercise to demonstrate the process used in performing the admission control system process quantitatively using the most complex protection case in the three-tier ecosystem.

This detailed analysis is included in this book for a number of reasons.

1. It serves as the computational exercise to demonstrate the process used in performing the admission control system process quantitatively using the most complex protection case in the three-tier ecosystem.
2. Satellite receivers are extremely complex, so they require a richer set of protection features than can be applied for protecting other systems.
3. The computational model is not specific to satellite protection, and aspects of it are applicable to the other protection requirements.
4. Since questions have been raised about the practicality of constructing the admission control system engine, this most stressing case is a good working example of the process used to protect all types of protected nodes.

The stressing nature of satellite receiver protection arises for a number of causes:

1. Satellite receivers are extremely sensitive, so they require a rich set of analytic features to assure protection.
2. They typically utilize highly directional antennas, and so provide significant spatial sharing opportunities, but also require complex consideration of each potentially interfering path.
3. Many of the uses of fixed satellite services are such that interference to a single downlink node or uplink channel would have significant impact to a large number of individuals, even if they did not utilize the satellite service directly.[2]

---

[1] Radio astronomy is even more sensitive to interference, but protection of radio astronomy is a unique challenge, and not addressed in this work, as it is typically performed through regulatory and administrative process, and is not as dynamic.

[2] An example is video distribution to individual cable providers, where all of the customers of the provider would be impacted by interference to the downlink site.

Questions have been raised about the practicality of constructing the admission control system engine. An exercise that emulates the processing of an admission control system therefore also serves to address the complexity concern. Since questions have been raised about the practicality of constructing the admission control system engine, an exercise that emulates the processing of an admission control system may also serve to address the complexity concern.

The reader more interested in the policy, commercial, or economic aspects of three-tier spectrum sharing can skip the detailed algorithms and equations without loss of context, with the assurance that they are available for implementation as needed.

## 9.2    Unique Aspects of Satellite Uplink and Downlink Protection

Satellite receivers pose unique technical challenges to protect:

1. Satellite receivers have a deep space thermal noise background at most times,[3] so the noise floor that they can achieve is much lower than the noise floor of terrestrial aligned devices, by at least a factor of ten.
2. The very low noise, high gain front end amplifiers, or Low Noise Amplifier (LNA) stages have a relatively low dynamic range. They were designed to receive 24 equal channels that are tightly balanced, so do not have the 70 dB or more dynamic range common in terrestrial applications.
3. The typical use is with high gain antennas. This creates very high gain towards a limited field of regard, and at least 10 dB of loss in most of the others. These patterns are also knowable a-priori, as the geometry from a given location on the earth to a given orbital slot is readily computable.

An extensive allocation of spectrum has been agreed to internationally, to support satellite operations and communications services. It is therefore an attractive candidate incumbent for spectrum sharing. It also poses some unique, and instructive, challenges in spectrum sharing. There have been a large number of proposals to share this spectrum, including a proposal at several of the World Radio Conference (WRC) conferences to allow International Mobile Telecommunications (IMT) operations in the band, and the ill-fated United States (US) initiative to share some spectrum between INMARSAT and Long-Term Evolution (LTE) operations by LightSquared.

It is worth examining the extent of the spectrum that is currently allocated exclusively or primarily to fixed geosynchronous satellite operations. Some of the major allocations to these services is shown in Table 9.1.[4]

It is obvious from Table 9.1 that a very significant amount of spectrum is used for satellite purposes, both below 6 GHz and below 35 GHz. Therefore, sharing of satellite spectrum is certainly an attractive opportunity, at least from the perspective of secondary

---

[3] Exceptions are solar or lunar backgrounds that occur occasionally, raising the noise temperature of receivers and the physical temperature of dishes.

[4] International military primary spectrum allocations are not shown.

**Table 9.1** Key international satellite spectrum allocations.

| Band Name | Use | Range (MHz) | Notes |
| --- | --- | --- | --- |
| C-Band | Downlink | 3700–4200 | |
| | Uplink | 5925–6425 | |
| Extended C-Band | Downlink | 3400–3700 | Grandfathered use only in USA |
| | Uplink | 6425–6725 | |
| Ku-Band | Downlink | 12200–12700 | |
| | Uplink | 17300–17800 | |
| Ka-Band | Downlink | 27500–31000 | |
| | Uplink | 17700–21200 | |

users. In this chapter, we will examine the unique protection requirements of satellite service protection, and then relate it to the opportunities for three-tier sharing.

## 9.3 Antenna Pointing and Gain Models

Geosynchronous satellites are all positioned over the equator, and are spaced in positions separated by (typically) two degrees of longitude. All satellites' ground terminals point to the equator. If they are in northern latitudes, or are located at significantly different longitudes, the patterns point very low on the horizon. If they are in more southern latitudes, and close in longitude to the satellite's slot, their elevation angle to point at the satellite is much higher. This is important, because it greatly influences the gain or loss of the antenna dish to terrestrial sources.

Computation of the elevation and look angles for a geosynchronous ground terminal located at latitude *lat* and longitude *long* to a geosynchronous satellite at longitude *slot* is given by Eq. (9.1) and (9.2).[5] The direction along the ground is given by *azimuth*, and elevation angle of the antenna is given by *elevation*. All units are in degrees and true north.

$$elevation = \arctan\left(\frac{\tan(slot - long)\cos(lat) - 0.1512}{\sqrt{1 - \cos^2(slot - long)\cos^2(lat)}}\right) \tag{9.1}$$

$$azimuth = 180 + \arctan\left(\frac{\tan(slot - long)}{\sin(lat)}\right) \tag{9.2}$$

The antenna gain/loss is a function of the angle to the antenna boresight. It is symmetric around the boresight, so a single angle represents this angular displacement. In most cases the emitter does not lie directly under the main beam, so spherical geometry is needed to determine the angle of the emitter to the boresight of the antenna. This uses the difference in the antenna *azimuth* and the bearing (*bearing*) from the

---

[5] A good derivation of these equations is provided by the US Federal Communications Commission (FCC) in their Appendix D supporting the sharing of 3650–3700 MHz [1].

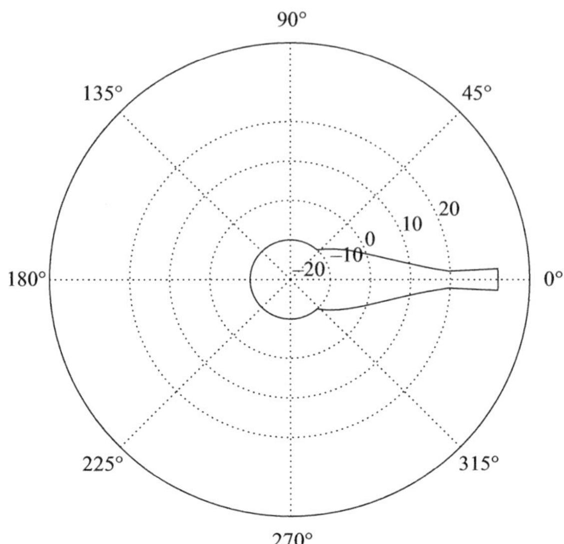


**Figure 9.1** Satellite antenna gain as a function of boresight angle ($\theta$) using US CBRS criteria (FCC rule 209(a)(4)).

satellite receiver to the emitter, and the elevation angle to compute the angle from the emitter to the antenna boresight. This conversion is shown in Eq. (9.3).

$$\theta = \arccos(\cos(elevation)\cos(azimuth - bearing)) \tag{9.3}$$

The $\theta$ value is used to derive the gain along any axis from the earth station. This gain (*gain*) is given by Eq. (9.4)[6] for protection of C-Band downlink receivers. This computation is band specific. This computation of antenna gain (in dB) is shown in Eq. (9.4).

$$G_{ant} = \begin{cases} 32 & \text{for } 0° \leq \theta \leq 3° \\ 32 - 25\log(\theta) & \text{for } 3° < \theta \leq 48° \\ -10 & \text{for } 48° < \theta \leq 180° \end{cases} \tag{9.4}$$

This gain is highest in the *azimuth* direction, if the elevation (or $\theta$) is less than 48°, the peak gain in the boresight of the antenna is given by $G_{int}$. If the elevation is more than 48°, then the antenna gain is modeled as constant in all directions, and is the worst case front-to-back ratio of the antenna, which is highly variable outside the main gain beam.[7]

Graphs of typical values of antenna performance are useful. These all utilize the internationally used C-Band downlink (3 700–4 200 MHz) band. Figure 9.1 illustrates the peak gain of the antenna as a function of the boresight angle ($\theta$). Antenna response

---

[6] There are several equations for this calculation. This is the version invoked in the US Citizens Broadband Radio Service (CBRS) regulations.

[7] Worst-case for interference is the least front-to-back ratio across the back side of the antenna.

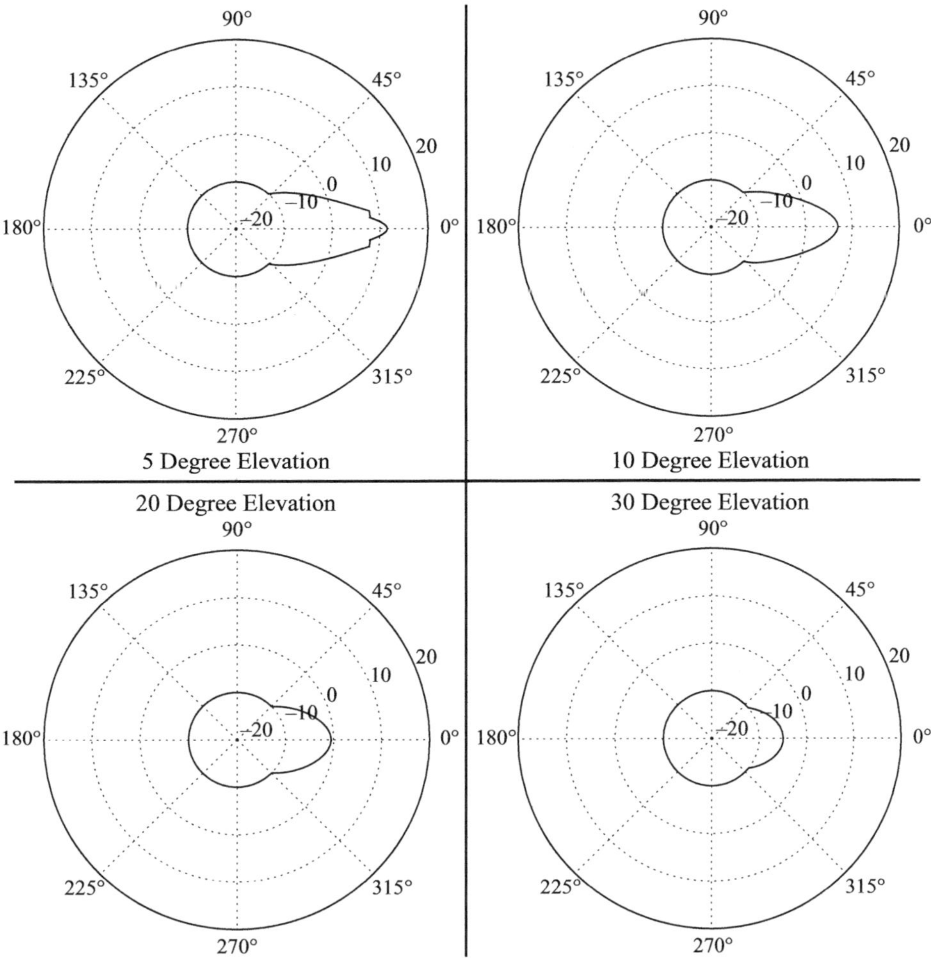



**Figure 9.2** Ground antenna pattern for elevation angles of 5°, 10°, 20°, and 30°.

must fit within this mask, so the mask represents the worst-case interference reception conditions.

The combination of elevation look angle and off-axis position plays an important role in the actual gain footprint of an antenna along the ground. Figure 9.1 illustrated the regulatory model of the direct gain of an antenna, essentially at a zero-degree elevation. Figure 9.2 illustrates the actual ground pattern of the same C-Band antenna when it has been elevated through a range of typical elevation angles; in this case 5°, 10°, 20°, and 30°. Even the 5° elevation angle, which is the lowest reasonable viewing geometry, has a gain reduction of almost 20 dB. As the elevation angle increases towards 30°, the gain rapidly is diminished towards the default front to back ratio of −10 dB.

It is important to understand and reflect the wide range of potential satellite geometries and visibility that can occur over a large land mass. This discussion uses the Continental United States of America (CONUS) land mass as an example.

Figure 9.3 is a map of the United States of America (USA) that depicts several locations in extreme north/south and east/west positions in CONUS. Table 9.2 provides the Fixed Satellite Service (FSS) visibility and geometry for the east- and western-most visible USA domestic satellites that are visible to these sites. The range of US domestic satellites licenses is from 72° to 139° west longitude.

Some general conclusions can be reached by inspection:

- Locations that are in the more southern areas have essentially the fixed front-to-back ratio gain (a loss) pointing at all satellite slots that are somewhat close to the ground site's longitude.
- Locations that are on the side of the ground station further from the equator operate in the front-to-back gain region. This is the northern side in the northern hemisphere, and the reverse in the southern hemisphere.
- The worse case is when a satellite is close to the horizon, regardless of the receiver's longitude or latitude.
- Not all domestic satellites are visible throughout the USA at reasonable (> 5°) look angles.

In the case of the USA, the fact that the domestic satellites are only present in the arc from 72° to 139° west longitude means that the lower latitude sites have full visibility to any of the orbital positions, and have look angles that are high enough to have their effective gain at, or near, the front-to-back ratio gain level for all domestic services. The site at the very tip of the State of Maine (far-northeastern corner of the USA) has an elevation angle of 4.3° when pointed at the far western domestic orbital position (139° west longitude), so the first viewable satellite (defined as above a 5° elevation angle) is the orbital slot at 137° west longitude. This is an almost unique case in the USA. It is important, but not the case that should drive spectrum protection.

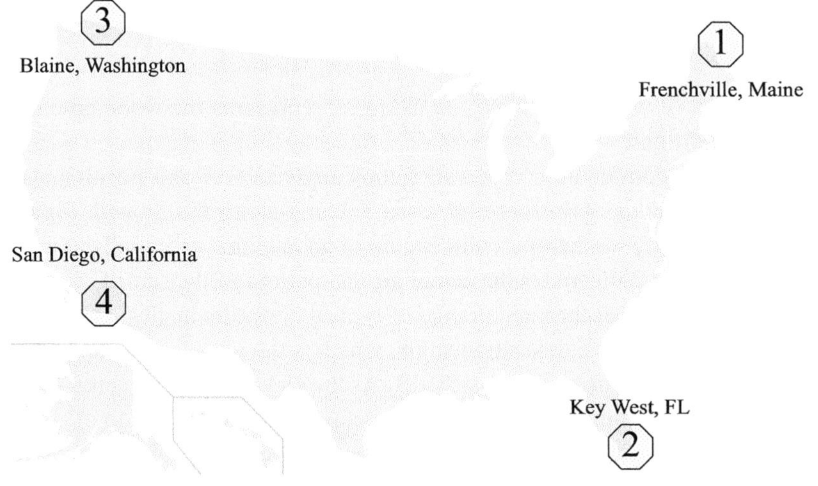


**Figure 9.3** US positions used for antenna gain and look angle analysis.

**Table 9.2** C-band satellite geometry in US corner positions (west longitudes are negative).

| Location | | Frenchville, ME | Key West, Fl | Blaine, WA | San Diego, CA |
|---|---|---|---|---|---|
| | Latitude (°) | 47.28 | 24.56 | 48.99 | 32.55 |
| | Longitude (°) | −68.36 | −81.71 | −122.74 | −117.06 |
| Eastern-most Slot | FSS Slot (°) | −72 | −72 | −72 | −72 |
| | Elevation (°) | 35.5 | 59.3 | 16.2 | 28.9 |
| | Azimuth (°) | 185 | 158 | 122 | 118 |
| | Gain (dB) | −6.77 | −10.00 | 1.77 | −4.54 |
| Southern-most Slot | FSS Slot (°) | −68 | −82 | −123 | −117 |
| | Elevation (°) | 35.7 | 61.3 | 33.8 | 52.1 |
| | Azimuth (°) | 180 | 180 | 180 | 180 |
| | Gain (dB) | −6.81 | −10.00 | −6.22 | −10.00 |
| Western-most Slot | FSS Slot (°) | −137 | −139 | −139 | −139 |
| | Elevation (°) | 5.7 | 21.3 | 31.6 | 45.3 |
| | Azimuth (°) | 254 | 255 | 201 | 217 |
| | Gain (dB) | 10.20 | −1.23 | −5.51 | −9.41 |

## 9.4 Satellite Receiver Characteristics

Satellite receivers also have unique characteristics. The spectral noise density ($N_0$) that is created by any passive radiating body was previously provided in Eq. (7.2). Typical terrestrial systems point at the horizon, and their antenna noise temperatures are generally driven by the equivalent black body radiation from the ground and objects in their field of view; an effective noise temperature that is in the range of room temperature, or about 290 K.

However, the antennas for satellite receivers point towards what is considered empty space. Arno Penzias and Robert Wilson won the 1978 Nobel prize for showing that the temperature of empty space (the Cosmic Microwave Background Radiation (CMBR)) in the cosmos was 2.8 K. Given the losses, internal noise, imperfect antenna illumination, and other causes, the typical satellite receiver (with highly directional antennas) has a system noise temperature around 78 K. This is approximately one third of the noise level of a typical receiver.

One other consideration in satellite receivers is the nature of satellite link dynamics, or lack of them. Because the normal causes of link uncertainty, such as weather, blockage, multi-path, and Fresnel effect impacts are not present, or as significant, on satellite links, they tend to be designed to operate very close to the link margin. For this reason, the interference specification for satellite receivers are very close to the noise floor level. In the International Telecommunication Union (ITU) recommendations, they are typically one tenth of the system noise temperature. The worst-case interference constraints compound a very low acceptable interference noise level, made more constraining by the possible high antenna gain values.

The large difference in allowable signal level between the best and worst case geometry demonstrated the advantage of situational and case-by-case admission control, as provided by the three-tier admission control system. Establishing a fixed level of power would have to be established based on the worst-case conditions, which are the low noise levels, and the high antenna gain at low angles of elevation. The range of worst case to best case is the difference between the lowest performance level of the antenna and its highest gain positioning. From Eq. (9.4), this is a range of 42 dB.

## 9.5    Frequency Domain Analysis

The previous chapter's discussion focused on the propagation of power across the environment. This is reasonably insensitive to frequency. We now consider aspects of the receiving and transmitting systems that are frequency dependent. We consider frequency-dependent characteristics in the following components of an end-to-end system:

**Transmitted Signal** This emission mask reflects the power spectral density of the conducted signal presented to the antenna.

**Transmit Antenna** This gain/loss mask reflects the gain and frequency-dependent performance of the antenna.

**Propagation Path** This analysis provides the path loss across the range of spectrum being managed or protected.

**Receive Antenna** Receive antenna characteristics reflect the gain/loss at the receiver location at each frequency.

**Receive Filter** In some applications, very high-performance filtering is essential to protect from adjacent band operation. The admission control system must reflect the performance of these filters to ensure protection of the receiver.

These receive chain blocks essentially correspond to an interference margin computation. One fundamental difference is in the selection of uncertainty regions. In communications link analysis, we select values deliberately reflecting the set of worst-case (highest-loss) conditions that the designer anticipated. In performing interference protection, the access control system must consider the best-case (least-loss) conditions that might occur.

One other distinction from traditional link interference analysis is that the admission control system has to concern itself with many types of protection, on many different frequencies. Therefore, the model is for the full range of the spectrum addressed by the admission control process. What is out-of-band emissions to one device is co-channel noise to another, and blocking energy to yet another. For this reason, the data needed to support the interference analysis is considerably more detailed than that used for single-channel link margin analysis.

The example analysis will use several typical small cell Access Point (AP) devices as the potentially interfering nodes. We will walk through the entire process of computing the aggregate interference, and applying protection criteria for in-band devices and a satellite receiver in an adjacent band. The interference conditions are shown in Figure 9.4.

This chapter will first develop the one-on-one case analysis shown in Figure 9.4, and then develop the aggregate response from the entire set of nodes. There are several nodes that aggregate to potentially cause interference to the downlink receiver. The example in this text will show the details of the calculations for *AP1* only, and then extend to determine the impacts of *AP1* through *AP8*.

For simplicity in demonstration, we will consider them as point sources, and not model a usage region for the User Equipment (UE) cloud. The protected receiver node is modeled as located at the upper northeastern corner of the USA. It is sited at location: 38.66° north latitude, 77.25° west longitude. The elevation is set to 5.2°, the minimum elevation angle necessary to view the western-most domestic FSS C-Band satellite slot (139° west longitude). For analytic convenience, the azimuth is set to point directly at *AP1*, to reflect the worst case geometry that can occur in the USA. If on an earth map, zero degrees in this graphic example would be approximately an azimuth of 254° true.

### 9.5.1    Interfering node transmission frequency response

Figure 9.5 illustrates a typical transmission frequency domain mask, in this case a specific LTE mode. Because what may be an out-of-band signal to the emitter may be co-channel to some other protected node, the emission mask must reflect worse-case emissions throughout the range of spectrum considered by the admission control system.

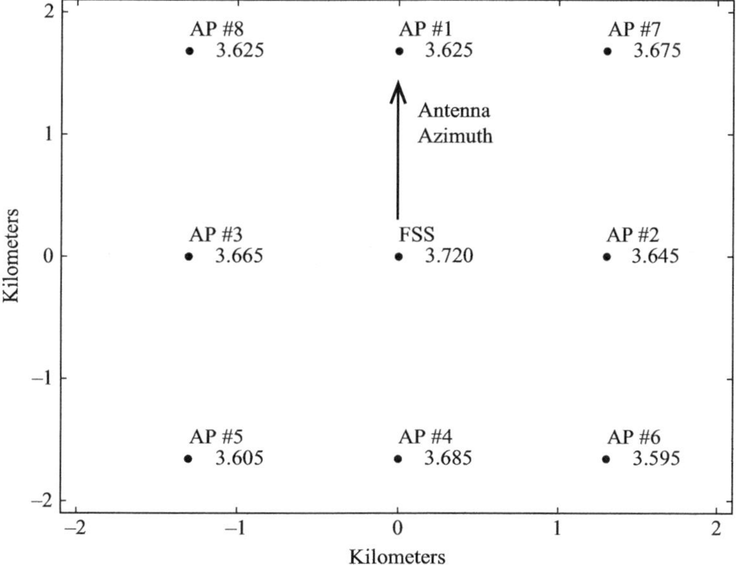


**Figure 9.4** Example node placement for computation of aggregated interference to satellite downlink receiver, oriented to the FSS antenna aperture.

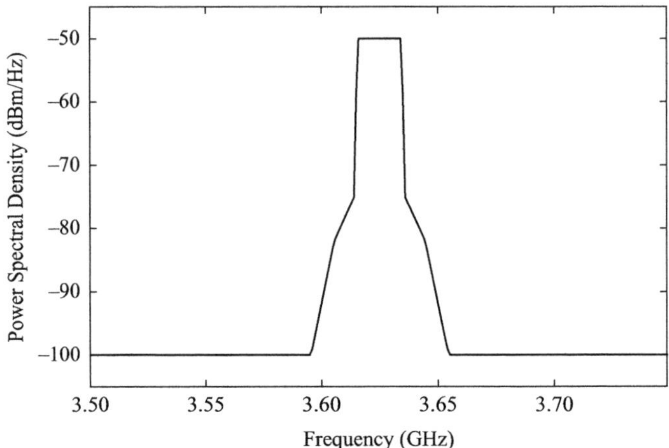

**Figure 9.5** Frequency domain characteristics of typical emitter (Wi-Max) in 3.55–3.7 GHz band.

This data is available for most waveforms and modem implementations. In the case when it is not available, the worst case, regulation-permitted emission mask can be used as a default, with any additional information improving the performance outside of the occupied channel.[8]

The model reflects both the channel-specific and fixed-frequency responses. Typically, the baseband element has a distribution around the center frequency, which translates with the center frequency. Many devices also have fixed bandpass, or high- and low-pass filters, whose response is not responsive to the center frequency. The admission control system should be provided with the composite of the channel-specific and fixed frequency response elements.

## 9.5.2    Transmit antenna frequency response

Figure 9.6 shows the frequency response of a typical, omnidirectional small cell AP antenna. Such an antenna is typically reasonably flat in its spectral response. However, antennas that have significant directionality are typically also more frequency selective, and this may be a factor in their performance, and interference, potential. Poor antenna performance at a frequency somewhat mitigates the interference potential at that frequency.

The important aspect of the antenna frequency response is that it can offer some additional filtering of out-of-band emissions when the devices being protected are far from the emitting node's own operating frequency. In the absence of spectrum response data, the admission control process can base its decisions on the assumption that antenna response is flat across all frequencies.

---

[8] We assume that use of the actual data would not make the performance worse, as it would no longer be compliant for operation.

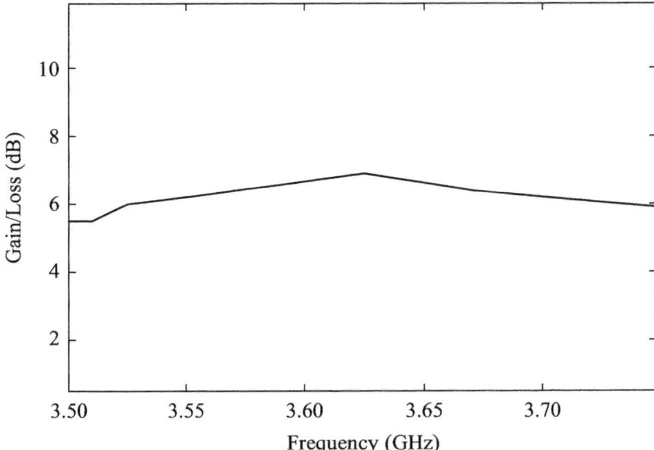

**Figure 9.6** Example of transmit frequency response of a typical access point antenna.

Devices may choose to provide the admission control system with a composite of the baseband and Radio Frequency (RF) sections of the device. In this case, a single mask can be provided, in lieu of individual ones. The admission control system can compute this given reference information provided prior to the grant request, as these two models are channel independent, or the admission control system can permit the device to compute it and include it with the grant request. The results are technically identical. A more complex device may change amplifier modes and linearity, in which case it would benefit from dynamic computation of this mask.

### 9.5.3    Propagation path frequency response

Figure 9.7 illustrates a sample path loss analysis. In this case, it is a 500 meter path that is modeled as free space.

Most path loss determinations are fairly frequency independent, at least over narrow bands of interest. The models adjust the results appropriately using the effective capture area ($\lambda^2$) of the reference isotropic receive antenna for which the path loss models normalize using the $\lambda^2$ term. The path loss increases 6 dB for each octave. Within a given small band, there is relatively little variation in path loss across an operating range.[9]

### 9.5.4    Receive antenna frequency response

Receive antenna response is an analog of the transmit frequency response. In AP applications, the antenna in the end user device is constrained in form factor and ability to control orientation. In the case of satellite installations, the dish itself is

---

[9] The path loss is particularly small when compared to the uncertainty of the estimate as a whole.

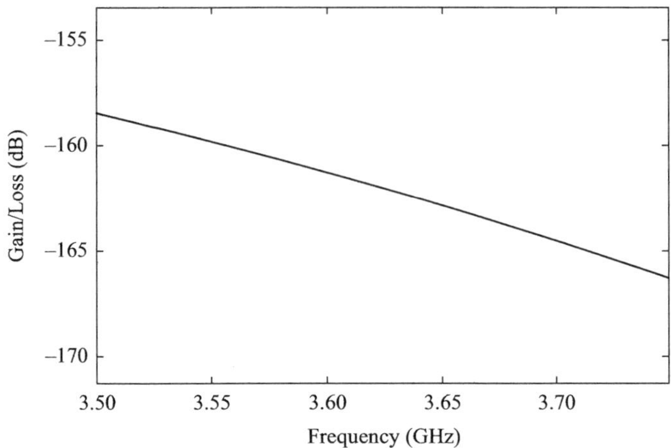

**Figure 9.7** Example of signal path frequency response for a simple free-space link.

essentially frequency independent, other than the $\lambda^2$ term. In other antenna designs, such as omni-directional antennas with gain, the receive antenna may have a significant impact on the performance of the receiving system.

The packaging of antennas in devices often results in highly variable spatial performance. Coupling into other antennas, circuit components, and blockage by closely located components, such as batteries and humans, creates highly variable and unpredictable spatial patterns. So the frequency mask must reflect the worse case response. In the case of interference protection, the worst case is actually the highest performance level (maximum gain at any frequency).

## 9.5.5    Receiver Frequency Response

High performance devices will typically include fixed or tunable filters that protect the receiver front end from overload conditions that would otherwise drive a non-linear response. As shown by Eq. (7.1), the non-linearity is responsive to voltage, not frequency, so limiting power input into the LNA/Low Noise Block Converter (LNB) stage is critical for low-linearity or low-noise receivers. The frequency response of a typical C-Band LNB, without any additional filtering, is shown in Figure 9.8.

Many of these devices are marketed as international, and thus have frequency response that go well beyond the domestic US band. This creates a high potential for blocking due to strong signals in this adjacent band, when it is not restricted to low signal strength satellite reception. In the Citizens Broadband Radio Service Devices (CBSD) proceeding in the US, LNA response data was presented [2] in opposition to sharing the spectrum, by Alion Science and Technology, on behalf of a filing by Content Interests. This data showed that C-Band LNA model had essentially flat response from 2 600 MHz to 4 900 MHz; almost an octave of susceptibility. Implicit in this data was an argument that any of this range of frequencies would block the C-Band receivers. This

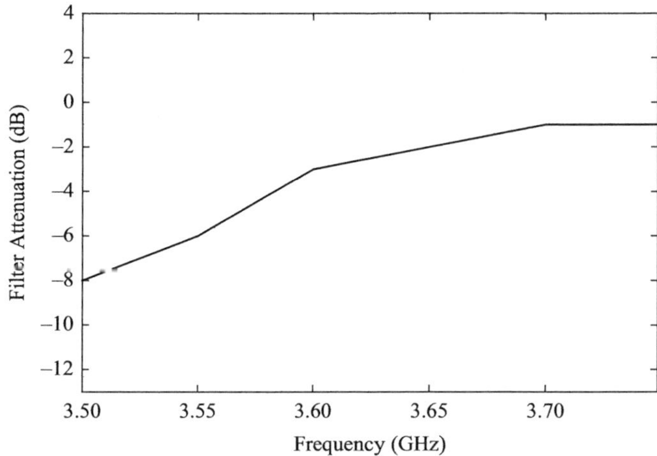

**Figure 9.8** Example of US C-Band receiver unfiltered frequency response.

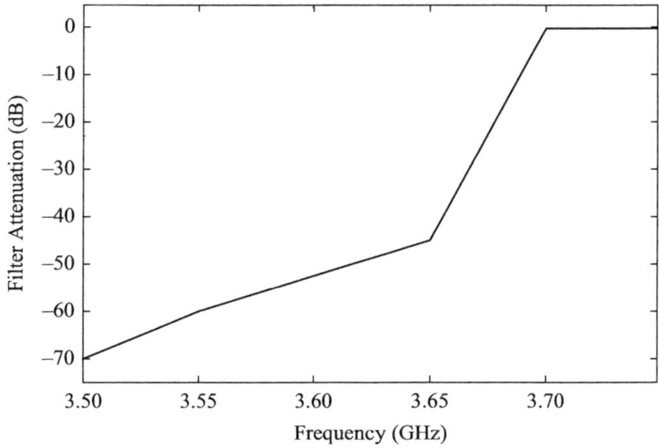

**Figure 9.9** Example of US C-Band receive high pass filter frequency response (example performance data from Microwave Filter Corporation).

is not an unusual claim by an incumbent; poor performance receivers are an effective tool for blocking other uses of spectrum.[10]

Due to this wideband response, many FSS sites add high pass filters to the receiving chain. An example of this filtering is the frequency response shown in Figure 9.9. This is a representative product already sold in the US to protect C-band operators from interference from a number of sources, such as radar altimeters. The example in this figure is the Microwave Filter Corporation (MFC) model 13961W.

---

[10]  The LightSquared conflict with the Global Positioning System (GPS) community is another example.

### 9.5.6     Composite received signal

The ultimate product of this process is to understand the Power Spectral Density (PSD) across the band of regard. This is the product of the masks described previously. The first mask (Transmission Frequency Response) has units of power per frequency range. The other masks are all gain/loss multipliers (or addition in dB units), so the resulting product has the units of a power spectral density. The spatial gain model from section 9.3 provides the directionality characteristics, and is also included in this computation.

Figure 9.10 illustrates the signal that arrives at the C-band receiver, and is presented by the antenna. All of the power below 3.7 GHz is considered to be blocking power. This power level is computed by integrating the PSD in this band of regard. In this case, the total power is -67.2 dBm, so the admission control system must conclude that the satellite receiver is considered to be effectively blocked, and thus would be interfered with if entry of the node was permitted.

Figure 9.11 illustrates the signal after it is processed through the high pass filter, (if present) and is presented to the LNA/LNB. Since all of the energy below 3.7 GHz is typically reduced by a value of almost 50 dB, the power is approximately this level below that shown in Figure 9.10. The filter effectively reduces the adjacent band energy to below the level at which intermodulation energy would significantly raise the noise figure of the satellite receiver. In this example, the third-order intermodulation noise level would be 150 dB lower than in the no-filter case, and far below any impact on receiver operation.

In this analysis we consider two protection criteria that the admission control system could be directed to implement. These two independent coexistence considerations are shown in Table 9.3. Each requires different treatment by the admission control system.

Table 9.4 illustrates the actual criteria measurements that would be created in our sample example. The 0.1 Interference to Noise (I/N) ratio is at a system noise (N) temperature of 78 K, or −179.7 dBm/Hz. This equates to a total interference power level

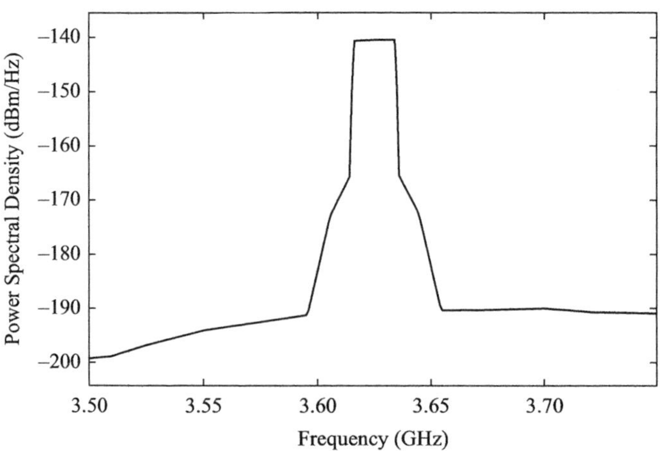

**Figure 9.10** Pre-filter received signal at a C-band downlink receiver.

**Table 9.3** Example satellite protection criteria.

| Criteria | Definition | Level |
| --- | --- | --- |
| Blocking Power | Total power outside of the operating band caused by devices controlled by the admission control system. | −80 dBm |
| Co-Channel Interference | Average power spectral density in any single channel in the protected service. | 7.8 K noise temperature |

**Table 9.4** Example satellite protection criteria satisfaction for a single possibly interfering node.

| Criteria | Level | Before Filter | After Filter |
| --- | --- | --- | --- |
| Blocking Power | −80 dBm | −67.6 dBm | −112.5 dBm |
| Co-Channel Interference | −103 dBm/Chan | −114.3 dBm | −113.8 dBm |

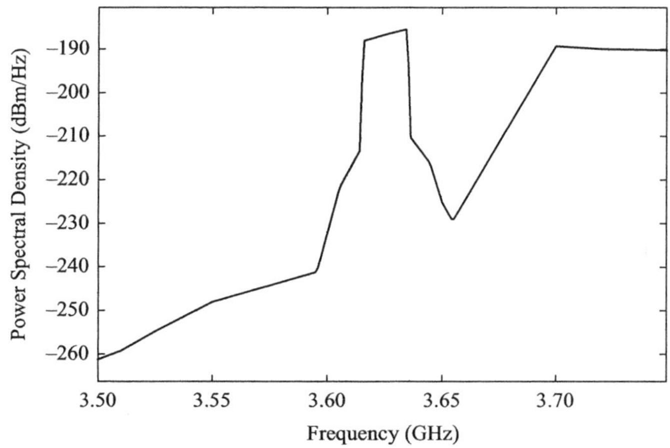

**Figure 9.11** Composite received signal at a C-band downlink receiver.

of −103.7 dBm for a single C-Band transponder channel. The slightly lower co-channel noise reflects the slight loss induced by the high-pass filter. In this particular scenario, the high-pass filtering is so effective in protecting from blocking power that 1 000 more devices could be deployed in similar geometries. On the other hand, there is only approximately a 10 dB margin in the co-channel interference margin, so the density of devices could only be increased by 10 times without exceeding this limit. More devices could be admitted if the out-of-band emissions level for these devices was reduced.

Without the high-pass filter, only very limited deployments could be achieved. This same constraint informs us about the issues in sharing spectrum with FSS sites that are co-channel with the sharing devices. Even when frequencies are deconflicted so that the secondary devices are not using the same frequencies that are in active use by the receivers (specific transponder channels), the total energy that reaches the receiver must be controlled. Unless the power of devices is to be greatly constrained, this implies that spatial directionality management, along with frequency management, is key to any such spectrum sharing regime.

## 9.6    Aggregate Emissions Into FSS Receivers

The same process can now be extended to address the aggregate emissions into the receiver from all of the devices shown in Figure 9.4. Figure 9.12 illustrates the aggregate emissions into a non-filtered receiver. In this case, there is a very significant amount of energy in the shared band below the FSS band, and the blocking energy is much higher than in the one-on-one case. However, the filtering is effective in eliminating even this high density of usage to below the blocking power limit.

The post-filter energy is shown in Figure 9.13. The co-channel noise is not significantly increased, as the one source (*AP1*) was in the main beam of the dish, and all other sources were angularly far from this beam.

Table 9.5 illustrates the numeric results of this analysis. Even though the number of nodes is greatly increased, the co-channel noise is still within the protection criteria.

Comparison of Figure 9.10 (just *AP1*) with Figure 9.12 illustrates that although seven additional nodes were introduced into the region around the FSS site, the signal at the receiver is dominated by *AP1*, which is located in the direct path of the antenna's maximum gain. The high directionality of FSS sites creates very significant differences in impact among devices at different locations and directions.

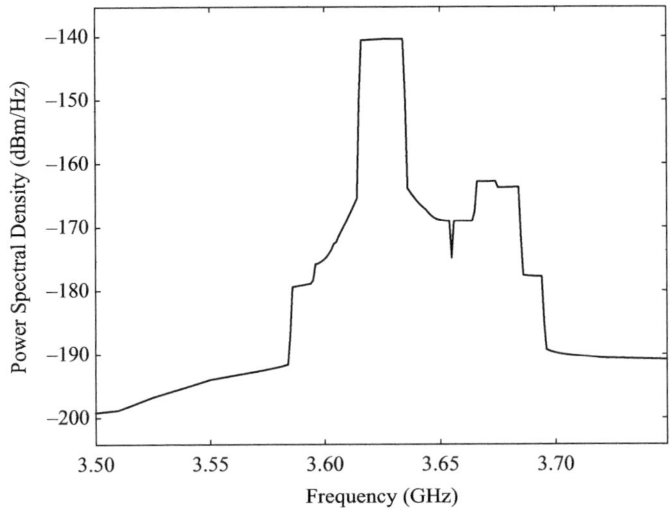

**Figure 9.12** Aggregate received signal at a C-band downlink receiver before high-pass filtering from AP1 to AP8.

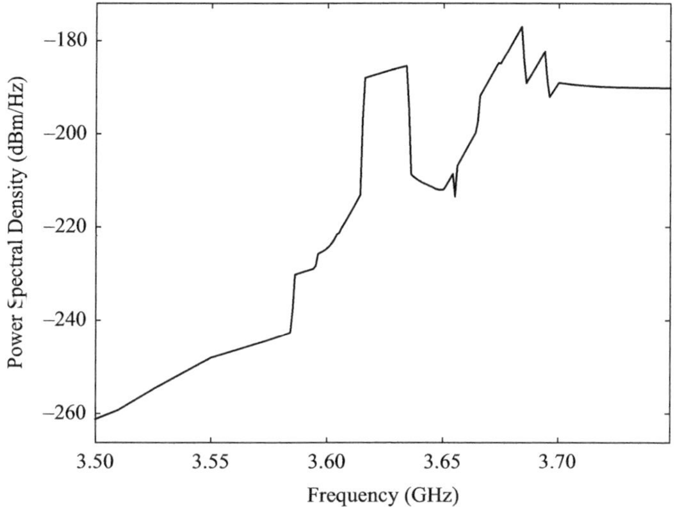

**Figure 9.13** Aggregate received signal at a C-band downlink receiver after filtering from AP1 to AP8.

**Table 9.5** Example satellite protection criteria satisfaction for multiple possibly interfering nodes.

| Criteria | Level | Before Filter | After Filter |
|---|---|---|---|
| Blocking Power | −80 dBm | −67.6 dBm | −107.0 dBm |
| Co-Channel Interference | −103 dBm/Chan | −114.4 dBm | −113.7 dBm |

In this case, the co-channel noise is within the acceptable criteria in both of the filtering cases (before and after front end filter). This is reasonable, as the filter has negligible impact within the satellite's operating spectrum. The blocking power criteria is significantly exceeded before the high-pass filter, but is well within the criteria after the filter. With this analysis, the admission control system can be assured that interference levels into a filter-protected receiver would exceed the protection criteria levels.

This process is computationally intense, as compared to a traditional link analysis, but it is not algorithmically any more complex. The data required is typically supported by device vendors, both in their own literature, and in the compliance documentation provided to regulators. The most flexible method would be to enable both preloaded device characteristics for devices that have limited control over their operating modes, but provide the flexibility for devices to include "replacement" data when the device can adjust these characteristics dynamically. For example, if power control was used in the device, it is likely that the Out of Band Emissions (OOBE) levels would be reduced if the channel power was reduced.

## 9.7     Other Possible Satellite Spectrum Sharing Scenarios

The previous example has focused on sharing of the FSS downlink. Other possible sharing scenarios that must be considered include:

- Sharing of the FSS uplink band
- Sharing with non fixed services
- Sharing with non-geosynchronous satellite services

Sharing with the uplink band of the satellite's transponder is constrained by several factors:

- The receiver antenna coverage is quite wide. In the case of C-Band, it is almost hemispheric; Ku- and Ka-bands have smaller spots, but they are still typically $1\,000\,\mathrm{km}^2$ or more in area. Therefore, any device in that area would be received at the satellite.
- The wide area of earth coverage, or spot beam coverage, results in many devices aggregating energy into the uplink.
- The operation of C-Band, Ku-Band and some Ka satellites is to function as a "bent pipe," in which the received energy at the transponder is transverted to the downlink frequency and re-radiated without processing. This noise on the uplink is additive to the noise introduced in the terrestrial downlink receiver.
- The link from the ground to the orbital position is line of sight, and has only $r^2$ propagation loss. For example, from the ground to a C-Band GEO satellite at $40\,000\,\mathrm{km}$ altitude has only a propagation loss of $200.3\,\mathrm{dB}$, and has typically around $28\,\mathrm{dB}$ of receive antenna gain on the satellite. Considering that the Google clutter loss measurements shown in Figure 7.3 performed at $3.65\,\mathrm{GHz}$ indicated that terrestrial links had measured losses of $180\,\mathrm{dB}$ over just $1\,\mathrm{km}$. The loss to a satellite is much less than might be thought. If these effects are aggregated across thousands, tens of thousands, or millions of devices within the uplink footprint, the aggregation increases of 30, 40 or $60\,\mathrm{dB}$ would raise the noise floor, even $40\,000\,\mathrm{km}$ away. Since a typical uplink operates under $60\,\mathrm{dBW}$, the aggregation of devices on this order could significantly impact the noise level at the satellite.
- The uplink spot sizes are typically larger than the control of any one National Regulatory Authority (NRA). In the downlink case, the possible interference was local to the ground site, and thus an appropriate decision on an NRA-by-NRA basis. However, interference to the uplink would cause disruption to ITU protected primary services, in violation of the United Nations (UN) ITU treaty obligations.

There is a considerable spectrum resource committed essentially exclusively to satellite uplink, as was shown in Table 9.1. However, the limited antenna performance of the in-orbit systems minimizes the ability to effectively share that spectrum without a risk of interference to the ongoing operations of the communications system. Reuse of this spectrum may be dependent on a policy decision regarding the transition of FSS from RF media to terrestrial mechanisms, such as the increasingly pervasive fiber and internet connectivity.

Sharing with non-fixed services, such as mobile services from both geosynchronous, as well as low and medium earth orbit, satellites provides several challenges:

- Mobile users are likely to be anywhere! Tracking them has privacy issues, operational complications, and would likely be disruptive to practical use of the band, unless there is some geographic segmentation of the services provided by the satellite. Fixed geographic segmentation protection of the downlink is similar to the previous discussion in this chapter. However, providing protection of a large number of mobile, partially connected, user terminals is a significant operational challenge.
- Mobile terminals have much less control over antenna gain, and will not provide the predictable and fixed antenna patterns which makes sharing with FSS terminals so practical.
- Non-geosynchronous satellites typically use orbits with significant inclination to the equator. As such, they may be located in any possible alignment to a user terminal. Except at attitudes above their inclination, they could be located to the north or south of the terminal, or to the east or west, as well.
- The protection of the FSS satellites that were in-band to the CBRS band in the USA, and the impact on devices above 3 700 MHz was one of the major, and most contentious, issues in the US FCC CBRS proceeding. The two FCC reports and orders [3, 4] have a considerable discussion of the FCC analysis of the discussion on all sides of the issue, and provide an extensive summary of all of the positions that were presented. The proceeding [1] to authorize Part 90 usage in 3 650–3 700 MHz also has an extensive discussion of the regulator's view of this sharing.
- The ITU has developed extensive material on the protection of satellite operation. As in most ITU processes, the recommendations are largely driven from the perspective of the incumbent users, and tend to be driven by worst-case interference possibilities, and do not reflect the views of other spectrum users, or the opportunity cost of the recommendations they impart [5].
- An example of spectrum sharing analysis is provided in the US FCC proceeding to share 3 650–3 700 MHz. The proceeding includes an appendix that provides the regulator's analysis on the shared spectrum in this range of spectrum. The analysis methods in Appendix D of this document is consistent with the ones in this chapter.

# References


1 Federal Communications Commission, *Report and Order and Memorandum Opinion and Order: In the Matter of Wireless Operations in the 3650–3700MHz Band*, FCC 05–56 (Washington, DC: 2005). http://apps.fcc.gov/edocs_public/attachmatch/FCC-05-56A1.pdf.

2 Alion Science and Technology, *Effects of the Proposed Citizens Broadband Service to CBand DOMSAT Earth Stations*. Attachment to filing by the Content Industries to the FCC on Proceeding 12–354 (CBRS), Consulting Report ESO-13-011-v3 (2013).

3 Federal Communications Commission, *Amendment of the Commission's Rules with Regard to Commercial Operation in the 3550–3650 MHz Band, GN Docket 12–354*, Report and Order

and Second Further Notice of Proposed Rulemaking and Order (2015). http://apps.fcc.gov/edocs_public/attachmatch/FCC-15-47A1.pdf.

4 ——, *Amendment of the Commission's Rules with Regard to Commercial Operation in the 3550–3650 MHz Band, GN Docket 12–354* (2016). https://apps.fcc.gov/edocs_public/attachmatch/FCC-16-55A1.pdf.

5 International Telecommunications Union, *Radio Regulations International Telecommunications Union Radio Regulations, Volume 1* (2012).

# 10 Protection to and From Terrestrial Access Points and Point-to-Point Networks

## 10.1 Unique Aspects of Terrestrial Protection

In this chapter, we consider the protection of terrestrial networks from other sources of interference and the reverse problem: how to characterize the aggregate ecosystem impact of terrestrial networks on other terrestrial, space, or sensor networks. Much of the necessary material has been developed in the prior chapters, so this chapter will only introduce, or further refine, the additional elements of interference analysis necessary to fully integrate terrestrial network protection into a fully functional three-tier spectrum regime.

We differentiate access point protection[1] from other point-to-point protection mechanisms because an access point implies that the access point node itself is not the sole protected entity; it also implies a potential cloud of user devices, or clients, that must be protected from interference. Similarly, if an Access Point (AP) is a potential source of interference, the AP is not the sole source of the network's interference, as the cloud of devices that surrounds it may be sources of interference, as well as the access point itself.

In the case of point-to-point networks, we assume a (typically) directional antenna at each endpoint, and protection driven by the angular susceptibility of the receive systems in use. Of course, these two are the notional, but not the only, cases, and some network architectures may make use of hybrids between these two alternatives which will require some integration of both of these "pure" cases.

There are a very wide range of technologies in use in terrestrial networks, and it is likely as networks move from 4th-Generation Wireless (4G) to Fifth-Generation Wireless Systems (5G) technology (whatever 5G turns out to be), and from networks typically below 6 GHz to networks above 20 GHz. There are a wide range of potential topologies that will have to be introduced into the model. Ten years from now, the simple model laid out in earlier chapters will have to be significantly extended, and made more complicated, to address these emerging topologies and the adaptive nature of many of these technologies.

---

[1] We consider these are typically going to be Local Area Networks (LAN) due to the lower power and higher frequencies typically in use in shared spectrum. However the principles apply equally to higher power, lower frequencies, such as in Wide Area Network (WAN) applications.

**Table 10.1** Terrestrial Network Characterization.

|  | Access Point | Point-to-Point |
| --- | --- | --- |
| Technical Characteristics |  |  |
| Directionality | Assumed to be omnidirectional in horizontal plane, or a large field of regard | Assumed to be highly directional in 2 or 3 dimensions |
| Protected Area Definition | Region around the AP, out to a service area | End points of the link, and locations within the transmitter and receiver antenna regard |
| Separation Methodology | Path loss (distance) from the AP and clients | Deconflicting antenna gain azimuth alignment |
| Regulatory Examples |  |  |
| Non-Part 96 | Television White Space (TVWS) | Microwave Coordination |
| Part 96 | Priority Access License Protection Area (PPA) Part 90 Wireless Service Protection Areas | Fixed Satellite Service (FSS) Protection(s) |

But in the near term, we have two widely deployed models of terrestrial networks: AP-based, and point-to-point. Table 10.1 illustrates the major differences between the pure access point model, and the pure point-to-point model, as they are defined for purposes of this chapter.

## 10.2    Implications of Sub-channeling of Access Point or its Clients

A reasonable assumption in the AP model is that the AP and the client have a symmetric spectrum usage, even if their power distribution might be different (as is the case in the United States (US) Citizens Broadband Radio Service (CBRS) regulations. Different power level ceilings are readily incorporated into this model.

A more complex situation is when the spectrum channel used by the AP is sub-channeled and assigned dynamically to one or more of the client devices associated with the AP. There are several interference conditions that can arise:

**Spectral Power** Even if only one device is operating as a client, if it only uses a portion of the AP channel, but uses its full power level, it will create a Power Spectral Density (PSD) that is much higher than if the whole channel was utilized. This is not an unlikely condition. The lower power limits of the client may make partial band operation similar in power with the AP. If these assignments are made dynamically, any PSD sensitive interference protections will have to be made highly restrictive around worst-case conditions.

**Total Power** Extending the example above, in this case the AP assigns sub-channels to multiple clients, but also allows the devices to transmit in the same slot. Each device might be limited by the power limit of client devices, but the aggregate emission level is the sum of all of the devices, and therefore is linearly increased.

When we consider both of these, the possibility is that both the total power in the band and the individual sub-channel PSD increases. For example, Long-Term Evolution (LTE) permits allocation of 50 individual resource blocks (sub-channels) per 10 MHz channel. Since the regulatory limits are on a per User Equipment (UE) basis, resource block scheduling enables multiple UEs to emit at their full power limits simultaneously. This increases the total aggregate emissions by 17 dB when the maximum of fifty individual devices are allocated their full power by the AP, each in a single resource block. The total aggregate channel emission level is increased by 17 dB compared to operation by a single UE.

The US CBRS regulations establish different power levels and PSD limits for clients.[2] UE devices are allowed only 10 dB less power than the AP in all cases, which comes close to addressing the 17 dB possible (but unlikely) impact of simultaneous sub-channel operations. Although it establishes limits on the PSD of the AP, it establishes no specific level for clients. In fact, the maximum PSD for a client of any category AP is actually 27 dBm/10 MHz, well above a "Category A" AP, which is limited to 20 dBm/10 MHz, but below the "Category B" limit of 37 dBm/10 MHz.

## 10.3    Implications of MAC layer, TDD, and FDD operation

The interference environment created by a network of nodes is determined by more than just the radios themselves, but also the impact of the upper layer processes, particularly the Media Access Layer (MAC), which controls access to the channel. How the MAC layer manages resources is key to the design of an interference management regime. The previous section described methods that a MAC or Radio Resource Management (RRM) agent might use to control access by multiple devices to a single channel simultaneously. In this section we consider distribution in time, or the allocation of channels into independent uplinks or downlinks. These processes address the scheduling and deconfliction of the clients and AP transmissions.

The simplest case to analyze is that of Frequency Division Duplex (FDD). FDD essentially is two independent radio systems, with one operating as an uplink (client to AP), and one operating as a downlink (AP to client). Spectrum used in this way is considered to be *"paired."* This has been the most common approach in modern commercial (cellular 2G, 3G, and 4G technology) wireless practice, but is becoming difficult to achieve, as pairable spectrum is exhausted. Unpaired operation in Time

---

[2] The regulations actually invoke the LTE "User Equipment" nomenclature for the clients.

Division Duplex (TDD) mode is becoming the only option to exploit available spectrum opportunities.

It should be noted that the use of FDD in three-tier spectrum is likely to be rare, if ever present. There are a number of engineering reasons for this:

1. FDD operation requires significant frequency separation between the uplink and downlink channels. A minimum of 30 to 100 MHz separation is typical. Locating uplink and downlink channels that are available for use, and appropriately separated, will be a challenge in a constrained, shared band with dynamic access.
2. The diplexer component is very frequency specific. The diplexer separates the antenna signal to, and from, the receive and transmit ports of the radio. The current technology has a fixed demarcation between these two ports, greatly constraining the choice of spectrum.
3. Pairing is typically predefined between the receive and transmit frequencies. In TDD, the client can recognize a beacon or other signature of its associated AP and then utilize the same channel. With FDD, the knowledge of the receive channel does not inform the device of what is expected on its uplink channel. It cannot be fixed in advance with certainty.

## 10.4    Impact of Network Client Device Clouds on External Networks

Section 6.2 provides a general analytic framework for defining the protected area of a network, which drives the protection of the network from external networks. In that chapter, the interest was in understanding the theoretical density implications of various network client protection strategies in a homogeneous network.

This was a suitable treatment to determine the protection of clients within the network, as only the most impacted protected device had to be considered. It did not develop a model for how the aggregation of this network should be considered as a source of interference to other networks, which we will now develop. In this development, we will consider all the non-theoretical aspects of real MAC layers, resource scheduling and other aspects of modern wireless design that were introduced in the previous sections.

Different regimes may choose to reflect, or abstract, these considerations differently. How detailed or abstract does not change the nature of the three-tier regime. It will impact either the density it can achieve, or the level of protection assurance it can provide.

Statistical assumptions and confidence levels can play an important role if there is the possibility of multiple client transmissions occurring. First, we examine the worst case for a single client transmission. For simplicity in expression, we imagine that we have a single propagation model that provides a loss value between any point separated by distance $d$ at frequency $f$ as $L_P(d)$.

Since all of these examples are co-channel, we will ignore the $f$ term. Client devices have a PSD of $P_C$ and AP PSD as $P_{AP}$. For convenience, all are in dBm. This simple lay-down is shown in Figure 10.1, showing the location of the client, the AP, and the

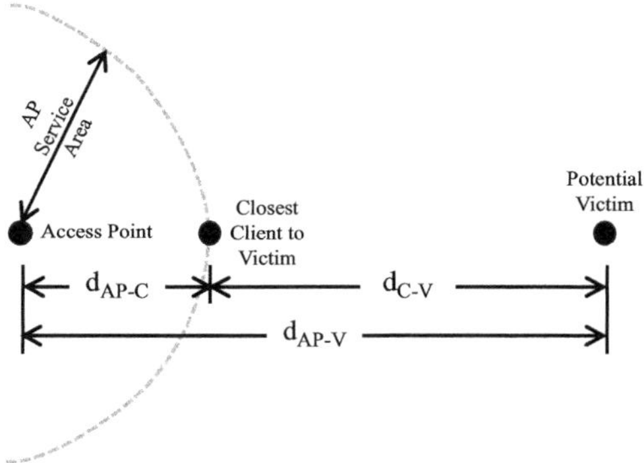


**Figure 10.1** Single client transmission worst case positioning.

potential victim of their interference. They are in the worst-case alignment, for a locally flat earth.

If we assume a Listen Before Talk (LBT) and/or a simple TDD model, then only one of these nodes can be transmitting in any single channel, so the interference at the victim is given by Eq. (10.1).

$$\max(P_{\text{AP}} - L_P(d_{\text{AP-V}}), P_C - L_P(d_{\text{C-V}})) \tag{10.1}$$

Since $L_P(d_{\text{C-V}}) \leq L_P(d_{\text{AP-V}})$, if $P_{\text{AP}} \approx P_C$ then the client dominates the worst case interference estimate external to the AP service area in simple TDD mode. The domination of the client or AP is a function of the relative power allowed to each, and the distance from the victim to the AP service area. Specifically, the AP dominates the interference when $P_{\text{AP}} - P_V > L_P(d_{\text{AP-V}}) - L_P(d_{\text{C-V}})$. So, if the victim is far from the service area, the two propagation loss estimates converge ($L_P(d_{\text{AP-V}}) \approx L_P(d_{\text{C-V}})$). This is due to the convergence of the limit of the ratio of the propagation losses, as shown in Eq. (10.2).

$$\lim_{d_{\text{C-V}} \to \infty} \left( \frac{d_{\text{C-V}}}{d_{C-V} + d_{\text{AP-C}}} \right)^{\alpha_p} = 1 \tag{10.2}$$

where $\alpha_p$ is the propagation exponent. At distance, the more powerful AP dominates the worst-case interference estimate. We can further characterize the ratio of these two distances by considering the relative signal level that defines the boundary of the AP service area, and the protection afforded the victim. Typically, the receiver requires that the $\frac{S}{N}$ ratio be significantly above 1 to support meaningful operations of the network. This defines the maximum size of the AP service area, and thus the maximum value of $d_{\text{AP-C}}$.

The allowable interference level at a victim is typically limited to some fraction of the thermal noise level ($N$) in order to ensure meaningful operation of the network links. We consider the definition of the AP service area boundary is the interior of a region defined by having a signal level of at least $P_{\text{SAB}}$. Similarly, we protect victims to an interference level of $P_{\text{int}}$. The ratio of $\frac{P_{\text{SAB}}}{P_{\text{int}}}$ establishes the ratio of $d_{\text{AP-C}}$ to $d_{\text{C-V}} + d_{\text{AP-C}}$.

We know that the relationship of the ranges, transmit power, and interference limits is given by Eqs. (10.3) and (10.4).

$$L_P(d_{\text{AP-C}}) = P_{\text{AP}}/P_{\text{SAB}} \tag{10.3}$$

$$L_P(d_{\text{C-V}}) = P_{\text{C}}/P_{\text{int}} \tag{10.4}$$

We know that $L_P \propto d^{-\alpha}$. In dB units, $L_P \propto -10\alpha \log_{10}(d)$. The ratio of these distances is therefore given by Eq. (10.5).

$$\frac{L_P(d_{\text{AP-C}})}{L_P(d_{\text{C-V}})} \propto \frac{P_{\text{AP}}/P_{\text{SAB}}}{P_{\text{C}}/P_{\text{int}}} = \frac{P_{\text{AP}}P_{\text{int}}}{P_{\text{SAB}}/P_{\text{C}}} \tag{10.5}$$

In dB terms, Eq. (10.5) becomes as shown in Eq. (10.6).

$$L_P(d_{\text{AP-C}}) - L_P(d_{\text{C-V}}) \propto P_{\text{AP}} - P_{\text{SAB}} - P_{\text{C}} + P_{\text{int}} \tag{10.6}$$

The difference in the path loss permitted between AP and client compared to the client and the victim is directly related to the differences in the size of the service area, and the protection limits of the devices.

Solving Eq. (10.6) for the ratio of these distances yields Eqs. (10.7) and (10.8), which provides the relative size of the service area and the excluded region around the service area. This is based on the fact that $L_P(d) \propto 10\alpha \log_{10}(d)$.

$$L_P(d_{\text{AP-C}}) - L_P(d_{\text{C-V}}) \propto P_{\text{AP}} - P_{\text{SAB}} - P_{\text{C}} + P_{\text{int}} \tag{10.7}$$

$$10\alpha \log_{10} \frac{L_P(d_{\text{AP-C}})}{L_P(d_{\text{C-V}})} \propto P_{\text{AP}} - P_{\text{SAB}} - P_{\text{C}} + P_{\text{int}} \tag{10.8}$$

We can consider this analysis using the specific values established in the US CBRS regulations. The CBRS equivalent of the AP service area is a PPA, which is defined as out to a distance where the signal level is −96 dBm/10 MHz. The protection of an adjoining PPA is at a level of −80 dBm/10 MHz. The Equivalent Isotropic Radiated Power (EIRP) permitted to an indoor access point is 30 dBm/10 MHz, and an outdoor one is permitted 47 dBm/10 MHz. Similarly, a typical client device (not requiring registration, but permitted to be mobile) is allowed an EIRP of 23 dBm/10 MHz.

The indoor and outdoor relative ranges of the service area and exclusion zone are shown in Table 10.2 for a propagation exponent ($\alpha$) of 2 indoors, and both 2 and 4 outdoors, as might be typical.

This model works for technologies that have either an AP transmitting or a single client. This includes Wireless Fidelity (Wi-Fi), Bluetooth and most LBT systems. It does not address the situation when the LTE network allocates resource blocks to multiple clients within a single time slot. In that case, we need to consider the much more complex conditions that multiple-clients transmission raise.

Some conclusions from this table include:

- The propagation model drives everything in considering the dimensions of the protected service areas and the protection region. With unrealistic propagation, the zones are extensive. With reasonable assumptions about outdoor propagation, the zones shrink dramatically. The indoor size is quite large, but a 30 dB loss in the

**Table 10.2** PA Service Area and Protection Zone relationships.

| Parameter | Symbol | Indoor $\alpha = 2$ | Outdoors $\alpha = 2$ | $\alpha = 3$ |
|---|---|---|---|---|
| Maximum AP EIRP (dBm/10 MHz) | $P_{AP}$ | 30 | 47 | 47 |
| Maximum Client EIRP (dBm/10 MHz) | $P_C$ | 23 | 23 | 23 |
| AP Protection Level (dBm/10 MHz) | $P_{int}$ | –80 | –80 | –80 |
| AP Service Area Level (dBm/10 MHz) | $P_{SAB}$ | –96 | –96 | –96 |
| Service Area Loss (dB) | $P_{AP} - P_{SAB}$ | 110 | 127 | 127 |
| AP-to-Victim Loss (dB) | $P_{AP} - P_{int}$ | 126 | 143 | 143 |
| Client-to-Victim Loss (dB) | $P_C - P_{int}$ | 119 | 119 | 119 |
| Radius of Service Area (m) | $d_{AP\text{-}C}$ | 2,08 | 14,763 | 112 |
| Distance: AP-Victim (m) | $d_{AP\text{-}V}$ | 13,158 | 93,154 | 385 |
| Distance: UE-Victim (m) | $d_{C\text{-}V}$ | 5,877 | 5,877 | 61 |

exterior wall would reduce this range by a factor of $2^5$, or a 32 times reduction in the range of all areas.

- Clients are an important consideration in interference to and from the access point service area only if the power levels are similar. With high power, or high EIRP, the AP dominates all of the interference considerations.
- The area tied up in service areas is a function of propagation model. With high loss models, the service area dominates, while with low power APs, and clear propagation, the protection zone around the service area is the dominant driver in density.

## 10.5 Point-to-Point Protection

Most of the methodology for point-to-point protection has already been developed in the more complex, three-dimensional case of FSS protection in Chapter 9, and will not be repeated. The computation of terrestrial point-to-point paths is a simpler process, since the projection of the gain pattern on the earth is identical to the gain pattern of the antenna, as if the elevation angle was zero.

The differences arise from the diversity of antenna performance and characteristics. While most of the FSS receivers utilize prime focus dishes, the antenna designs deployed above 2 GHz typically have a wider range of technologies, including the conventional dishes and horns, but also planar arrays, phased arrays, increasingly beam-forming, and soon, massive Multiple Input/Multiple Output (MIMO). Each of these antenna types poses unique challenges in performing effective and spectrally efficient point-to-point protection.

**Focused Dish** The model for dish antennas is similar to that outlined in the FSS case. In the case of FSS operations, the antenna pattern, and thus gain, is dictated by the necessity to separate adjacent orbital slots. They are regulated by most, if not all, National Regulatory Authority (NRA) regulations. An admission control system can assume the worst-case (maximum possible gain) perimeter of the three-dimensional pattern. However, the antennas used in terrestrial point-to-point links are dictated by the economics of the deployment and the need for antenna gain and pattern control. Therefore, the admission control system cannot assume any specific de facto antenna characteristics, and these must be uniquely provided and validated for each receiver and uplink deployment, and maintained throughout the lifetime of that deployment. Technically, this is not a challenge for the control system operator, but it is a significant operational challenge to assure this discipline is maintained in the field.

**Planar Array** Planar arrays are similar in their challenges to dish antennas, but introduce the additional consideration that planar arrays often have relatively high, almost random, angles of relatively high gain, referred to as grating lobes. The grating lobe angular positions can be highly variable by frequency, so they are hard to characterize in advance. They impact both the transmitted pattern and the receiver susceptibility to interference. Since the grating lobes are highly frequency dependent, they may be difficult to map for interference analysis, so the antenna pattern may have to be an enclosing perimeter, that may significantly overstate antenna gain by tens of dB in most directions.

**Phased Arrays** Phased array antennas introduce all of the issues associated with the planar antenna case, but introduce their own, as well. A phased array is presumably deployed because the antenna will serve devices on more than one azimuth. Therefore, the admission control system must either be aware of all of the node locations (and assume they are fixed), or assume that the phased array could be in any setting within the field of regard of the array, typically from $\pm 30°$ to $\pm 60°$ from the angle perpendicular to the array face. Phased arrays have known (and specific) directionally/directionalities, and must be explicitly pointed by varying the time delay in the antenna elements. A system operator should be able to inform the admission control system of the specific angles, or range of angles, over which the antenna will operate.

**Beam-forming** Beam-forming antennas function similarly to the phased array antennas, except instead of being programmed to specific directions, they adapt the beam to optimize the signal by controlling gain and nulling across the field of regard. Many of the uses of beam-forming antennas are distinct from phased arrays. Whereas phased arrays are effective in line-of-sight, beam-forming is often effective in exploiting non-line-of-sight paths through dynamically focusing on the multi-path, and simultaneously nulling other interfering signals. Therefore, it is unreasonable to make any assumptions regarding the actual antenna pattern of such devices, and they have to be considered to have emissions at peak focus in any azimuth in their field of regard, with an understanding that the beam direction cannot be predicted, even if the endpoint is known. In some applications, beam-forming is utilized to provide multiple independent beams, in the same spectrum, but in different directions. In this case, the peak gain will be reduced from the maximum focused EIPR, but the angle of regard will be maximized. The total emission model from such antennas may exceed unity gain when integrated spherically.

**Massive MIMO** Massive MIMO or, space-time coding, is increasingly viewed as one of the key technologies in expanding wireless capacity within very constrained spectrum availability. However, although it includes a large number of antennas, it focuses information, not energy, so the far field of the device is not predictably focused in the same manner as a beam-formed device. We must consider the gain pattern as typically Gaussian around the field of regard of the array or MIMO elements.

The previous description of FSS protection is applicable to point-to-point network outbound and inbound interference, once the adjustments for the antenna characteristics at both ends of the link are reflected in the analysis. Typically, isolation is less of an issue for terrestrial applications than for FSS, due to a number of factors:

- The antenna gain achieved by point-to-point antennas is typically less than that generated by the 3.3 to 8 m diameter dishes in use for C-Band, as an example. They therefore are a less stressing case.
- The noise floor of the terrestrial applications is at least 6 dB higher, due to the earth temperature rather than deep-space background.
- Terrestrial receivers are typically more robust to blocking power due to higher Third-Order Input Intercept Point (IIP3) dynamic range of signals in mixed-usage bands, with a wide range of emitter power, EIRP and path losses.

## 10.6 Three-Dimensional Considerations

The traditional network planning has treated propagation analysis as a set of points along a two-dimensional surface. The distance between two points was considered to be the surface difference between their latitude and longitude points. This was certainly a valid approach for frequencies below 1 GHz. It might even have been correct for high-power elevated base stations at frequencies below 2 GHz. However, in lower-powered, indoor or outdoor, networks, the two-dimensional assumption is worth examining.

The One World Trade Center building (without considering antennas) is 417 m above the ground. Although two *lat, long* points may be almost adjacent, the actual distance could be greater than 417 m. The free-space path loss for points 10 m (at 3.62 GHz) is not 64 dB, but over 96 dB including the vertical distance. There is a 33 dB additional path loss. If the path is through building materials, the additional loss could be literally hundreds of dB more, considering that the per floor loss can exceed 50 dB.

Concepts such as a client service area, the US CBRS regulations PPA, or census tracts are not meaningful in a three-dimensional environment. There is no technical reason that devices on floors above a protected network cannot share the same spectrum without interference to the protected network. This three-dimensional environment is not addressed by extending traditional methods of spectrum management.

An interesting consideration is what is actually provided by a protected license. The protection level of the US CBRS regulations PPA is defined at a height of 1.5 m. So, a node sufficiently high may be quite close in horizontal distance, but with sufficient path loss (due to vertical distance and clutter loss) to not violate the Priority Access License (PAL) criteria. Even if the building was not as tall as the World Trade Center, the loss through numerous concrete floors with up to 50 dB loss per floor,[3] would provide sufficient protection even a few floors away.

It is a matter for the policy community to determine how to address the vertical aspect of protection rights. There are no clearly absolutely correct engineering options that do not have some regrets, or opportunity for "gaming."

- If the protected network can define the elevation at which the protection is provided, it can create excessive protected regions. Higher elevations would transition the path loss from terrain-driven to free-space, greatly expanding the size of the exclusion zone. The temptation will be for PA users to claim these higher elevations as the basis for protection.
- If a fixed elevation is established by default for the second tier (as is provided in the US CBRS regulations), then they will be inappropriate to many of the use cases that appear.

The need for more advanced concepts than protected regions is driven by several considerations. These include the high material loss at frequencies above 2–3 GHz, the transition from outdoor to indoor coverage, and the low-powered nature of likely shared-spectrum deployments.


## 10.7     Suggested Reading

There are a wide range of readings that can contribute to an understanding of current and future technologies in terrestrial networking.

- An understanding of the antennas in use in point-to-point systems is essential to development of the wide range of models needed to represent the wide range of architectures used in this application.

  - MIMO systems are likely to emerge in both point-to-point and AP architectures. The original papers on this technology [1] have been extended to much more complex, highly scaled applications. Examples of critical work in massive MIMO have a fundamentally different impact on architectures than the original work did.
  - It is likely that adaptive beam-forming will also play a big part in future wireless systems.

---

[3] A table showing a range of loss through building materials is provided in Table 12.1.

- It is desirable to abstract the specific details of the wireless protocols in developing an interference protection regime. Technologies in use do change, and inclusion of technology specific features can tend to create technology favoritism in the regulations. However, the subtleties of these technologies are important considerations. The prevalence of both Wi-Fi and LTE in current usage certainly implies that any regime reflect their subtleties, as a minimum. There are a wide range of works describing both of these physical and MAC layer designs.
- The US CBRS regulations [2] include an appendix that describes the PPA protection (the CBRS implementation of AP service areas) with drawings of each scenario and how protection is to be afforded. Similarly, a Public Notice (PN) was issued by the Federal Communications Commission (FCC) [3] to define the protection rights of the grandfathered Part 90 devices; primarily wireless Internet service provider (WISP) and utility networks.

## References


1 G. J. Foschini, Layered space time architecture for wireless communications in a fading environment using multiple antennas. *Bell Labs Technical Journal*, **1**/2, (1996) 41–59.
2 Federal Communications Commission, *Amendment of the Commission's Rules with Regard to Commercial Operation in the 3550–3650 MHz Band, GN Docket 12–354* (2016). https://apps.fcc.gov/edocs_public/attachmatch/FCC-16-55A1.pdf.
3 ——, *Wireless Telecommunications Bureau and Office of Engineering and Technology announce Methodology for Determining the Protected Contours for Grandfathered 3650–3700 MHz Band Licenses GN Docket No. 12-354*. Public notice (2016).

# 11 Protection to and From Radar Systems

## 11.1 Unique Aspects of Radar Systems

Although most spectrum sharing discussion revolves around communications services sharing with communications services, radars consume massive amounts of spectrum, and meaningful spectrum sharing will require that this spectrum be meaningfully shared. Technical and operational issues impacting consideration and implementation of radar and communications spectrum sharing are many, including:

- Radars have very high peak power, and also very high gain antennas, so their interference potential could appear to be quite significant over an extensive region.
- Radars are often used for safety of life (aviation, maritime) or national security (military) related applications, so the tolerance for interference is low or non-existent.
- The emission characteristics of some radars are not publicly available from the Government or military users, and thus sensing design for detecting some radars is difficult. Others are so variable in pulse structure and scanning patterns that they are essentially random to the sensor designer. Similarly, the interference susceptibility of these radars is also not publicly available for use in determining protection modalities.

- Military radar users are concerned about the detection of some of their operations and characteristics.
- Although the transmission duty cycle for a radar is very low, the radar is "using" the spectrum while it is listening for a return signal, and is impacted by interference for a period twice the time it takes light to reach the outer radius of the radar.
- Bi-static radars have receivers that are not located at the same location as the transmitters (the illuminator), and therefore the angle or distance to these radar receivers cannot be determined by a sensing regime.

For these reasons, other than the Dynamic Frequency Selection (DFS) band at 5 GHz, there are few success stories for radar spectrum sharing.

## 11.2 A General Model of Radar System Design and Performance

This section introduces some basic parameters of radar operation, with only the factors related to the detection, protection, and mitigation of interference from radars considered. In most cases, the models are fairly simple radar modes of operation. Specialized military and space-borne radars have typically much more complex operational modes, but still have the basic characteristics provided in this discussion.

The challenges in radar spectrum sensing derive from their high power and high antenna gain. These features are necessary to overcome the very challenging link budget, which is expressed as the radar range equation, shown in Eq. (11.1). $R_{\max}$ is the idealized range of the radar, while $P_S$ is the transmitted power, $G_{\mathrm{ant}}$ is the antenna gain (absolute, not dB), $\lambda$ is the wavelength, $\sigma$ is the target's effective radar cross section (typically in $m^2$), and $P_{E_{min}}$ is the minimum power to declare a detection.

$$R_{\max} = \left( \frac{P_S G_{\mathrm{ant}}^2 \lambda^2 \sigma}{P_{E_{min}} (4\pi)^3} \right)^{1/4} \tag{11.1}$$

Several factors that drive radar design are apparent in this equation. Communications in free space follows the $r^2$ law. However, because radar signals must traverse the path twice, they follow an $r^4$ relationship. For a radar to double its range requires 16 times more power, rather than 4 times, as in the case of communications. Also, antenna gain is squared for radars, as the antenna functions at both ends of the link.[1] These factors drive the fundamental differences between communications systems and radar systems.

Radar systems have very unique emission characteristics compared to communications systems, and equally so, compared to each other. There are a number of characteristics of radar emissions that are fundamental to the sensing and protection of radars, and the mitigation of radar interference into the communication systems.

The most significant ones from the spectrum sharing perspective include:

**Pulse Width** The Pulse Width (PW) is the time interval in which the radar actively emits the pulse. Typically measured in nano-, micro-, or milli-seconds.

**Pulse Repetition** Pulse Repetition Frequency (PRF) is the rate at which the radar emits pulses. Typically measured in Hz.

**Pulse Bandwidth** The spectral bandwidth of an individual pulse. Radar range resolution is directly related to the pulse bandwidth, so accurate ranging requires significant bandwidth.

**Operating Range** The frequency range of the radar within which the pulse frequency can be selected.

**Peak Power** The peak power during the transmission of the pulse. Measured in watts, or dBW.

**Frequency Agility** Some radars will vary the frequency of the radar, possibly on a pulse-by-pulse basis.

**Beam Agility** Electronically Steerable Array/Antenna (ESA) provides the radar with the capability to radiate each pulse in arbitrary directions, rather than a fixed rotation. This makes the radar appear to a sensor as if it was a random impulsive noise event, rather than an intentional emission.

---

[1] This is applicable to monostatic radars only.

**Rotational Period** For radars that have rotating, mechanical antennas, this is the time it takes for the antenna to make one full rotation. Rotation is measured in seconds.

**Beamwidth (H)** Horizontal beam width proving the angular range of maximal gain. Most radars have asymmetric patterns to maximize resolution in the intended dimension. Typically measured in degrees.

**Beamwidth (V)** Same as above for the vertical dimension.

**Main Beam Gain** Maximum gain within the main beam of the radar. Typically measured in dB.

**Front to Back Ratio** The difference in antenna gain between the main beam (front) and the least gain region (backside) of the antenna pattern.[2]

**Elevation Angle** Aircraft tracking radars may slightly elevate their beam to avoid illuminating excessive ground terrain, and deduce the stationary ground return component.

**Chirping** Some radar waveforms sweep through a set of frequencies within a single pulse. The apparent pulse on any one sub-channel is therefore shorter than the actual pulse width.

Some generalizations can be made. Solid-state and vacuum-tube radars deliver similar energy per pulse, but solid-state radars tend to have lower power and longer pulses than similar vacuum-tube designs, thus often have higher duty cycles. Modern military radars typically have agile beam directionality (phased arrays), while civil radars remain mechanical. Military radars are typically agile in frequency, while civil radars are typically fixed. There are more radical deployments of radars, such as bistatic radars, where the transmitter and receiver are at different locations,[3] passive radars, which use environmental energy for detection, and Ultra Wideband (UWB) radars that do not have conventional frequency assignments.

The impact of the radar on communications systems is driven by both the peak and average power. The peak Equivalent Isotropic Radiated Power (EIRP) of the radar is given by: *Main Beam Gain + Peak Power* in dB units. The average EIRP is computed by Eq. (11.2).

$$Average\ EIRP = EIRP + 10_{Log_{10}}(Pulse\ Width * Pulse\ Repetition). \qquad (11.2)$$

A simplified version of the time signature of a traditional mechanical rotating radar is shown in Figure 11.1. Some military radars have multiple modes of operation (such as search, acquisition, fire control, counter-countermeasures, etc.) that may modify these characteristics rapidly; even on a pulse-by-pulse basis.

A representative set of values for some common radars, and radar types, is shown in Table 11.1. The first radar (USA AN/SPN-43C) is the radar that is incumbent in the band the United States (US) Citizens Broadband Radio Service (CBRS) is sharing. Although

---

[2] Small grating nulls that create very low gain nulls in the pattern are generally not considered in this measure.

[3] Bistatic radars are primarily for military applications to preclude using the direction of arrival of the transmission to jam the receiver.

**Table 11.1** Example radar characteristics.

| Measure | USA AN/SPN-43C[a] | 3 GHz Maritime Radar [1] | Meterological ITU Radar G [2] |
|---|---|---|---|
| Pulse Width ($\mu$sec) | 0.95 | 0.05–1.2 | 1.6, 4.7 |
| PRF (kHz) | 1 | 0.375–4.0 | 318–1 304 |
| Pulse Bandwidth (MHz) | 1.6 | Unspecified | 0.6 |
| Peak Pulse Power (dBW) | 60 | $\approx$45–48 | 57 |
| Frequency Agility | Tunable, Fixed | Tunable, Fixed | Tunable, Fixed |
| Operating Range (MHz) | 3 500–3 700 | 3 020–3 080 | 2 700–3 000 |
| Beam Agility | No | No | No |
| Rotational Period (sec) | 4 | 1.0–3.0 | 20 |
| Beam Width (H) ($^\circ$) | 1.75 | 1.0–4.0 | 0.92 |
| Beam Width (V) ($^\circ$) | 4.4 | 24–30 | 0.92 |
| Main Beam Gain (dB) | 43 | 26–28 | 45.7 |
| Elevation Angle ($^\circ$) | 3 | Not Specified | 0.5–20 |
| Chirped | No | No | No |

[a] Source: http://spectrumwiki.com/wiki/DisplayEntry.aspx?DisplyId=225

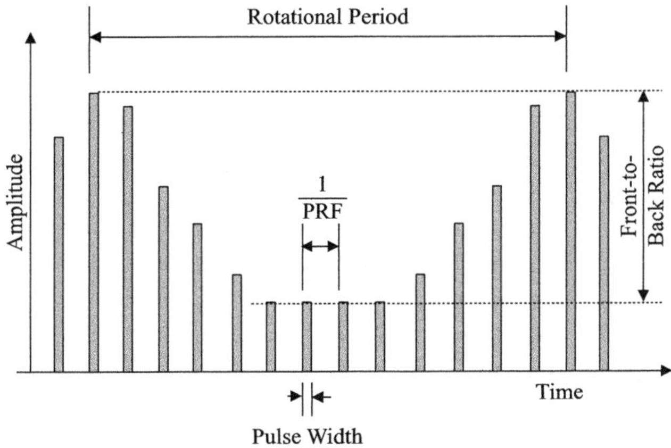


**Figure 11.1** Simplified time-sequence of a radar signal.

operated on modern naval carriers for air traffic control purposes, it is more typical of a civil air traffic control radar than a tactical military one.

## 11.3    Interference to Radar and Space Sensor Systems

Demonstrating that a given spectrum sharing regime will avoid interference to radar system operation is a nontrivial challenge. The communities that operate radars are generally extremely conservative, and point to the safety of flight, national security and other protection requirements that they can argue mandate a zero-risk approach to any spectrum sharing regime.

There are several radar modes of operation we can consider:

**Ground-to-Air** Radars that search for airborne or weather targets often have slightly elevated antenna orientation. This is intended to reduce the amount of ground clutter, but also reduces the susceptibility of the radar to terrestrial sources somewhat.

**Air-to-Ground** Air-to-ground radars typically can scan large areas and have relatively small apertures, so have large spot sizes when projected onto the ground. An additional complexity is that the sources move rapidly, making the detection of the radars harder, since listening to them cannot preclude the sudden appearance and need for protection.

**Air-to-Air** Air-to-air radars typically have one of two purposes. Military users use these radars for search and weapons fire control, and both civil and military users utilize these radars for in-flight weather detection. The radar waveforms for search, fire-control, and weather are very different, and must be addressed individually.

**Ground-to-Ground** Terrestrial-to-terrestrial radars are typically short range, and often Continuous Wave (CW). The principles of this radar are similar to the communications sharing examples in other chapters, rather than a typical pulse radar.

**Space-to-Ground** This category includes look down radars, as well as passive sensing using the Radio Frequency (RF) spectrum. The spot size of a receiving antenna from reasonable orbits is quite large, so local protection is less practical than with ground- or air-based radars. Instead, orbital mechanics can inform users of the potentially interfering conditions based on projecting satellite positions. This is well suited for implementation by a three-tier admission control system.

The CW case we will not consider further. CW transmissions do not have the pulse characteristic, and appear to be, and are detectable by, the same techniques as for detecting communications systems, with the possible exception of high chirp rates and ranges. This has been extensively researched in the Dynamic Spectrum Access (DSA) community.

We will consider the impact of interference on the radars' effective search area. Search area is proportional to the square of range, so it is much more sensitive to interference than the range metric, but reflects the use of radars more effectively, even if it magnifies the impact of interference.

Air-to-ground and air-to-air are challenging cases due to a combination of location ambiguity, the reaction time to sensing requirements, and their long range due to the height at which they may operate. Consider an aircraft flying at a ground speed of 270 m/sec (972 km/hr or 600 mph). If flying at 10 km altitude, it has a range to the horizon of 357 km, or a total line-of-sight area (ignoring antenna height) of approximately 400 000 km$^2$. Just one aircraft has line of sight to four percent of the Continental United States of America (CONUS), or around nine percent of the European Union (EU). And these paths are line of sight, so the path loss from the ground to the aircraft may be less than the path loss in 0.5 km over an urban environments! Of course, this range is for a zero degree look angle from the ground, which is unlikely,[4] but

---

[4] Possible if the antenna is highly elevated compared to the local environment.

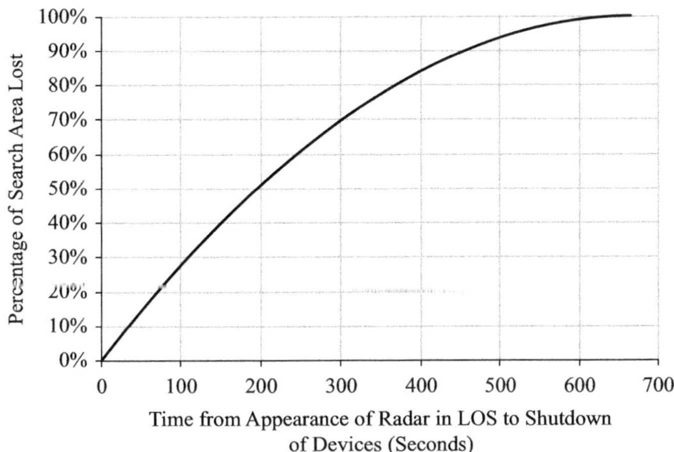


**Figure 11.2** Shutdown time impact on airborne radar search area.

it is the type of "corner case" which all interference issues appear to end up revolving around.

A more realistic scenario is that the slant angle limits the effective horizon to half of the distance (180 km) of the previous example. The aircraft is moving at 0.27 km/sec. If (as the radar community will argue) a device emitting interferes with the radar, then the search radius, and thus search area of the radar, will be reduced due to the interference, until interfering source(s) are shutdown. Figure 11.2 illustrates the worst-case relationship of the shutdown time to the search area from the aircraft.

It is clear that even a minimal delay in detection and shutdown would have significant impact on the possible search area. The 300 sec shutdown time in the US CBRS regulations would potentially cause interference to 30 percent of the search area. To reduce the potentially lost search area to five percent would require a shutdown interval from appearance on the horizon to device termination, of less than 17 seconds. "Fast Movers" are a challenge!

The opportunity in ground-to-air is not symmetric with air-to-ground. Ignoring any possible reflection onto the airborne platform, the interference path is all terrestrial, so it is subject to all of the losses along the ground, and becomes a diffractive path rapidly, with path loss approaching $r^4$. It is subject to additional clutter loss along the terrestrial path.

We should consider the impact of interference on the range the sensing range radar can achieve. We derive this by differentiating the radar range equation in Eq. (11.1)[5] in terms of changes in the detection threshold. Since the threshold is in terms of noise, changes in noise level are linear with changes in this threshold. Doubling of the noise floor essentially doubles the received pulse energy $PE_{min}$ required to achieve the same

---

[5] Note: differentiation of the radar range equation provides the impact on the range due to a change in the energy level required to identify a radar pulse.

target signal to noise ratio. The derivative of Eq. (11.1) with respect to $P_{E_{min}}$ is given by Eq. (11.3).

$$\frac{\delta R_{\max}}{\delta P_{E_{min}}} = k\frac{1}{4}(P_{E_{min}})^{-3/4} \tag{11.3}$$

The impact of an increase of the typical threshold level of 0.1 interference to noise ratio is a reduction in the radar's range of $\frac{1}{4}(0.1)$, or about 2.35 percent. The impact of doubling of the noise level is a reduction in range of $\frac{1}{4}1^{-3/4}$, or about 16 percent. In terms of search area, the impacts are essentially double this, in other words 4.5 and 29 percent respectively.

## 11.4     Detection of Radar System Presence

Detection of radar spectrum usage would appear to be a relatively simple task. Radars have a very high power signal, they have detectable pulses, and in many cases, they are not mobile. So detecting radar signals would appear to be a solid basis for spectrum sharing. In fact, detection of a single, simple radar waveform is a straightforward engineering challenge. However, there are a number of problems when designing a generalized radar detection capability, or to detect more complex, or variable waveforms.

- Radar detection algorithms must be designed to detect specific ranges of radar characteristics, and may have very poor performance when attempting to detect other characteristics. In some bands, there is a wide range of radar pulse characteristics in use, each of which must be considered individually.
- High probability of detection necessarily and inherently implies a higher probability of false alarms, as well as the converse. A "perfect" sensor is impossible to implement, so compromise in protection assurance, or in usage assurance, is inevitable.
- Radars have a very low duty cycle, so detection methods must have a high sensing presence on any possible channel.

In considering the performance of a radar detection regime, there are three overall considerations that compound to determine the effectiveness of the regime, and its impact on sharing users:

- The ability of the detector to detect the radar in the absence of any other signal.
- The ability to not report radar detections in a typically crowded band, in the absence of any radar signal.
- The ability to detect the radar at low signal levels, in the presence of typical, crowded band, non-radar signals.

There is an inherent relationship between the likelihood of detection ($P_D$), and the corresponding likelihood that the sensor will declare a radar is detected, when one is not actually present ($P_{FA}$). The different curves represent different thresholds of detection. Lower thresholds increase the probability of detection, but create false alarms. Higher detection thresholds reduce the false alarm rate, but also decrease the probability of detection.

This is a fundamental constraint from detection theory that must be addressed in designing detection approaches. There is no closed-form model for these relationships, but they can be readily computed numerically. There exists a fundamental conceptual relationship of signal level and the probability of false alarm ($P_{FA}$) and of detection ($P_D$). This is depicted as a Receiver Operating Characteristics (ROC).

For a given radar signal to noise level, there is a defined curve for the achievable $P_D$ and $P_{FA}$. The designer has the flexibility to improve one or the other, but improvement in one must result in reduction in the quality of the other. At the extreme, a system with no signal input could achieve $P_D = 1.0$ by always indicating a detection, although it would also have a $P_{FA} = 1.0$. Alternatively, it could never report any detection, and achieve a perfect false alarm rate ($P_{FA} = 0.0$), but at the cost of a $P_D = 0.0$. Detection energy improves these trades, but finite energy can never achieve a sensor that achieves $P_{FA} = 0.0$ and $P_D = 1.0$ simultaneously. Therefore any sensor-based spectrum sharing regime must be resilient to some sensing errors. It can shift this burden between disruption of secondary usage or incumbent radars (or other incumbent uses), but it cannot avoid some compromises.

As an example, a signal level of almost 40 times (15 dB) higher than the noise level is required to obtain high confidence detection at a reasonable false alarm rate ($P_{FA} = 10^{-8}$, $P_D = 0.99$). It is important to recognize that the $P_{FA}$ value is in terms of sensing epochs, which for radars is quite short; on the order of several the inverse of the PRF. A $P_{FA}$ of $10^{-8}$ might appear to be minimally disruptive, but for a PRF of $10^3$ per second, and single pulse detection, it equates to a false alarm every $10^5$ seconds, or about one false event every day.

The noise signal seen by the detector is Additive White Gaussian Noise (AWGN) only if it is thermal, such as natural background or equipment front end temperature. If the noise level is from signals, then there will be some correlation within this noise, and the AWGN assumption will be invalid, typically raising $P_{FA}$ and/or lowering $P_D$. Any cyclo-stationary behavior in the waveform will resemble a fixed PRF radar signal, for example. For example, the Long-Term Evolution (LTE) waveform has slotting periods that are identical to the SPN-43C radar that must be detected in the US CBRS band.

In discussing the considerations that are practical issues in the design of a generalized sensing regime, we will refer back to some of the examples in Table 11.1 to quantify the impacts on realizable systems.

**Integration Time** How long the sensor integrates the signals and noise in a given sensing channel.

**Sensing Presence** The ratio of time the sensor device is sensitive to signals in a monitored channel. This reflects time processing, and receiving other channels.

$P_{FA}$ **and** $P_D$ **Trades** Threshold settings, and criteria for declaring a detection involve making trades between high probability of detection and low false alarm rates. At a given signal level, in a Gaussian environment, these two measures are directly linked.

If a sensor integrates a sensed channel for $t_{\text{sense}}$ and a radar has a pulse width of $t_{\text{pulse}}$, then the effective noise elevation ($N_{\text{integ}}$ in dB) is given by Eq. (11.4). Noise is integrated linearly with time, but the radar signal has fixed energy.

$$N_{\text{integ}} = \begin{cases} 0 & \text{for } t_{\text{sense}} \leq t_{\text{pulse}} \\ 10\log_{10}\left(\dfrac{t_{\text{sense}}}{t_{\text{pulse}}}\right) & \text{for } t_{\text{sense}} > t_{\text{pulse}} \end{cases} \tag{11.4}$$

For example, we consider a sensor with an integration time that was a compromise among many radars of $1\,000\,\mu\text{sec}$. When applied to sense a radar with a pulse width of $100\,\mu\text{sec}$, it would integrate $10\,\text{dB}$ of additional noise above the actual noise level (assuming AWGN noise) when the radar actually transmitted.

Another consideration of integration time is how the timing in the sensor impacts the detection probability of a single pulse. If our sensor has an integration time of $t_{\text{sense}}$ out of a total sense scan time (sense start time to the next sense start time) interval of $t_{\text{scan}}$, then Eq. (11.5) shows the sensor system's probability of sensing on the radar's channel ($P_{\text{Dscan}}$) during any given pulse, for the full duration of the pulse.

$$P_{\text{Dscan}} = \begin{cases} 1 & \text{for } t_{\text{sense}} = t_{\text{scan}} \text{ and } t_{\text{sense}} << t_{\text{pulse}} \\ \left(\dfrac{t_{\text{sense}}}{t_{\text{scan}}}\right) & \text{for } t_{\text{sense}} < t_{\text{scan}} \text{ and } t_{\text{sense}} << t_{\text{pulse}} \end{cases} \tag{11.5}$$

The first case in Eq. (11.5) represents a staring receiver that is continually scanning the channel(s). The second case could represent a scanning receiver, or a Fast Fourier Transform (FFT) based sampling, with a period for processing.

The last trade space to be introduced is in the channel scanning and bandwidth coverage. The previous two equations showed that a sensor needs to be present on a channel with a high probability, and that the scan time has to be closely matched to the pulse width in time. That would motivate design decisions towards use of a wideband detector if the noise integration was acceptable.

Eq. (11.6) shows how the noise ($N_{\text{bandwidth}}$) is impacted by integrating across wider bandwidths than that occupied by the radar. The relationship of the radar bandwidth ($B_{\text{radar}}$) and the sensed bandwidth ($B_{\text{sensed}}$) is critical in establishing the effective noise floor of the detection system.

$$N_{\text{bandwidth}} = \begin{cases} 0 & \text{for } B_{\text{radar}} \geq B_{\text{sensed}} \\ 10\log_{10}\left(\dfrac{B_{\text{sensed}}}{B_{\text{radar}}}\right) & \text{for } B_{\text{radar}} < B_{\text{sensed}} \end{cases} \tag{11.6}$$

These equations are quite idealistic, as in many cases the channel noise is not AWGN, but is highly correlated as the band achieves usage. In the case of the US CBRS band, the radars are mostly at sea, so the sensors can use directional antennas to restrict their field of regard to the ocean. In a more typical setting, where radars could be located on any azimuth, the noise background will not be the thermal noise of the receiver, but the interference from other users of the band. This greatly increases the false alarm probability over the Gaussian statistics.

Eqs. (11.4), (11.5), and (11.6) constrain choices. Narrow sensing bandwidth reduces noise, but cuts down the scan rate. Wideband sensing provides more scan time, but at a higher noise floor level. Additional noise levels reduce the probability of detection. Since most sharing regimes will have a stressing requirement to detect the incumbent radar, the likelihood is that the shortfall in noise reduction will have to come at the expense of the false alarm rate, which is at least an inconvenience to users, if not a negative factor in band adoption.

It is likely that the criteria for radar sensor performance will establish minimum probability of detection in a given time, at a specified signal level. The users of the band also have an implicit false alarm rate constraint. Using the AN/SPN-43C example, the sensor has $10^3$ detection opportunities every second. If a maximum of one false alarm event is allowed per 24 hours (86 400 seconds), the maximum false alarm rate is 1 per $8.64 \cdot 10^7$ seconds, or a rate of $1.1 \cdot 10^{-8}$.

The detection threshold has to be extremely high to achieve this extremely low rate of false alarms. In practice, most sensors have criteria for declaring a detection, such as two consecutive pulses, or two out of three predicted pulse times.[6] This shifts the stress onto the detection algorithm, as it must now achieve a much higher single pulse probability of detection in order to meet this criteria. Requiring $n$ successive pulse detections effectively lowers $P_D$ to $P_D{}^n$. The $P_D$ can only be determined for the intervals when the radar is actually within the sensor's field of regard and when radiating towards the sensor, while the $P_{FA}$ is applicable to all times at which the sensor is working.

One other factor is that for mechanically steered radars, the pulse occurrence is far from Gaussian. The radar swings through the field of regard of the sensor periodically, and only a few pulses are emitted in the direction of the sensor.[7] If the radar is not detected during that sweep, the next opportunity is during the next sweep of the antenna. So the time to detect is granular, not continuous. Again, using the AN/SPN-43C as an example, approximately 20 pulses are sent to a sensor during each sweep, and if the sensor does not declare a detection, it is 4 seconds until the next opportunity to detect occurs.

The $P_D$ can only be determined for the intervals when the radar is actually within the sensor's field of regard, and when radiating towards the sensor, while the $P_{FA}$ is applicable to all times at which the sensor is working. A radar with a 1.75° beamwidth (such as the AN/SPY-43C) has an opportunity to detect the radar at only 0.5% of the time the sensor operates, essentially reducing the quality metric of $\dfrac{P_D}{P_{FA}}$ by 200 times.

Table 11.2 illustrates some sensor designs that could be considered for a radar sensing system. The wideband sensing approach implements a single receiver that covers the entire band simultaneously. The stepped channel detector has a narrowband receiver that samples each possible radar channel individually. The parallel FFT version has one

---

[6] This approach only works with known and fixed PRF radars, and when the antenna is not agile between pulses.

[7] These are non-Gaussian, since if there is one pulse at a given time, the likelihood of one immediately before or after it is much higher than average, and if none was received, the likelihood of one in the next slot is lower than average. Rotating radar pulses are not independent in time.

**Table 11.2** Radar sensor implementation approaches.

| Design Approach | Advantages | Disadvantages |
| --- | --- | --- |
| Wideband scanning | Simple, and potentially inexpensive. Could have rapid time to detect. | Reduced sensitivity due to noise integration across such a wide band. Potential impact of dense secondary sharing signals throughout the wide band of coverage. May also be vulnerable to overload by any signal in the sensed range. |
| Stepped Channel | Can achieve maximum sensitivity. | Very reduced detection probability in a given time window. |
| Parallel FFT | Can achieve maximum sensitivity. | FFT processing likely means the spectrum is not being monitored at all times ($t_{scan}$), which implies reduced probability of detection, similar to the stepped model. |
| Parallel Channels | Achieves maximum sensitivity, probability of detection, and time to detect. Simplicity of operation and high performance may allow some of the analog and digital requirements to be relaxed, decreasing cost of implementation. | Highly duplicated and redundant hardware may increase cost unless per channel performance can be relaxed significantly. |

wideband FFT that is subsampled for each possible radar channel. The parallel channel detector has multiple receiver chains that operate on each channel independently.

The point of this table is not to argue for or against any sensor design. They each have significant implementation advantages and performance disadvantages. Even sensing a simple, legacy design, non-tactical radar, such as the AN/SPY-43C is a significant challenge, even though its waveform is simple, and well known; it has no frequency or beam agility; and its operator is cooperative in the effort to deploy detection capabilities.

## 11.5      Interference From Radar Systems to Communications Systems

Some in the radar community have stated that even if it was shown that communications devices would not interfere with radars, the high power radars would be highly disruptive to the communications service. By this view, they are, in fact, doing a favor for the communications community in avoiding disappointment, or even disaster, if investment in such communications devices was contemplated. An example of this argument is the exclusion zones that are shown in Figure 4.2. These limitations were intended to protect the communications usage from the gigawatt EIRP AN/SPN-43C radar, and are provided in the National Telecommunications and Information Agency (NTIA) Quick Look Report [4].

Before addressing this topic analytically, an anecdotal piece of evidence is useful. In the United States of America (USA) CBRS proceeding, Google staff reported that they decided to experiment with the actual service quality impact of interference from the incumbent CBRS band radar into LTE. This was reported in two ex parte filings by Google, and in more detail in a paper coauthored with Virginia Tech [5]. In the experiment they reported, that they rented time on a test AN/SPN-43C at Naval Air Station, Patuxent River in Maryland, USA. The TD-LTE devices operated co-channel with the radar with no detectable degradation at a distance of only 4 km from the radar, and in the adjacent channel of the radar at a distance of 1 km. This is approximately 100 times less protection distance than in the NTIA analysis shown in Figure 4.2.

The values from Table 11.1 are the key to understanding the 100-fold reduction in the actual in-bound protection operation distances. They provide the basis for analysis of how a communications system can operate co-channel with a gigawatt radar! We consider two broad categories of impacts: time and power. In this discussion, we consider a specific interference target, TD-LTE. The numerical aspects of the conclusion are specific to the AN/SPN-43C Radar and LTE, but the general principles are applicable to pulse (non-CW) radars and packet radios with reasonable error handling, in general.

**Time** Table 11.1 shows that the AN/SPC-43C radar PW is only 95 $\mu$sec in duration, and has a duty cycle less than 1 percent. An LTE subframe (an uplink or downlink slot) is 1 000 $\mu$sec long. A subframe can only be interfered with by at most one radar pulse, which could be present for 9.5 percent of the subframe. LTE operates at a variable code rate, but is typically in the neighborhood of 0.4, so correction of a bursty subframe block is within its capacity. Also, the number of blocks impacted is quite small. Table 11.1 shows that the horizontal dimension of the main beam is 1.75° wide. The main beam points at the victim for only 0.49 percent ($4.9 \cdot 10^{-3}$) of the time. The victim is not in the main beam for the remaining 99.5 percent of time, and receives much less impact. The radar transmission duty cycle is 0.095% $\left( \dfrac{0.95\,\mu s}{1\,000\,\mu s} \right) = 9.5 \cdot 10^{-4}$, so the actual duty cycle of the radar at the receiver is $4.6x10^{-6}$. A few subframes might be impacted during each radar rotation, out of a total of 40 000.

**Power** Table 11.1 shows that the vertical half beam width is 2.2°. However, the radar has an elevation angle of 3°, so the main beam is above the ground height at all locations around the radar site. Added to this consideration, the radar is mounted 150 feet above the water line, so the main beam encroaches even less on the ground.

Adding to the interference tolerance is the contribution of Media Access Layer (MAC) and transport layer error control. Even if the subframes are uncorrectable, both the LTE and the upper layer of the transport layer (if using Transport Control

Protocol (TCP), or an application that is packet loss tolerant using User Datagram Protocol (UDP)) would address the lost packet at very low system cost.

Although it is hard to generalize interference analysis and policies, a baseline assumption should be that communications systems should be afforded the right to share spectrum with radar systems, without restrictions due to concerns over interference to the communications applications. Most commercially viable communications technologies can probably inherently operate reasonably in the presence of pulse radars. In the worst case, tailoring of PHY layer features, such as interleaving, coding, and Error Detection and Control (EDAC) strategies can likely assure operation even in highly stressing environments.

Yet another reason for reducing concern about interference from radars, at least in urban, small cell deployment, is that measured clutter losses are quite high. Figure 7.3 showed an additional 40 dB loss over 1 km in even moderate density commercial areas. Although a 90 dBW EIRP, such as in the AN/SPN-43C radar, would appear to be intimidating, when you adjust to reflect operation below the antenna main beam (43 dB), add a typical minimum of 40 dB clutter loss per kilometer, then the effective power left is only 7 dBW EIRP at 1 km. The average effective power is reduced a further 57 dB due to the duty cycle at the receiver, The average effective EIRP is in the microwatt range!

## 11.6    Future Technology to Increase Radar and Communications System Coexistence

The Defense Advanced Research Projects Agency (DARPA) Shared Spectrum Access for Radar and Communications (SSPARC) program performed extensive research on improving coexistence and spectrum sharing between radars and communications systems. The DARPA web site states its goals as:[8]

- Spectrum sharing between military radars and military communications systems ("military/military sharing") increases both capabilities simultaneously when operating in congested and contested spectral environments.
- Spectrum sharing between military radars and commercial communications systems ("military/commercial sharing") preserves radar capability, while meeting national and international needs for increased commercial communications spectrum, without incurring the high cost of relocating radars to new frequency bands.

Everything in this chapter changes if, and when, radars coordinate with communication users at very granular levels of time. Certainly, civil radars could inform a coordination agent of their frequency and azimuth continuously, which could allow users in the vicinity of the radar to subscribe to the feed, and be agile in their use of the spectrum. Coordination of pulse epochs could even be established based on common Global Positioning System (GPS) time references, similar to how transmit and receive slots are coordinated in TD-LTE.

---

[8] From: www.darpa.mil/program/shared-spectrum-access-for-radar-and-communications.

These sharing techniques are not ones that are likely to be retrofitted into existing radars, but they are not included in the design of any future radars either. The problem is not technical, but a matter of incentives. If spectrum protection is granted to users for exclusive use for one application, there is little incentive to share that spectrum with others. Creation of that incentive is a challenge for policy makers, not engineers.

## 11.7 Radars in a Three-Tier Framework

Sharing radar bands is certainly consistent with the principles of the three-tier framework presented in this book. The first application, the US CBRS band has the objective of sharing with a single type of radar, which has absolute priority. An interesting extension of this is to operate at least private, and non-safety-of-life, government radars, as tenants within the three-tier framework, rather than as primary incumbents. This approach has several advantages:

- It encourages the radar community to share the burden of spectrum scarcity by reducing the extent of the protection it seeks from the spectrum management regime.
- It encourages, and monetizes the implementation of the coordination features discussed in the last section, and in the DARPA SSPARC program, as examples. By coordinating with other users, the radars reduce their footprint, and thus spectrum protection costs.
- It encourages future radar designs to be interference tolerant, rather than depending on regulator protection. Radar designs often have a zero budget for interference, which is unrealistic for any participant using such a scarce resource as spectrum.

It should be clear that sharing of radar spectrum is an issue that will require some additional communications technology, but more so, cooperation on the side of the radar community. It is likely that this may require some changes in the incentive structure of the spectrum ecosystem in regard to the protection of radars, and their obligations to share the spectrum on reasonable terms.

## 11.8 Suggested Reading

The technology and engineering of radars is addressed extensively in the literature, and other than suggesting Skolnik's classic work on radars, the reader will find a rich set of sources at all levels of discussion. The subject of sharing with radars is addressed much less in the literature, so the reader will find that a more varied source of material will have to be sought.

- The essential reference standard of radar technology is by Skolnik [6], whose work has been updated may times over the decades.
- Some DARPA SSPARC material has appeared in academic publications [7, 8, 9]. Information on this program is available at: www.darpa.mil/program/shared-spectrum-access-for-radar-and-communications.

- There is an evolving field of cognitive radar [10] that is addressing many of the challenges in this chapter, and in recent publications on cooperative communications/radar operation, such as [11].
- Google reported experimentation with the interference caused to TD-LTE operation in the proximity of the AN-SPN-43C. The US Naval Academy and Virginia Polytechnic University collaborated in these experiments [5]. This is anecdotal evidence, but instructive in the impact of real radar operation on real TD-LTE operation (and likely be extrapolated, to some degree, to other similar wireless protocols). The experiment showed only a minimal impact on TD-LTE error rates, even in close proximity to the radar.

## References


1 International Telecommunications Union, *Technical Characteristics of Maritime Radionavigation RADARS, Recommendation ITU-R M.1313* (1997).

2 ——, *Characteristics of Radiolocation Radars, and Characteristics and Protection Criteria for Sharing Studies for Aeronautical Radionavigation and Meteorological Radars in the Radiodetermination Service Operating in the Frequency Band 2700–2900 MHz, Recommendation ITU-R M.1464-1* (2003).

3 G. Brooker, *Sensors and Signals* (Sydney: Australian Centre for Field Robotics, 2009).

4 US Department of Commerce, National Telecommunications and Information Agency (NTIA), *An Assessment of the Near Term Viability of Accommodating Wireless Broadband Systems in the 1675–1710 MHz, 1755–1780 MHz, 3500–3650 MHz, and 4200–4220 MHz, 4380–4400 MHz bands.* Technical report (2010).

5 J. H. Reed, A.W. Clegg, A. V. Padaki, T. Yang, R. Nealy, C. Dietrich, C. R. Anderson, and D. M. Mearns, On the co-existence of TD-LTE and Radar over 3.5 GHz band: An experimental study. *IEEE Wireless Communications Letters*, **5**/4 (2016) 368–371.

6 M. Skolnik, *Introduction to Radar Systems,* 2nd edn (New York: McGraw-Hill, 1980).

7 A. Khawar, A. Abdelhadi, and T. C. Clancy, Coexistence analysis between radar and cellular system in LoS channel. *IEEE Antennas and wireless propagation letters*, **15** (2016).

8 G. M. Jacyna, B. Fell, and D. McLemore, A high-level overview of fundamental limits studies for the DARPA SSPARC program. *2016 IEEE Radar Conference* (2016), 1–6.

9 M. P. Fitz, T. R. Halford, I. Hossain, and S. W. Enserink, Towards simultaneous radar and spectral sensing. *2014 IEEE International Symposium on Dynamic Spectrum Access Networks* (2014), 15–19.

10 J. Guerci, *Cognitive Radar: The Knowledge-Aided Fully Adaptive Approach* (Norwood, MA: Artech House, 2010).

11 J. R. Guerci, R. M. Guerci, A. Lackpour, and D. Moskowitz, Joint design and operation of shared spectrum access for radar and communications. *2015 IEEE Radar Conference* (2015) 761–766.

# 12 Coexistence Between Unprotected Peer Devices

## 12.1 The Role of Coexistence in Spectrum Protection Systems

One of the desirable features of the three-tier regime is that it offers the opportunity for network operators to acquire protection of their nodes from interference. This feature is judged to be essential to creating confidence in the investment in these bands. However, it is not clear that paying for protection in small segments of the spectrum is a superior solution to other methods of assuring a dependable level of service. In this chapter, we will investigate the conceptual basis of alternatives to protected mid-tier status, and investigate methods of implementing them.

Use of protected status has a number of disadvantages:

- The amount of spectrum that can be acquired in a protected status may be relatively small compared to the spectral bandwidth that is available in the sharable band.
- The cost of the protection may be high compared to the access devices, and may create significant disadvantages to the economics of the deployment.
- The use of a specific frequency that is accorded protection might limit the flexibility of the network to deal with conditions such as strong adjacent channel signals, primary spectrum reclaiming events, and network spectrum reuse strategies.
- Protection of large areas is inappropriate for many small/femto cell deployment cases, such as indoor and three-dimensional deployments. It adds an unnecessary cost burden to the second tier, and limits access to spectrum by the third tier.

For these reasons, this chapter will investigate methods of achieving high levels of predictability in spectrum access effectiveness that do not require explicit grants of protection or elevation in rights, and minimize the extent of spectrum that is committed to elevated rights usage.

## 12.2 Model for Coexistence Services

Our model for coexistence has two possible levels of operation.

**Partitioning** Partitioned use divides the available spectrum resources in a given location between a number of networks that desire spectrum, and are willing to enter into an agreement to coexist with each other. This sharing makes no assumptions on the technology used, or on the operation of the technology within the allocated channels.

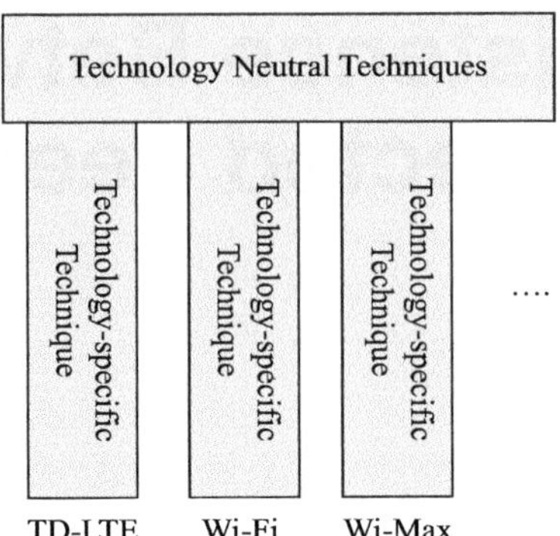


**Figure 12.1**  Relationships of technology-neutral and specific coexistence techniques.

**Integrated Sharing** Integrated sharing enables multiple users of the spectrum, as above, to share more extensive spectrum resources, but with some control over network operation and technology selection.

The first level (partitioning) is technology neutral; it divides the available resources in some previously agreed to framework, and assures separation of the operation of networks. In the second level (integrated sharing), the specific nature of the technology common to contending users is considered and utilized to maximize the access by each of the cooperating parties.

The hierarchal relationship among these techniques is shown in Figure 12.1.

## 12.3   Indoor Coexistence Analysis with Realistic Environments

It is relatively easy to become concerned about coexistence in unprotected spectrum. Many locations have the unlicensed spectrum crowded with tens of individual Wireless Fidelity (Wi-Fi) Service Set Identifier (SSID) broadcasts; rapidly scrolling off the screen when attempting to select one. Wi-Fi is a very "polite" protocol, and one Wi-Fi device will not transmit when it detects that the channel is occupied, even if the signal is very weak, and would not interfere with its own operation. The threshold for a channel to be declared busy by the Listen Before Talk (LBT) algorithm is -62 dBm. Because of the LBT algorithms, all users of a Wi-Fi frequency essentially share one common channel, and its aggregate network bandwidth is at best shared by all of these users. In reality, more users reduce the aggregate channel capacity. Wi-Fi therefore attempts to handle coexistence by ensuring that interference does not occur, but does so at the cost of reducing the possibility of dense spectrum reuse when strong signal conditions are present.[1]

---

[1] We consider spectrum reuse in this context to be simultaneous use of a frequency, not time-shared access. LBT time-shares, but does not reuse. Operation at high interference to noise ($I/N$) levels has the potential to enable reuse.

**Table 12.1** Measured propagation clutter loss from building materials at 2, 3, 3.5 and 5 GHz (from [1]).

| Material | Propogation Loss (dB) | | | |
| --- | --- | --- | --- | --- |
|  | 2.0 GHz | 3.0 GHz | 3.5 GHz | 5.0 GHz |
| No rebar, concrete wall, 203 mm | 34.6 | 50.0 | 51.0 | 53.0 |
| 1% Steel rebar concrete wall, 203 mm | 32.0 | 51.2 | 51.5 | 55.0 |
| 2% Steel rebar concrete wall, 203 mm | 36.8 | 52.6 | 54.0 | 57.2 |
| Wallboard Wall | | Negligible | | |
| Brick Wall, 89 mm | 5.4 | 15.0 | 15.2 | 15.2 |
| Brick Wall, 178 mm | 7.6 | 15.9 | 20.2 | 30.6 |
| Brick Wall, 267 mm | 10.6 | 26.8 | 29.2 | 31.4 |
| Masonry Wall, 203 mm | 11.2 | 12.5 | 12.5 | 12.5 |
| Masonry Wall, 406 mm | 17.9 | 24.2 | 25.2 | 28.5 |
| Masonry Wall, 610 mm | 30.0 | 34.2 | 34.0 | 39.2 |
| Brick Faced Concrete Wall, 90 mm brick plus 102 mm of concrete | 16.8 | 32.8 | 33.4 | 40.5 |
| Brick Faced Concrete Wall, 90 mm brick plus 203 mm of concrete | 32.8 | 58.4 | 60.3 | 69.7 |
| Brick Faced Masonry Wall, 90 mm brick plus 194 mm block | 11.6 | 28.9 | 30.2 | 32.8 |
| Glass Panels, 6 mm thickness | 1.4 | 0.5 | 0.2 | 0.1 |
| Glass Panels, 13 mm thickness | 3.4 | 1.5 | 1.1 | 0.5 |
| Glass Panels, 19 mm thickness | 3.9 | 1.9 | 1.5 | 1.0 |

Long-Term Evolution (LTE) is designed to operate with high levels of spectrum reuse, and therefore high $I/N$ conditions. The criteria for LTE operation is that there be a sufficient ratio of signal-to-noise-plus-interference $\frac{S}{I+N}$ to ensure reliable link operation. This approach places no absolute limit on the value of $I$, so long as $S$ is significantly higher than the sum of $I$ and noise $N$, referred to as Signal to Interference and Noise Ratio (SINR).

A significant factor in managing true spectrum reuse is understanding the impact of clutter losses. A number of organizations have reported the collection of external and interior clutter data. Figure 7.3 illustrated the exterior clutter losses reported by Google. These losses were at 3.62 GHz center frequency, but are similar to those in the range of spectrum that would be considered in spectrum sharing.

Unfortunately, there is no single, curated, source for building loss measurements and models. National Institute of Standards and Technology (NIST) performed an extensive study of the losses in some typical building materials and construction methods in 1997, which covered a wide range of frequencies [1]. Some of the common materials in use in commercial and Multi-Dwelling Unit (MDU) construction are shown in Table 12.1, which illustrates the propagation loss of interior and exterior building features in both a residential MDU and commercial building construction. The values shown are for 2.0, 3.0, 3.5, and 5 GHz, and estimated from the graphic plots in the report, which include this frequency range. The 3.5 GHz is included to be directly applicable to the US Citizens Broadband Radio Service (CBRS) band coexistence analysis.

**Table 12.2** Measured propagation clutter loss from window designs at 2, 3, 3.5 and 5 GHz (from [2]).

| Material | Propagation Loss (dB) | | | |
| --- | --- | --- | --- | --- |
| | 2.0 GHz | 3.0 GHz | 3.5 GHz | 5.0 GHz |
| 2-Layer Energy Efficient Window, Metal Frame | 18 | 23 | 25 | 27 |
| 3-Layer Energy Efficient Door, Metal Frame | 30 | 34 | 33 | 33 |
| 2-Layer Energy Efficient Window, Wooden Frame | 20 | 19 | 23 | 31 |
| 3-Layer Energy Efficient Window (Finland) | 33 | 35 | 36 | 41 |
| 1-Layer Glass Window, Wooden Frame | 3 | 2.5 | 2.4 | 4 |

The baseline values of loss function as viable lower bounds of the estimated loss, which is sufficient for use in estimating worst case interference conditions, if not for communications link availability. The internal structure of the wall can apparently create very frequency dependent loss peaks, but these are all perturbations with additional loss on top of a baseline loss value. The same behaviors are evident in concrete masonry blocks, which show high frequency dependence when constructed in multiple block depth walls. Materials with minimal internal structure (such as concrete) generally show monotonic behavior, with loss increasing with frequency (with the exception of glass).

The data provided by the NIST report, summarized in Table 12.1, considers only traditional plain silica window materials. In modern commercial and MDU structures, window treatment typically utilizes leaded or thermal materials, rather than pure silica. Another source of information specific to the loss that occurs in modern window construction is provided by Rodrigues, et al., [2]. The window-related material is provided in Table 12.2 in a similar format as Table 12.1. Note that the advanced window design (the line titled "3-Layer Energy Efficient Window (Finland)" has losses that are close to the equivalent masonry construction.

With these measurements, one can examine the likelihood of interference to a given absolute power, and the impact on the required $\frac{S}{I+N}$ ratio, which is the more significant criterion, from the perspective of network operation. To perform this analysis, we consider several floors of a commercial facility, all with access to the same extent of spectrum, and determine where coexistence is required and where it is not, and the impact on spectrum availability when it is required. Propagation within the facilities is mostly free-space, with the additional clutter loss of any intervening structures on the path. Each location will be assessed for both the noise floor elevation, and the likely signal to external network interference plus noise of a network within that space. These discussions are anecdotal, but are intended to provide a framework from which to identify the most critical challenges in achieving indoor peer coexistence.

There are some immediate hypotheses that are plausible by inspection:

1. The through-floor loss of a concrete, or concrete plus steel reinforcing, bar is 50+ dB of additional path loss. It is reasonable to assume that in most cases, frequencies can be reused vertically in a commercial or MDU residential building. The latitude and longitude of the nodes may be identical, but the path loss between them is at least

50 dB, if not much more. Three-dimensional reuse should be feasible, even without coexistence methods for buildings with commercial concrete floors.
2. Building exterior walls have high loss. The exterior energy penetration will likely be driven by the construction and the area of the windows. For buildings with any appreciable window area, the through-window signal entry path will be the dominant source of external interference. For plate glass, the loss is insignificant. However, for multilayer energy-efficient windows, as in most modern construction, the loss through the windows is high enough to often enable operation without coexistence features, even in a 100 percent reuse of spectrum environment.
3. Networks on the same floor of a commercial building can likely coexist on the same frequencies if they are isolated through mineral materials, such as concrete block, and most certainly if they are isolated by solid concrete walls. If they are isolated by only wallboard or wooden construction, then coexistence in at least the boundary locations will be required.
4. In a shared-floor situation, the worst-case partitioning of the spectrum will be the number of tenants where the additional range-driven path loss, plus any obstructions, is under 20–30 dB. In most cases, this will still leave full carrier aggregation channels available even without coexistence features, and considerably more with them.
5. Single-family home construction methods, such as wood frame, plywood, wallboard, and brick veneer do not provide sufficient attenuation to enable operation in a shared-frequency environment without coexistence. If these networks are not spatially separated, they will require the use of coexistence features.

In the previous discussion, the most obvious, and simplest, coexistence feature is the simple assignment of separate channels of operation.

To verify these hypotheses empirically and quantitatively, consider three floors of a building, as shown in Figure 12.2. There are so many possible arrangements that a general proof is not practical, so we instead provide an analysis of a typical indoor enterprise deployment. The building has an external access point, with 10 dB more EIRP, in front of the building that has emissions that enter the building through thermally treated glass windows in the adjoining side. The loss in any paths through the wall material is ignored in this analysis, and all of the energy is assumed to enter through the window path, which is assumed to be visible to all locations on the interior.

Using these locations, free-space propagation, and the obstacle clutter losses from Table 12.1, the values of interference ($I$), signal ($S$), signal-to-noise-plus-interference ratio $\left(\dfrac{S}{I+N}\right)$, and signal-to-noise ratio $\left(\dfrac{S}{N}\right)$ are shown in Table 12.3 for a range of locations within the space.[2] Some of the analytic assumptions in this analysis are:

- Deconfliction between the interior networks devices of a single tenant is addressed by the management system managing these devices. Each floor has a unique tenant, and the third floor has two tenants that adjoin through wallboard.
- Center frequency of 3.62 GHz.

---

[2] Interference within the private network is not considered, as it is likely that the network handles its own coexistence, such as provided by LTE.

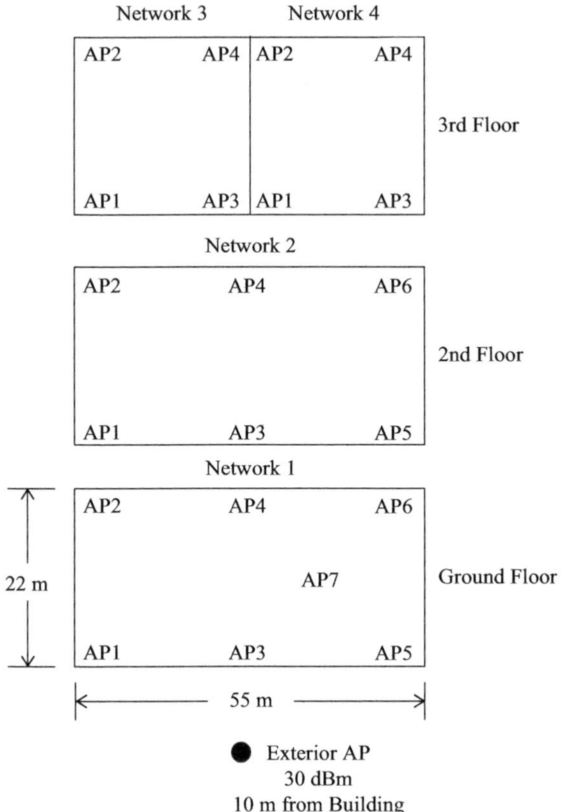


**Figure 12.2** Example interior layout for coexistence analysis.

- Free-space propagation dominates internal to the structure.
- No consideration of ducting and diffractive effects.
- Multiple enterprises are present, and they manage interference within their networks, such as with LTE Radio Resource Management (RRM). We consider them as all sharing one set of channels, with a Power Spectral Density (PSD) of 10 dBm/MHz, and perfect isotropic antennas. Since the example is interference limited, the absolute value of the power is not relevant.
- The outdoor small cell (30 dBm/10 MHz) propagated indoors to the second floor through a two-pane energy-efficient window, with 23 dB of loss in the least loss case. In most cases it will have higher loss, but the worst case is used for the example. The performance of this node is not modeled, so there are no signal strength estimates.
- Interference is computed by aggregating the 100% duty cycle of all emitters from other networks. This is the worst case assumption that an admission control system would have to make.
- The system noise temperature of the receiver is computed at 720 K.

**Table 12.3** Interference and signal to noise ratio in sample enterprise building at 3.62 GHz.

| Network | Node | Interference (dBm/MHz) | Signal (dBm/MHz) | SINR (dB) | SNR (dB) |
|---|---|---|---|---|---|
| Net 1 | AP 1 | −83.23 | −59.20 | 24.04 | 50.83 |
| | AP 2 | −86.16 | −59.20 | 26.96 | 50.83 |
| | AP 3 | −75.89 | −57.37 | 18.52 | 52.65 |
| | AP 4 | −84.25 | −57.71 | 26.54 | 52.32 |
| | AP 5 | −83.51 | −57.37 | 26.13 | 52.65 |
| | AP 6 | −86.28 | −57.71 | 28.57 | 52.32 |
| | AP 7 | −82.53 | −57.37 | 25.15 | 52.65 |
| Net 2 | AP 1 | −91.62 | −59.20 | 32.42 | 50.83 |
| | AP 2 | −91.62 | −59.20 | 32.42 | 50.83 |
| | AP 3 | −90.11 | −59.20 | 30.91 | 50.83 |
| | AP 4 | −90.11 | −59.20 | 30.91 | 50.83 |
| | AP 5 | −91.56 | −59.20 | 32.36 | 50.83 |
| | AP 6 | −91.57 | −59.20 | 32.37 | 50.83 |
| Net 3 | AP 1 | −58.32 | −59.20 | −0.88 | 50.83 |
| | AP 2 | −58.32 | −59.20 | −0.88 | 50.83 |
| | AP 3 | −39.55 | −59.20 | −19.64 | 50.83 |
| | AP 4 | −39.55 | −59.20 | −19.64 | 50.83 |
| Net 4 | AP 1 | −39.55 | −59.20 | −19.65 | 50.83 |
| | AP 2 | −39.55 | −59.20 | −19.65 | 50.83 |
| | AP 3 | −58.54 | −59.20 | −0.66 | 50.83 |
| | AP 4 | −58.54 | −59.20 | −0.66 | 50.83 |
| Outdoor | 30 dBm unit | −83.28 | N/A | −N/A | N/A |

The results of this example lead to several conclusions, that are example-specific, but probably have more general applicability:

- The free-space communications paths dominate over any of the interference paths through major attentive materials, such as exterior walls. The aggregate interference may rise to the level of −75 dBm in this example, but even in this worst case on the second floor (AP 3, Net 1), the SINR was over 18 dB, so the interference would have little impact on the operation of the deployment.
- Shared spaces which are not isolated have essentially zero chance of operating without interference. Active coexistence methods, such as frequency separation, timing, resource block scheduling or LBT would be required in these cases.
- The operation of an outdoor, higher-power Access Point (AP) had impact on the networks operating in the interior (AP3, Net 1) and did increase the interference level, but the modern, energy-efficient windows reduced this signal level sufficiently to not cause a significant reduction in the link throughput due to reduced SINR. The impact on higher floors was greatly reduced by the added loss of the through floor attenuation, and the reduced penetration due to the higher arrival angles in the through-window attenuation.

The message from this analysis is that coexistence may not be a major consideration in many deployments that are isolated by distance or highly attenuating materials. However, for networks without such isolation, coexistence will be essential to effective operations, unless spectrum is simply partitioned. Such coexistence may be trivial, such as isolated frequencies, and will still not consume all spectrum resources. Or, it will be more effective through tighter integration of the coexisting networks, up to the level of integration that is present within carrier LTE networks operating in high spectrum reuse conditions, requiring high levels of awareness and control over all member node status and operation. The technology to provide this exists; it is a business and policy issue about how much information and control is shared, and therefore how much sovereignty is compromised in the operation of the jointly managed systems.

## 12.4    Listen Before Talk MAC Layer Coexistence

The LBT Media Access Layer (MAC) layer creates a time partitioning of a single channel for use by multiple, uncoordinated networks that share it. LBT is the mechanism for several of the popular Local Area Networks (LAN) technologies, such as Wi-Fi and Bluetooth. The advantages of LBT include:

- There is no coordination or collaboration required between the sharing devices or networks. No interconnect between devices in the band or any third party agent is required to provide coexistence between users and networks. The only coordination between nodes is the sensing of energy in the channel.
- There is no requirement for any common technology, channel framework, cryptography, or other limits on the scope of devices that can participate. The framework is inherently technology neutral, once basic channel boundaries are established. An example of this is the Multefire technology that plans to provide LBT for Time Division Long Term Evolution (TD-LTE) in order to coexist with the dominant Wi-Fi use in the unlicensed bands.[3]

The disadvantages to LBT revolve around how conditions of channel congestion are determined and managed. As the density of channel usage increases, devices are required to increase the time that they wait to reenter the channel through an exponential back-off principle.[4] This impacts both throughput and latency of the data stream. Depending on the mixture of devices and traffic loads, the throughput efficiency of LBT varies, but the aggregate throughput is typically well below that of the channel if there was no device sharing occurring. This is the system cost of the simplicity of the LBT mechanism.

The distinction between an absolute measure of interference and a signal referenced one is significant in examining the channel usage characteristics of LBT and non-LBT MAC layers. LBT has a fixed threshold at which all devices must consider the channel

---

[3] Multefire™ is a registered trademark of the Multefire™ Alliance.

[4] This requires an increasing delay in reentering the channel as congestion is perceived to increase.

as occupied. For Wi-Fi, the standards state that all devices must consider the channel occupied if the device detects a signal level typically more than 10 dB above the noise floor. For a 20 MHz channel, this is –75 dBm.

In free space, this is a separation of around 500 m at 2.4 GHz and a typical antenna. So, any Wi-Fi access point at this distance will cause another device to consider the channel occupied and back-off.[5] A Wi-Fi access point at 6 m distance will have a signal strength of around –43 dBm, 32 dB above the occupied threshold. Several devices could be present and transmitting at this range with Signal to Noise Ratio (SNR) of at least 43 dB, which is a strong Wi-Fi signal. However, the distant node at the 500 m distance would cause the device to consider the channel busy, and not utilize an otherwise highly usable channel. Of course, real propagation loss is much greater than free-space propagation loss, but the relationships described here are maintained.

The use of LBT divides the aggregate capacity of the channel among any nodes with less than this implicit path loss, even if many of these networks would have sufficient signal to operate simultaneously, even in the presence of each other. If the network wanted to operate at an SNR above 12 dB, and at range of 6 m, the networks could be spaced at a range of 30 m. Idealized positioning could enable a 1 000 times increase in density over LBT.

## 12.5 Lessons Learned From LTE-U/LAA and Wi-Fi Conflicts

The proposal to introduce LTE-based technologies into spectrum that has traditionally been primarily utilized for Wi-Fi and Bluetooth technologies has created extensive controversy, largely due to the initial lack of coexistence of these technologies and potential impact on the existing uses, even though these uses were not entitled to protection. The technologies proposed included a variant of LTE, Long Term Evolution Unlicensed (LTE-U). Significant studies, analyses, and experiments were reported, typically by advocates for, or against, the proposed use. Advocates on any side are not generally the best source of impartial science, but some broad principles can be gleaned from the record.

- LBT MAC layers must detect the channel being busy at very low levels of signal, so the LBT devices will go into exponential back-off[6] and delay their access to the channel. Operation of a continuous source, such as a non-LBT protocol, even at low levels, may cause the LBT systems to not have opportunities to enter the channel at all, even though the level of signal would be quite low, and might not have caused interference to the LBT network.
- Continuously operating systems have very significant negative impacts on each other, when they operate at similar levels. Their performance is much worse than an LBT MAC in the same conditions.

---

[5] For a client with 0 dB gain, the range would be more like 200 m.
[6] The exponential back-off behavior increases the time between channel access on the assumption that the high rate of channel busy sensing indicates congestion on the channel.

Because of these issues, the LTE community appeared to decide that it could not (technically or politically) prevail in using unmodified LTE in the same bands that incumbent usage was dominated by LBT systems. Various proposals, such as Multefire and Licensed Assisted Access (LAA) are being proposed to introduce LBT into the baseline LTE protocols to coexist with the currently dominant LBT usage in the unlicensed bands.

## 12.6    Hybrid Strategies that Integrate Protected and Unprotected Spectrum

The previous discussion has treated the alternatives of protected access and some level of coexistence as discrete and exclusive options. It is important to recognize that this need not be so, and that network strategies can utilize both protected and unprotected spectrum access to operate networks effectively. Hybrid strategies that utilize some amount of protected spectrum and some amount of unprotected spectrum are a viable, and attractive alternative.

A specific example of this is the evolving LAA standard for LTE. This was, to some extent, a follow-on from the initial proposals for LTE-U. The initial concept for LAA was LTE operation in typical exclusive licensed carrier bands, and a supplemental downlink operation in 5 GHz unlicensed spectrum that was compatible with LBT. Thus, one set of channels was protected, and the unlicensed one was not.

Such strategies reduce the required reliability and predictability of the secondary channel, as it is offload traffic, and the network can allocate high reliability traffic to the assured access spectrum, and use the contended spectrum for less time critical or Quality of Service (QoS)-sensitive traffic. This spectrum is a good candidate for integrated sharing, so long as the technology can support this systems engineering framework.

## 12.7    An Integrated Framework for Technology-Neutral and Technology-Specific Coexistence

Technology-neutral coexistence is constrained to the very "crude" aspects of the channel, in which the orthogonal aspects of the spectrum usage, such as time and frequency can be used to isolate users from each other. LBT does this inherently for devices that share this methodology. We assume that a three-tier framework must provide some mandatory or optional mechanism to deconflict devices from each other, without any awareness of, or constraints on, their specific technology capabilities. We consider frequency allocation as either static or dynamic in time. In this context, static-frequency coexistence provides spectrum for longer term usage, such as the unitary (not subdivided) channel in use.

Dynamic frequency coexistence manages specific sub-channel usage on a highly variable manner, on a short time-scale, such as LTE resource blocks, which can be adjusted on a slot by slot basis. The coexistence can manage time at the level of a channel grant, which is a long term commitment in a technology-independent manner. Alternatively, a technology such as LTE can allocate time slots on a millisecond-by-millisecond basis. Longer-term time and frequency assignments can be made in a technology-neutral basis, while short-term allocations require technology assumptions, and thus are technology-specific.

Using the framework of Figure 12.1, the top-tier technology-neutral layer partitions the spectrum resource using the orthogonal dimensions of time and frequency. In most optimization problems, partitioning does not create an efficient use of resources. There is no reason to believe spectrum is any different. The internal design of LTE reflects detailed understanding of very efficient use of spectrum resources, and its approach to dense spectrum resource is a very tight, technology-specific integration of adjoining devices.

In the operational concept, devices register with a coexistence service and express the technologies that they have available and are willing to utilize to achieve coexistence. The coexistence service determines the common elements between adjoining networks, and utilizes the lowest common level of techniques available to each, and for which they have expressed a willingness to employ them.

It is useful to consider some use-case scenarios. In these scenarios, we consider the unit of management is networks, not devices. We assume networks provide control over the coexistence of devices within the individual networks. Therefore, we think of coexistence as an inter-network service, not an inter-device one. The responsibility for interference management among cooperating devices is provided by the system managing that cooperation, such as the components of an Extended Packet Core (EPC), cluster controller, or Self-Organizing Network (SON).

- A set of devices that have no coexistence features register with the coexistence service. The only choice the coexistence service has is to partition the available spectrum resources between the devices or networks using some predetermined ratio. Each device has access to interference-free and uncontended spectrum, although the extent of this spectrum is reduced proportional to the number of devices or networks accessing it.
- A set of devices that all have, and are willing to use, LBT MAC layers for coexistence with other LBT devices register with the coexistence service. The coexistence server can allocate the maximum extent of available spectrum to all of the devices in common. Each device would have access to a large extent of spectrum, although this access would be contended. Assuming that there was some statistical variation in demand, each device can achieve much higher peak access to the spectrum, at least in relation to the previous example. These devices need not have a common technology, but could, for example, be a mix of Bluetooth, Wi-Fi, and Multefire.
- A set of TD-LTE devices register with the coexistence service. Again, the service can allocate a large extent of spectrum to these devices/networks, or could partition it as if they had no common coexistence technology. The difference is the degree to which they are willing to exchange information, such as timing, configuration, aggregate demand data, and the topology of the networks, and comply with coexistence manager direction. For example, if the networks are significant in extent, and coexistence is

only needed in small boundary regions, then a large spectrum extent, with time-slot or resource-block sharing on the edges would be highly effective in optimizing capacity in the interior, and avoid interference at the edges. Even if the allocation of time slots and resource blocks was fixed, and not demand-based, the benefits of a large spectrum extent in the interior of the network would be beneficial to all users.

- A number of devices with a mix of LBT and LTE register with the service. In this case, a hybrid strategy has to be invoked. The LBT and LTE users are partitioned into separate spectrum, and the LTE-specific sharing methods are further applied within that partition, similar to illustration in Figure 12.1.
- Other technologies are likely to emerge and create new coexistence methods that will create new segmentations of this coexistence service. For example, a technology could develop that shared spectrum via a Code Division Multiple Access (CDMA)-like framework using Pseudo Noise Spread Spectrum (PNSS) to isolate users. In that case, the coexistence service would have to be extended to manage sets of codes, and maximize the orthogonality of neighboring networks' code assignments.

## 12.8    Suggested Reading

- The proposals for enhancing the compatibility of LTE with LBT systems have resulted in several proposals that are instructive in their approach to coexistence. The consortium is developing a technology, Multefire, that provides LBT functionality using the LBT MAC layer framework. Intended for compatibility and spectrum sharing with Wi-Fi dominated bands, it is proposed by some as a method to create coexistence among users in the lower tier, without any necessity to have any coordination, other than the in-channel sensing. Detailed technical material on Multefire is not available outside of the Multefire community at this time, but will likely be more publicly available in the future as its proponents move to adoption.
- LTE internal features are critical to the implementation of many of the coexistence features that will be required in rich three-tier spectrum regimes. There are a number of good reference sources on LTE [3, 4].
- The performance of LBT as a spectrum-sharing and coexistence mechanism is central to the choice of three-tier spectrum regimes. There are a number of readings that present some of the performance characteristics of LBT [5, 6, 7].
- There has not been very much research on the coexistence of widely different protocols. One example where this issue was studied was when the use of LTE in the unlicensed 5 GHz band was proposed in LTE-U. A number of experiments were reported in public filings with the US Federal Communications Commission (FCC). These results were typically arguing for, or against, permitting LTE in the traditional Wi-Fi band. Therefore, the results have to be viewed with some discretion. But, even with that caveat, they remain instructive in the issues of such coexistence.[7]

---

[7] In full disclosure, the author was one of the opponents of permitting unmodified LTE to be permitted in this band.

# References


1 National Institute of Standards and Technology, *NIST Construction Automation Program Report No. 3 Electromagnetic Signal Attenuation in Construction Materials,* NISTIR 6055 (Gaithersburg, MD, 1997). fire.nist.gov/bfrlpubs/build97/PDF/b97123.pdf.

2 I. N. Rodriguez, H. Cong, N. T. Jørgensen, T. Sørensen, B. Mogensen, and P. Elgaard, Radio propagation into modern buildings. *IEEE 80th Vehicular Technology Conference* (Aalborg Universitet, 2014), 1–5.

3 E. Dahlman, S. Parkvall, and J. Sköld, *4G: LTE/LTE-Advanced for Mobile Broadband* (Amsterdam: Academic Press/Elsevier, 2011).

4 A. Ghosh, J. Zhang, J. G. Andrews, and R. Muhamed, *Fundamentals of LTE* (NJ: Prentice Hall, 2011).

5 C. Chen, R. Ratasuk, and A. Ghosh, Downlink performance analysis of LTE and WiFi coexistence in unlicensed bands with a simple listen-before-talk scheme. *IEEE 81st Vehicular Technology Conference* (2015), 1–5.

6 A. Mukherjee, J. F. Cheng, S. Falahati, L. Falconetti, A. Furuskär, B. Godana, D. H. Kang, H. Koorapaty, D. Larsson, and Y. Yang, System architecture and coexistence evaluation of licensed-assisted access LTE with IEEE 802.11. *2015 IEEE International Conference on Communication Workshop* (2015), 2350–2355.

7 N. Rupasinghe and İ Güvenç, Licensed-assisted access for WiFi-LTE coexistence in the unlicensed spectrum. *IEEE Globecom Workshops* (2014), 894–899.

# Example Use of Three-Tier Spectrum: Use of the 3.5 GHz CBRS Band in the USA

# 13 Device and Network Interaction with the Spectrum Access System

## 13.1 Introduction

This chapter summarizes the design of the United States (US) Citizens Broadband Radio Service (CBRS) device to Spectrum Access System (SAS) interaction. The US standards have been developed by the Wireless Innovation Forum (WinnForum) through collaboration between its membership, including Mobile Network Operators (MNO), Mobile System Operator (MSO), equipment suppliers, SAS operators, and incumbent system users.

This discussion is not intended as a recipe for how three-tier regimes must be implemented. Instead, it is presented as a starting point for other implementations to consider this design, and, hopefully, further extend it, or adapt it to meet different objectives. Certainly, there are cost advantages if, all things being equal, there is a degree of harmonization of these mechanisms internationally. It will also serve to make the more general or abstract discussions more concrete and explicit.

This chapter will summarize the design, and the rationale for this design. However, it is not intended to be sufficient for CBRS design or development, and is intended to familiarize the reader with the frameworks selected for this first implementation of the three-tier concept. The complete hierarchy of objects and containers is not represented fully, but abstracted to discuss the information content of the exchanges, rather than their implementation.

It is also worth noting that all of these interfaces are defined consistent with Internet Engineering Task Force (IETF) internet standards, including the Internet Protocol (IP) standard. Rather than utilize any specific wireless technology standard, such as the Third-Generation Partnership (3GPP) Long-Term Evolution (LTE) layer 2 interfaces, the three-tier devices, or their proxies, operate on IP network backhaul. This supports operation over any public or private IP network. Similarly, the upper layer security protocols are specific to IP Transport Layer Security (TLS)/Secure Socket Layer (SSL). This is a significant departure from past telecommunications practice, which typically deployed access point networks over unique layer 2 interfaces.

As the design evolved, the CBRS community implicitly established a consensus around the following list of design objectives:

1. Assured device shutdown, or relocation, when directed, was essential to assure protected users of the integrity of the protection methodology.

2. SAS interaction must be technology neutral, and capable of supporting any reasonable technology in the band, and not limit future innovation.
3. Avoid requiring any substantial changes in the design of the Radio Access Network (RAN) or upper layers, in order to fully exploit the existing high-volume commercial products.
4. Be extensible to enable technology-specific optimizations and coexistence features. The immediate focus of this was for Time Division Long Term Evolution (TD-LTE)-specific extension to provide additional coexistence features.
5. Support multiple competitive, but cooperative, SAS suppliers, with no loss of protection methodology integrity.
6. Provide authentication and non-repudiation to ensure that devices, SAS, and installers were appropriately certified and accountable.
7. Enable SAS interaction on behalf of CBSDs by network management functions, such as those provided within the LTE Extended Packet Core (EPC).

## 13.2    CBSD Shutdown and Relocation Assurance

The key requirement for the Citizens Broadband Radio Service Devices (CBSD) ecosystem from a protected user's perspective is assured shutdown and relocation of devices that could cause them interference. Since there was (and still is) incumbent resistance to this regime, the shutdown mechanism had to be demonstrably reliable despite a wide range of conditions, such as:

- The SAS could fail and leave devices authorized to emit with no ongoing protection mechanism.
- The communications path between the SAS and the CBSD could fail, leaving the devices authorized to emit.
- Many enterprise and consumer devices would be positioned behind firewalls, where a Network Address Translation (NAT) service is used with Dynamic Host Control Protocol (DHCP) to dynamically assign IP addresses, and this address could change during operation, leaving the device unreachable by the SAS, and therefore not capable of receiving shutdown messages.
- Firewall policies, or overloaded DHCP networks, may preclude incoming connections to devices located within perimeter defenses.

There are two modes of operation between the SAS and the individual CBSD. A CBSD can access the SAS directly, or through a proxy, such as the control system used in LTE. The proxy interface is both more complex, and is technology specific, so the discussion in this chapter will focus on the direct, standalone SAS to CBSD interface. This interface implements a "pull" from the SAS by the CBSD for all interaction between them.

The CBSD is required to continuously request permission to continue operating under a spectrum grant. It does this by requesting a "heartbeat" periodically in order to continue to emit. If for any reason it is not granted permission, or cannot contact the SAS, the CBSD must terminate operation when the current authorization expires.

**Table 13.1** Typical shutdown/relocation time intervals that establish heartbeat intervals.

| Shutdown Event | Time Allowed | Applicable in Locations |
| --- | --- | --- |
| Naval Radar Operation | 300 seconds | Naval Protection Zones |
| FSS TT&C Site Registration | Unspecified | All CBSDs |
| PAL Licensee Deployment | Unspecified | In PAL-licensed census tract and on PAL frequency |
| Government Directed Shutdown | Under Discussion | All CBSDs |

Why have continual operation require a short time interval authorization when the spectrum grant may be valid for a much longer period? Spectrum grants provide stability and the reasonable expectation of the right to continue operation under the grant until its expiration. Grants provide stability in spectrum assignments over long periods of time. Systems that have complex spectrum management strategies can therefore perform planning and management.

Failure to receive continued authorization in response to a heartbeat request is an exceptional condition. It should be uncommon for most devices. The heartbeat process operates independently of the grant process, and addresses only the requirement to provide protection when protected devices enter the spectrum. The grant and authorization process maximizes long-term frequency stability, while still assuring that arbitrary spectrum clearing times can be achieved with 100 percent confidence.

The SAS computes a heartbeat interval dynamically based on the most constraining spectrum recovery time for any possible protected node that might register or be detected. The lowest applicable exit time is used to set the reauthorization time when the grant is initially issued, and may be updated in subsequent heartbeat messages, if needed. Table 13.1 illustrates the exit, or clearing, time for possible entries into the CBRS band.

The SAS response to a heartbeat request can consist of:

- Continuation of the authorization to use the grant, with no change in the grant conditions.
- Continuation of the authorization to use the grant, with a change in the valid period, power or other grant conditions.
- A message that informs the device that the authorization is not provided.
- No response at all, which essentially has the same effect as informing the device that the authorization is not to be provided.

The CBSD is free to request reauthorization at any time it decides is appropriate. It is aware of the quality of the path from it to the SAS, so it can balance the transaction rate and the risk of not obtaining reauthorization.

This interface may appear to be overly burdensome. Certainly for devices that are located in frequencies and regions where there are short clearing times, there will be a relatively high heartbeat rate. The virtue of this interface is that it ensures that devices do not exceed the shutdown time allowed, under any possible failure condition. Regardless of what component of the ecosystem fails, the device will behave properly, if inconveniently. Spectrum sharing is about incumbent trust, and creating a high assurance shutdown was believed to be important in this first implementation of three-tier spectrum.

## 13.3    General Principles of Device Interaction with the SAS

Figure 13.1 illustrates the valid CBSD states, the messages that cause transitions between these states. Note that in the US CBRS implementation, the first state is divided into two substates, *"Not Registered"* and *"Registered."* This is a convenience for making the registration transaction independent of the grant, so that location information and other invariant data need only be provided once. This is particularly convenient for professional installation, where the installer can cryptographically sign the initial registration. However, from the perspective of interference management, these two states are not distinguishable, or significant. In the later sections, the transactions specific to each of these two states will be described.

**No Grant** The device has no valid spectrum grant (and therefore no authorization to radiate). This is the initial state of a device. It may or may not be registered with the SAS in this state. Registered, or not registered, could be considered sub-states of this state.

**Granted** The device has received a spectrum grant. The grant is typically good for a long period. The device has not been authorized to actually radiate.

**Authorized** The device has a valid grant, and has been authorized to radiate.

The transitions between these states are described in more detail in Table 13.2.

## 13.4    SAS, CBSD, and Installer Authentication

Trust in the spectrum-sharing regime implies that the system has several attributes:

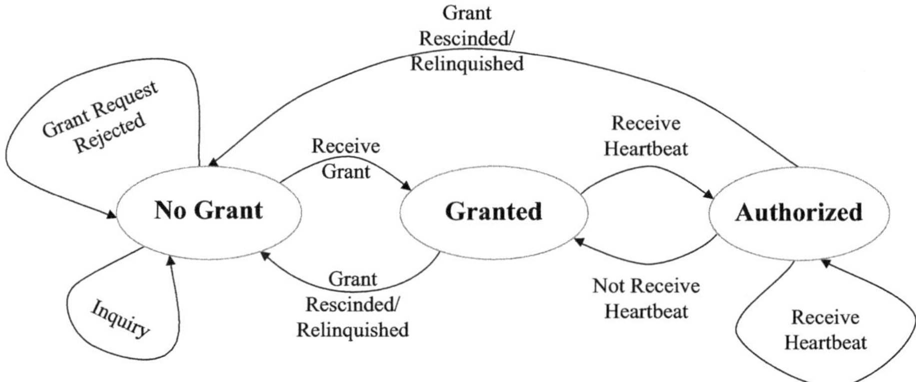


**Figure 13.1** CBSD authorization states and permitted transitions.

**Table 13.2** Valid CBSD states and actions.

| State | Valid Actions | Received from SAS | Next State |
|---|---|---|---|
| No Grant | Inquiry Request | Inquiry Response | No Grant |
| | Grant Request | Grant Issuance | Granted |
| | | Grant Rejection | No Grant |
| Granted | Request Heartbeat | Receive Continued Authorization | Authorized |
| | | Not Receive Authorization | Granted |
| | Relinquish Grant | Relinquishment Confirmation | No Grant |
| Authorized | Request Heartbeat | Receive Continued Authorization | Authorized |
| | | Not Receive Authorization | Granted |
| | Relinquish Grant | Relinquishment Confirmation | No Grant |

- Participants in the ecosystem are accountable for their actions, and the system can assure knowledge of the identity of all parties involved in decisions or actions.
- The only participants in the ecosystem are ones that have agreed to the terms of use, been validated through an independent review of their performance, or qualifications.
- No party can impersonate any other entity in the ecosystem without the explicit support of the impersonated party.

The authentication process verifies the identity of any party (device, person, or cloud service, as examples) that asserts identity or compliance with the rules. The WinnForum industry group decided to adopt the current public/private key structure that is used for many computer authentication applications, most commonly in the identity of web sites through their SSL certificates.

The optimal method would be to have a single root Certificate Authority (CA) that was the sole issuer of all of the certificates used in the ecosystem. However, any sole source, non-competitive role certainly violates many of the CBRS principles, so it was decided to have a set of certificate issuers that were "trusted" to issue each of the specific certificate types. This also integrated more efficiently with the manufacturing process used by some of the ecosystem suppliers. They issued their own certificates that were not tied to a root that would likely be the one jointly selected. This enabled Part 96 manufacturers to issue their own device certificates.

The other advantage of modeling the authentication on SSL/TLS certificates was that any participant in the community could validate these certificates. Devices, installers, and Priority Access License (PAL) owners could therefore move between SAS providers without having to re-establish their credentials with the new SAS supplier. This approach made trust a property between the CBRS community, as a whole, and the SAS providers, not just bilateral with a between a single SAS operator and a single device manufacturer or operator.

The range of authenticated identity information required for CBRS included the following categories:

**Device** The Federal Communications Commission (FCC) regulations require that the SAS verify that any device that receives a spectrum grant is authorized to operate in the band. This certificate would be only issued to manufacturers that held at least one Part 96 device certification.

**SAS** SAS operations have several authentication needs. SAS interfaces must be able to assure CBSD accesses that they are, in fact, a SAS. Otherwise, SAS emulators could impersonate SAS operators and permit unlawful entry of devices, shut down networks, and generally wreak havoc on the operations in the ecosystem. Also, the exchange of SAS data requires that the receiving SAS be assured that the information that it is receiving is from a true SAS, and not an attempt to corrupt the regime.

**PAL Owner** A PAL grant request asserts that operator of the device is the PAL licensee. The SAS needs to be able to validate that the node requesting PAL protection is entitled to ask for that.

**Trusted Installer** Trusted installers are allowed to state the location of devices in conditions where Global Positioning System (GPS) location cannot be determined, or where the functionality was not present. These installers are the only individuals authorized to provide the SAS registration data on outdoor CBSDs. This ensures that only installers who have been trained in the installation and applicable regulations are certificated. A CA issuing this certificate would have to verify the completion of a program that was recognized as having an acceptable syllabus and completion criteria.

In addition, it is likely that the SAS operators will utilize certificates for internal uses, such as validating that sensor and health and status reports are coming from the intended device.

## 13.5    Device Interaction with the SAS

### 13.5.1    Device and proxy interaction

The simplified model for a SAS-controlled CBSD ecosystem is direct interaction between a SAS and a CBSD. For standalone devices, such as a typical Wireless Fidelity (Wi-Fi) Access Point (AP) this is the typical model for deployment without SAS services, and can be directly carried over into the SAS-controlled model.

However, for more complex, network-oriented technologies, such as small cell LTE and its variants, there is an intermediate layer of management. LTE refers to this as the various elements within the EPC, including the Radio Resource Management (RRM) and Mobility Management Entity (MME), which directly manage the resource (spectrum and time resources) use by the controlled eNodeB devices. For technology generality, the SAS refers to these as *"proxies,"* as they interact on behalf of their controlled devices.

Interaction with the SAS by proxies is similar to the interaction of an individual CBSD. The message designs provide for request messages to provide arrays of one or more request objects to the SAS, and for the SAS to reply with an array of response objects, in the identical sequence as provided by the request. This enables a single proxy to interact on behalf of a network of CBSDs. This is not the most sophisticated implementation of a proxy in that much of the material may be common across the devices managed by the proxies. It has the virtue of a common structure with the standalone version of the interface, and its certification need make no assumption about the reliability of the proxy or any of the intervening communications systems. Note that the proxy now takes on the responsibility to assure CBSD shutdown through its own mechanisms.

### 13.5.2 Spectrum Measurement by CBSDs

It is obvious that the SAS can make better coexistence decisions when it is provided with measurements of the environment. These measurements can serve a number of purposes within the CBRS ecosystem, including:

**Coexistence** The measurements indicate the actual conditions at a site. While the SAS may not know the duty cycle of potential interferers, the use of measurements overcomes many of the challenges due to the lack of knowledge of building materials, foliage and other factors that might cause greatly increased path loss.

**Building Specifics** Indoor propagation in a given building is probably quite similar between floors and across major areas of the building. As soon as a few deployments in a building occur, the SAS will be in a position to develop building-specific models of the paths within the structure and more aggressively pack devices, with high confidence in achieving acceptable interference levels.

**Diagnostics** If the measurements reported exceed the worst case estimate prepared by the SAS, then there is a serious error in its algorithms, or there is some emitters that were not accounted for, or not accounted for properly.

**Enforcement** If the additional interference levels are not explainable by errors in the SAS computations, or by mis-registered CBSDs or User Equipment (UE) devices, then the logical conclusion is that there are devices present that are not operating in accordance with the regulations, and some enforcement action may be justified. The measurements may support such an action, or assist in its execution.

The SAS can request measurements by including a measurement request in many of its responses. The device then includes the measurement data in its next message to the SAS. The US CBSD community has currently only defined one type of measurement, which is an LTE Received Signal Strength Indication (RSSI) measurement. The set of measurements is anticipated to expand as different technologies enter the band, and CBRS-specific implementations emerge. The LTE RSSI measurement type is shown in the tables and figures as an example.

### 13.5.3 Transactions between the SAS and CBSDs

The following subsections describe the major SAS to CBSD transactions in some detail. They describe the current US CBSD and SAS interaction. This design is still very early in its evolution; there has been little experience with it in actual operation; and it has features that were included to facilitate certification, rather than because they were optimal. Nonetheless, it is a useful point of departure that any subsequent implementation is likely to draw from.

Each transaction is characterized by a drawing of the message hierarchy, indicating which items are arrays in which a list of one or more values can be provided. Each object or data item is noted as being *Required, Optional*, or *Conditional*. Conditional items are required to be included in the message depending on the success of the transaction, or the other content in the message, or the preceding one. A list of the message objects and data items is provided in Table 13.3.

### 13.5.4    Device Inquiry Requests

Device software must be aware of the rules of the national regulatory regime in which they operate, but they cannot be aware of local conditions and spectrum availability until they are deployed. The inquiry transaction provides this awareness, but does not provide any spectrum rights. The device can examine the options provided, sense the spectrum to detect the levels of interference to which it would be subject, and trade the possibility of disruption against the local environment present in each channel.

Knowledge of the other uses, and the regulatory constraints on a given band, is highly useful to the device in performing channel selection. A device that is examining spectrum for use as General Authorized Access (GAA) might interpret the band status impact on its operations as follows:

**GAA**  No constraints, and not likely to be disrupted once a GAA grant is received.
**PAL**  A PAL is licensed in this channel at this location, although it may not be present at this time. Deployment of this PAL might force relocation of GAA devices, even if the GAA assignment was possible at the GAA device entry time.

The SAS inquiry transaction processing does not examine the possibility of the device to cause interference to other protected entities, such as wireless Internet service provider (WISP) or Fixed Satellite Service (FSS) operation. Therefore the results of an inquiry are not an assurance of a spectrum grant, they are only indicative of the status of the channel.

Note that this inquiry object is very parochial to the US CBRS implementation. It does not provide many of the features that would be required in a much more extensive three-tier framework, including more bands, more nations, and more rule sets, such as will be described in Chapter 18. This function will need to be significantly extended when additional countries and bands are added to the spectrum regime. A device might have multiple bands available at a location. The inquiry function would let it inquire at a band level before addressing the specifics of any given band. This would provide a strategic level of band selection interaction to supplement the tactical nature of channel selection.

Similarly, a three-tier device should be able to roam the world and discover the specific rules in whatever location it found itself. This would include the list of bands, and also the certification that the device would have to meet in order to be authorized to use each one. The device could then select its credentials for any band to which it held

**Table 13.3** CBSD to SAS messaging data and object items.

| Item | Object/ Data | Content |
| --- | --- | --- |
| Air Inter. | Object | Description of the air interface to used by the CBSD. |
| Ant. Gain | Data | Maximum antenna main beam gain, in dB units. Required for category B CBSD registrations. |
| Ant. Model | Data | Name of a model that provides a more detailed antenna pattern characterization.[a] |
| Ant. Azimuth | Data | Boresight azimuth, 0 is true north. Required for Category B CBSD registrations. |
| Ant. Downtilt | Data | In degrees. A negative value indicates the antenna is pointed upwards from the horizontal. |
| Avail. Channel | Object | List of channel (frequency range), their type, and the rule base that authorized them. |
| Beamwidth | Data | Angle (in degrees) between the antenna's 3 dB horizontal pattern points. |
| Call Sign | Data | Identifier assigned by the FCC.[b] |
| CBSD Category | Data | Device category, implying limitations on conducted power, EIRP, location and installation authentication. Part 96 establishes two categories, *A* and *B*. |
| CBSD ID | Data | String providing the globally-unique identifier of a CBSD. Assigned by the SAS after successful registration. |
| CBSD Info | Object | Data describing the CBSD model and configuration. |
| CBSD Serial No. | Data | Unique identifier within a given FCC ID, assigned by manufacturer. |
| Channel Type | Data | *GAA* or *PAL* showing channel status. |
| EIRP Cap. | Data | Maximum EIRP, required for category B CBSD registration. |
| EutraRssiReport | Object | A list of LTE RSSI reports, as defined by 3GPP. |
| FCC ID | Data | The US FCC identifier assigned during certification. Required to be validated by the SAS before a grant is issued. |
| Firmware Ver. | Data | Firmware software version. |
| Freq. Range | Object | Provides an upper and lower frequency limit for an inquiry, channel, or grant. |
| Grant Expire | Data | Grant expiration date/time in UTC. |
| Grant ID | Data | Unique identifier of each SAS grant. |
| Grant Request | Object | Provides all of the information required to specify a grant request by a CBSD to the SAS. |
| Grant Renew | Data | Logical; true, if the request is to renew the current grant. |
| Group ID | Data | Group ID establishes membership in a particular CBSD group. The SAS can make assumptions about the management of the members of a single Group ID, but not across Group IDs. |
| Group Params | Object | Array of CBSD grouping parameters. |
| Group Type | Data | Defines the type of group that the associated Group ID belongs. Currently, only an *interference-coordination* group is defined, but additional group types are likely to be established to reflect different coordination schemes to which the SAS must be aware in making deconfliction decisions. |

**Table 13.3** (*cont.*)

| Item | Object/Data | Content |
| --- | --- | --- |
| Hardware | Data | Applicable revision or version of the FCC ID. |
| HB Interval | Data | Time in seconds after which an authorization is invalid. This is essentially the maximum heartbeat interval the CBSD should use. |
| Height Type | Data | Type of altitude: Above Ground Level (AGL) or Above Mean Sea Level (AMSL). |
| Height | Data | Altitude of the CBSD in meters. The reference for this is provided in *Height Type*. |
| High Freq. | Data | Highest Frequency in Range. |
| Horiz. Accuracy | Data | Accuracy of the measurement horizontal position. |
| Indoor | Data | Logical; true if the CBSD is deployed indoors. |
| InquiredSpectrum | Data | A list of frequency range objects that enumerate the range of spectrum that is being inquired. |
| Install Params. | Object | Provides all of the information regarding the actual placement of the CBSD. |
| Latitude | Data | CBSD location. Latitude in WGS84 reference. |
| Longitude | Data | CBSD location. Longitude in WGS84 reference. |
| Low Freq. | Data | Lowest Frequency in Range. |
| Max. EIRP | Data | Maximum EIRP requested by the CBSD, or granted by the SAS. |
| Measure Rpt. | Object | List of individual measurements. |
| Measure Rpt. Config. | Data | Array of strings that enumerate the measurement capabilities of the CBSD. |
| Measured Bandwidth | Data | Bandwidth in Hz above the value provided in the field *Measured Frequency*. |
| Measurement Cap. | Data | Array containing a list of measurement capabilities. |
| Measured Frequency | Data | Lowest Frequency in Range. |
| Measured RSSI | Data | Carrier RSSI as measured in accordance with 3GPP criteria. |
| Model | Data | Model number/identification. |
| Oper. Params. | Object | Provides the specific radiation request to the SAS, or grant from the SAS. |
| OperationState | Data | Response to the previous heartbeat response's request for CBSD operational state. Has values of "Authorized" or "Granted." |
| Radio Tech. | Data | Specifies the technology the CBSD can support. Currently only an LTE value is defined. |
| Resp. Code | Data | Success/Failure code. |
| Resp. Data | Data | Undefined contents. Will contain information provided by the SAS to assist the CBSD in responding to the status code. |
| Resp. Msg. | Data | Textual description of the result status. |

**Table 13.3** (*cont.*)

| Item | Object/ Data | Content |
| --- | --- | --- |
| Response | Object | SAS message success response, providing status code and optional description. |
| Rule Applied | Data | The FCC Rule used to generate the response. For CBSD devices, this is the literal string *FCC Part 96*. |
| Software Ver. | Data | Software revision level. |
| Support Spec. | Data | Which modes of the technology specified in the *Radio Technology* field. |
| User ID | Data | Identity of a specific CBSD user organization, pre-established with the SAS. |
| User Key | Data | Optional, pre-established credential from the SAS to be used to verify the association of CBSDs and the user. |
| Vendor | MData | Manufacturer name. |
| Vert. Accuracy | Data | Accuracy of the measurement vertical position. |
| Xmit. Expire | Data | Transmit authorization date/time in UTC. |

[a] A logical extension of this field is to enable the antenna pattern to be computed by the CBSD and provided dynamically to the SAS through this interface. This would accommodate electronically steerable and multiple beam antennas that would not have a single, static pattern.

[b] The meaning of a call sign in the context of an automated licensing regime is not apparent, or the mechanism for this assignment to be made. It may have been required for legacy tracking purposes.

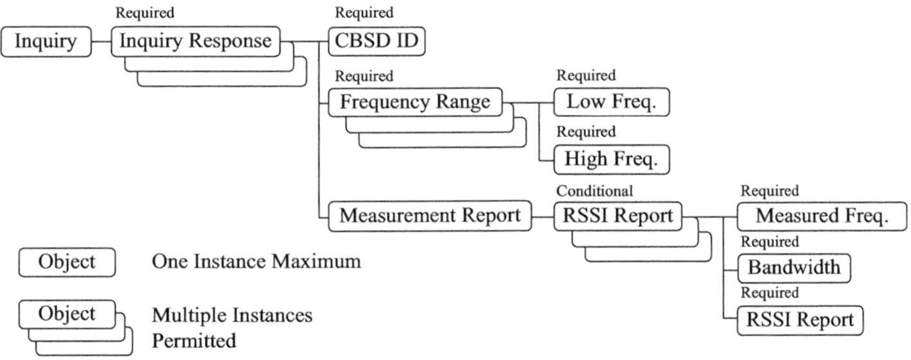


**Figure 13.2** CBSD SAS inquiry request object.

a matching certificate, and proceed to operate. This would enable an internationally harmonized framework for sharing spectrum, even in the absence of internationally harmonized band choices and rule structures.

A spectrum inquiry transaction consists of an array of *SpectrumInquiryRequest* objects. Each of these objects includes the information in Figure 13.2.

The inquiry transaction includes not only the status of the SAS perception of the interference analysis, but also provides the ability of the device to respond to measurement requests with channel measurement data.

The response is a channelized list of candidate bands and the regulatory authority the SAS used to make each determination. The response object is shown in Figure 13.3.

### 13.5.5    Registration Requests

Grant requests can only be issued following successful device registration. A device can be registered and the information can remain dormant until the device is operated. For example, a professional installer might make a network deployment without activating the boxes, but would register each one as it was deployed. Alternatively, the node can be registered, before immediately requesting a grant. A registration request consists of an array of *GrantRequest* objects. Each of these objects includes the information in Figure 13.4, with the detail on the very important CBSD deployment data provided in Figure 13.5.

The response to a registration request basically indicates that the request was successful or not, as well as an optional measurement request from the SAS. The SAS also provides the *CBSD ID*, which is a unique identifier that is used in all subsequent interaction with the SAS. This response message is shown in Figure 13.6.

### 13.5.6    Grant requests

Grant requests typically follow CBSD registration. They can be made at any time, and might be useful if a GAA channel became congested, or the device was forced off an existing grant. A given request message can include multiple *GrantRequest* objects. Each requested grant is for a specific *CBSD ID* and contiguous frequency range. There is no constraint on a given CBSD requesting multiple grants in order to obtain the right to use non-contiguous blocks of spectrum.

The grant request message structure is shown in Figure 13.7. The CBSD can respond to a previous SAS measurement request in this response message.

A grant response provides a set of *GrantResponse* objects in the same sequence as provided in the grant request message. Successful grant issuance is provided with an

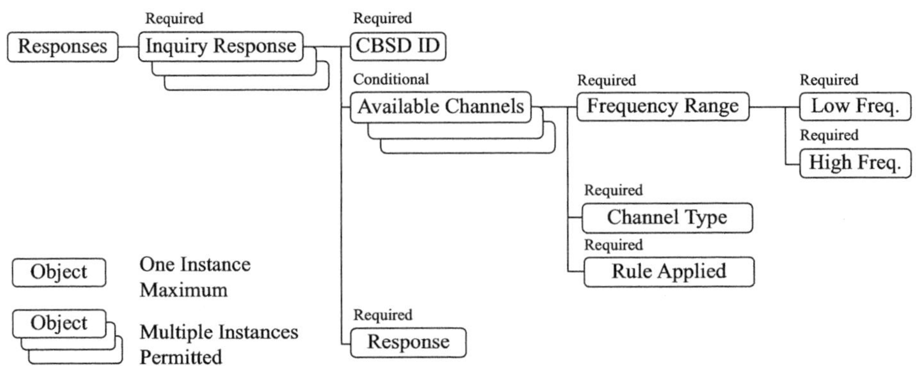


**Figure 13.3**  CBSD SAS inquiry response object.

expiration of the grant, as well as the heartbeat interval that is required to maintain current authorization to use the grant. The message also provides for the provision of a modified set of operational parameters if the SAS cannot provide a grant for the requested parameters, but could provide a subset, or alternative, to the requested grant.

The grant request response message is shown in Figure 13.8. As in all SAS to CBSD messages, this one can include a measurement request.

## 13.5.7    Continued Authorization to Operate Requests

Spectrum grants are for an extended period of time, and it may be necessary to terminate or suspend the right to transmit at some period in the life of the grant. This can be due to the deployment of a higher priority device, or by direct command of the government authorities. Therefore, devices are required to periodically request continued authorizations, or heartbeats, during the use of a grant. If a device holds more than one grant, it must renew each grant individually, although the requests can be included in a single message. This message can also include a measurement report, if one was requested in the last SAS interaction. The SAS can also request the device to provide its status and operational parameters to the SAS in the next heartbeat request. The request for continued authorization to operate message is shown in Figure 13.9.

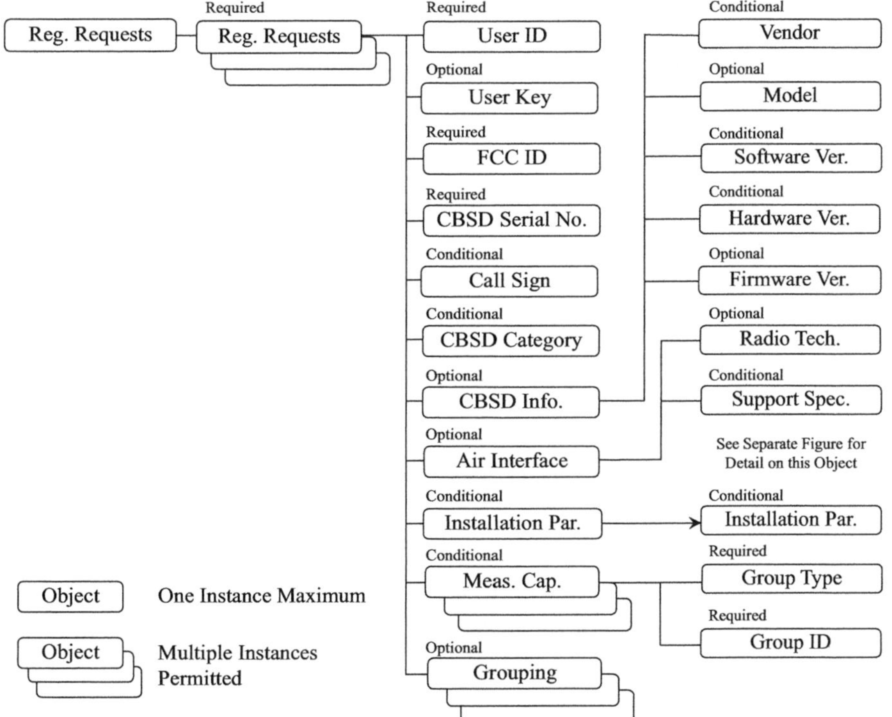


**Figure 13.4** CBSD SAS registration request object (part 1).

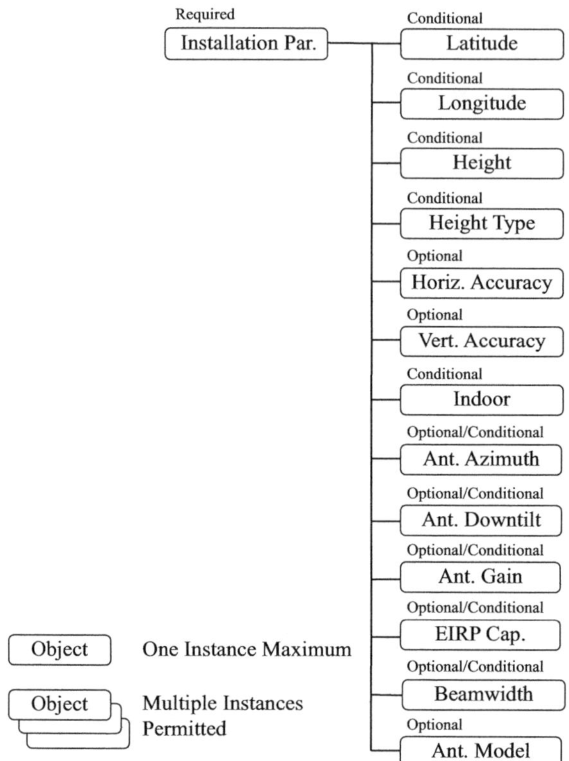



**Figure 13.5** CBSD SAS registration request object (part 2).

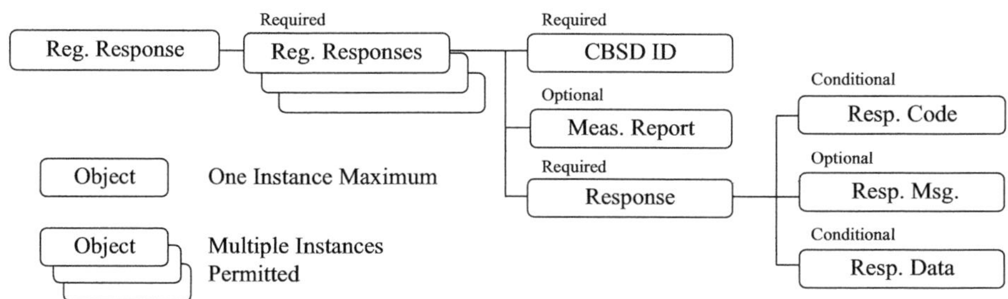



**Figure 13.6** CBSD SAS registration request response object.

The response to a heartbeat renewal request is a new expiration time, assuming the authorization is continued. The SAS can also use this response to change the heartbeat interval or request a measurement be included in the next heartbeat request. The SAS can also provide a new set of operational parameters to the device if it is necessary to modify power or frequency. The continued authorization message is shown in Figure 13.10.

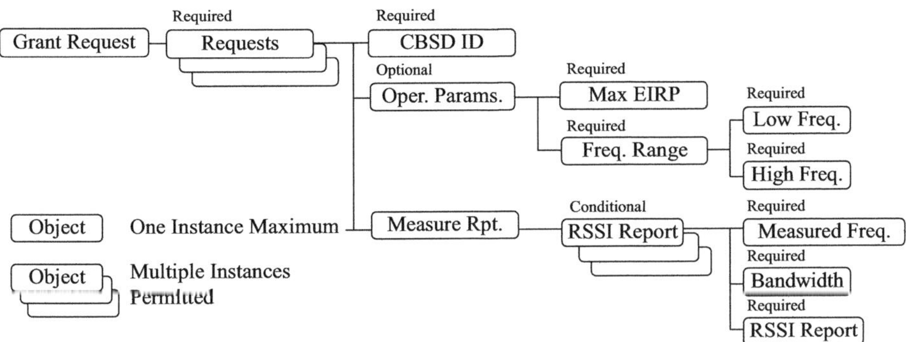

**Figure 13.7**  CBSD SAS grant request.

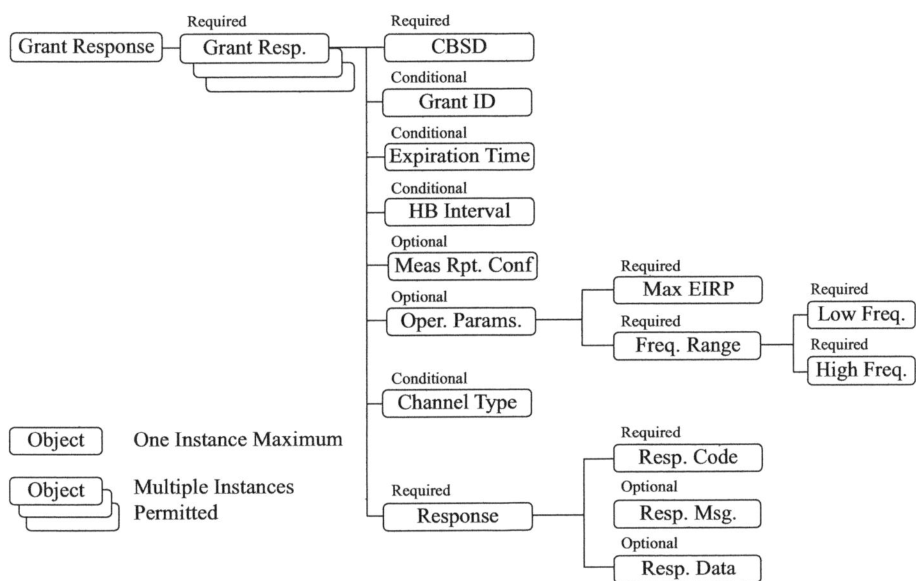

**Figure 13.8**  CBSD SAS grant response.

## 13.5.8   Grant relinquishment requests

One of the principles of three-tier spectrum is avoid spectrum warehousing. A device accumulating spectrum grants is a violation of this principle. Even if the authorizations are GAA, holding spectrum grants can impact other users. One of the protection modalities provided by a SAS is to protect higher tiers by estimating the aggregate interference from all sources of interference from the sharing devices into the protected receiver. The SAS cannot be aware of the instantaneous usage of any of these grants, so is forced to assume that all grants are being used fully. A device that holds unneeded grants is therefore improperly raising the estimate of aggregate emissions, and may therefore preclude other users from entry into the band.

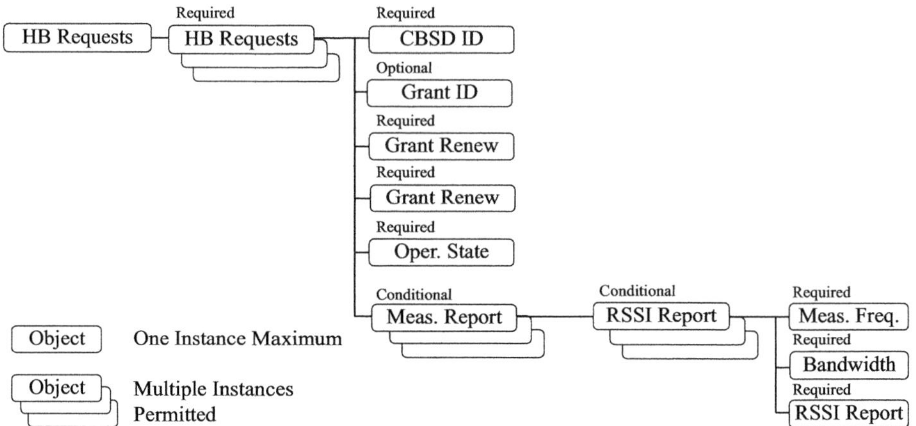

**Figure 13.9** CBSD SAS heartbeat request.

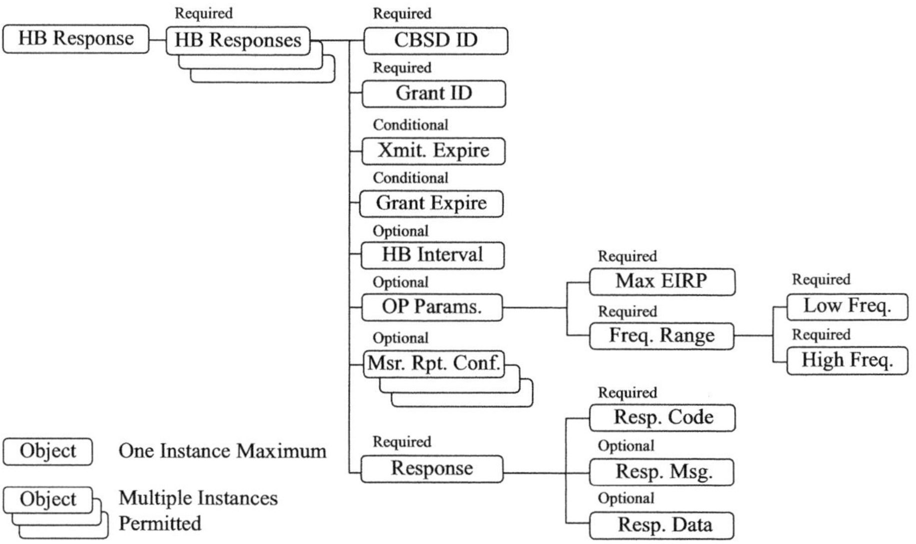

**Figure 13.10** CBSD SAS heartbeat response.

The grant relinquishment is therefore imposed as a precursor to any subsequent grant request. A device can only have one grant at a time. A grant could be for multiple channels, and even discontinuous spectrum, but it is a single grant.[1] The relinquishment is a precursor to requesting new channels. The relinquishment request message is simply the *CBSD ID* and the *Grant ID*. The grant relinquishment request message is shown in Figure 13.11.

---

[1] This does not preclude one grant that is noncontiguous, nor having more than one CBSD contained in a single "box."

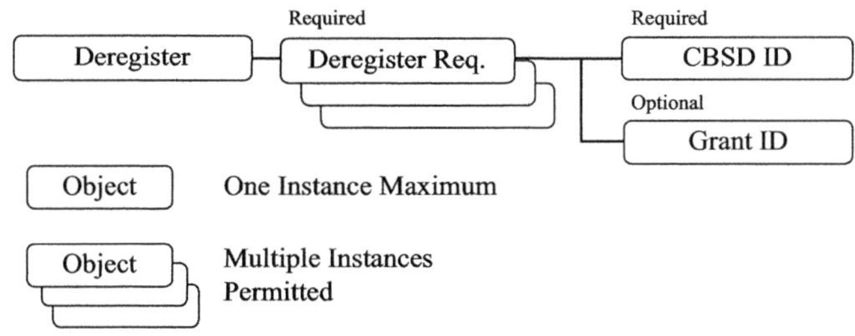

**Figure 13.11**  CBSD SAS grant relinquish request.

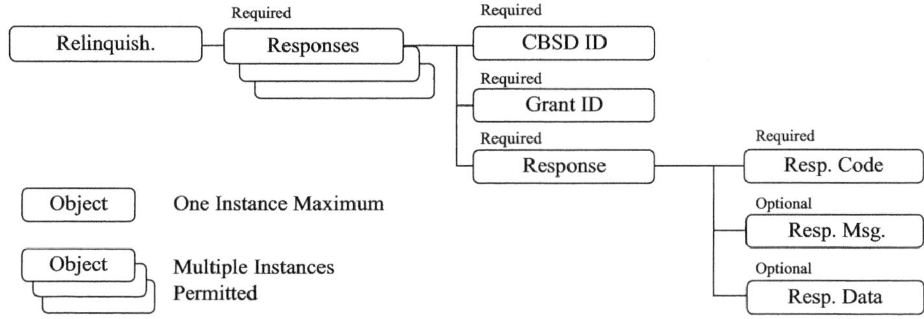

**Figure 13.12**  CBSD SAS grant relinquish response.

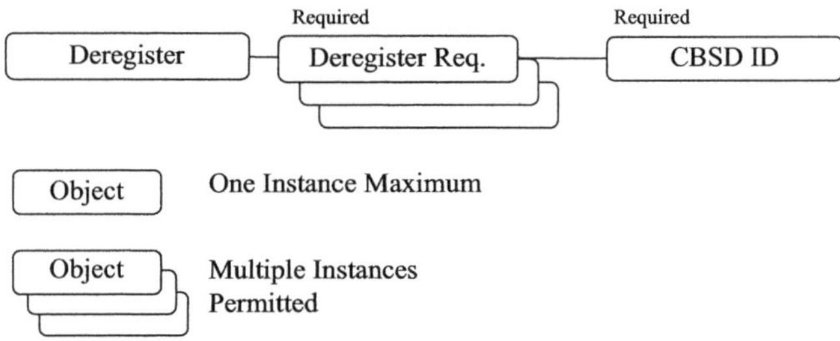

**Figure 13.13**  CBSD SAS deregistration request.

The grant relinquishment request response message is similarly simple, and provides the status of each relinquishment request. This message is shown in Figure 13.12.

## 13.5.9  CBSD Deregistration Requests

Devices can be removed from SAS registration by deregistering them from the SAS. This is a simple transaction of providing one or more *CBSD ID*s. The deregistration request message structure is shown in Figure 13.13.

The response is a set of response codes for each of the *CBSD ID*s that were deregistered. The deregistration request response message structure is shown in Figure 13.14.

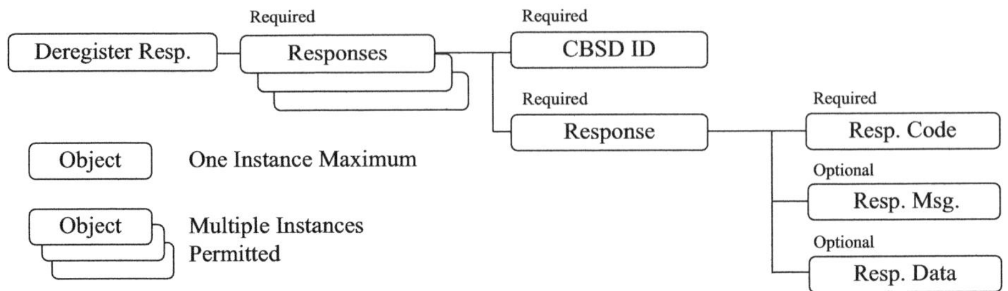

**Figure 13.14**  CBSD SAS deregistration response.


## 13.6     Extensibility of the Interface

There are likely to be several motivations for extensions of this interface framework.

- There are going to be significant *"lessons learned"* from trials and early deployment experience. These will have to be introduced into an ecosystem that will already be populated with devices.
- There will be technology-specific extensions needed to support technology-specific capabilities, such as coexistence methods and measurement capabilities.
- Technology-specific coexistence services will likely overload these messages with fields providing the technology-specific coexistence parameters from CBSDs to the service, and from the service to the CBSDs.
- Extension of the CBRS framework to additional spectrum bands and nations will require unique information be provided to the SAS, and may require unique information from the SAS.

As examples of the second of these drivers, at the time of this writing, only one measurement request for the measurement request object (*SpectrumInquiryRequest*) there is only one defined measurement type, which is an LTE-specific *MeasReportConfig*, which is defined as the 3GPP RSSI measurement, and currently only has one possible value, *EutraCarrierRssi*. A number of new measurement types are anticipated to be proposed to measure path losses, temporal characteristics of interference, and other vendor-specific methods. Similarly, the technology-specific response object, *EutraCarrierRssiRpt* could have parallel response objects to carry different measurement results.


## 13.7     Suggested Reading

- The details on the message design for these interactions is provided in the WinnForum Technical Specification for the SAS-to-SAS Protocol. It provides the specific message formats and semantics for these exchanges, and extends the general concepts of the

Technical Requirements A and B Level documents to the specific implementation. The details of the interface are provided in a technical specification [1].

## References

1 Spectrum Sharing Committee (SSC), *SAS to CBSD Technical Specification*, WINNF-16-S-0016-V1.0.0 (Wireless Innovation Forum, 2016).

# 14 CBRS SAS Requirements

## 14.1 Introduction to the SAS Design

This chapter will discuss the design of a specific implementation of a three-tier admission control system; in this example, the United States (US) Citizens Broadband Radio Service (CBRS) Spectrum Access System (SAS). The SAS discussed in this chapter is very much an image of the Citizens Broadband Radio Service Devices (CBSD) interface design described in the previous chapter. That chapter outlines the operation of the SAS in responding to an access point or point-to-point end node under the CBRS regulations. Much of this chapter parallels the interface discussion in the previous chapter, except from the perspective of the SAS side of the transaction.

The US implementation of a three-tier admission control system is the first ever specified through regulation, or implemented. The emphasis is therefore heavily biased towards protection of the current incumbents in the band, and the assurance of no bad outcomes as initial experience is gained in the band. It can be anticipated to mature, and relax, some of the extremely low-risk control specifications that are currently in place. Nevertheless, it is the best current example of such a system, and worth examination in detail.

For this reason, the US design of the CBRS SAS should not be considered an example that is universal. It is the best point of departure to examine the design and functionality consequences of one set of regulations for three-tier spectrum. Where specific features might be unique to the US regulations, they are noted in the text.

Although it is not explicit in the regulations, the presumed model of the SAS implementation is an application on one of the large-scale public or private cloud computing infrastructures, such as operated by Amazon Web Services (AWS), Microsoft Azure, or Google Cloud Platform (GCP). Such services provide redundant processing and communications capacity, internet exchange services, and failover. Therefore, the computer science related aspects of the design, such as backup, reliability, database management, and the like will not be further discussed here, as they are specific to the cloud service that would be selected, have been addressed in much more complex and stressed applications, and are not unique to the SAS functionality.

## 14.2　CBRS SAS Departures from the Reference Three-Tier Model

The major difference between the baseline three-tier regime and the US SAS is the lack of a dynamic spectrum marketplace for spectrum rights. The CBRS regime provides for annual auctions that are executed similarly to how traditional spectrum auctions have been held. These are executed independently of the SAS, on an annual basis. The results of this offline auction, and renewals of previous licenses, are sent to the SAS for use over the next annual period. The SAS does implement a number of features to support a more dynamic secondary market, which are discussed in a section later in this chapter.

## 14.3　Authentication and Protection of Data from Corruption

Spectrum sharing is dependent on trust between all of the parties in the sharing regime, particularly by participants in the protected tiers. This trust has to extend to the individual devices and systems that provide whatever protection methods are afforded. The Federal Communications Commission (FCC) requires that the SAS systems verify the certification status of CBSDs. The choice of this mechanism and other authentication requirements was implicitly delegated to industry definition. The Wireless Innovation Forum (WinnForum) established industry requirements for communications security in a Communications Security (COMSEC) [1] document that established common methods for authentication that would be implemented by all SAS operators and CBSD devices so that SASs, devices, and installers could interoperate across the CBRS ecosystem.

The WinnForum established a process to approve specific Certificate Authority (CA) organizations to issue one or more of the CBRS certificates. Individual CAs would provide an approved process used to validate each certificate type, similar to the extended Secure Socket Layer (SSL) certificate process. Certificates would be provided through the Transport Layer Security (TLS) protocol. The certificates types that provide the mandatory authentication are shown in Table 14.1.

In addition to these ecosystem-wide certificates, individual SAS operators who employed Environmental Sensing Capability (ESC) networks were also required by the WinnForum practices to issue and validate certificates. Certificates between the ESC sensors and the SAS ensures both that the ESC status and reporting were received from the certified ESCs, without corruption, and that the ESC only reported band occupancy data to an authorized SAS. This precludes the ESC being used by imposters in order to detect naval operations, or to cause interference to the naval radar by falsely reporting no detections. There is no requirement that ESC certificates be globally recognized, or provided through the same CA as the ecosystem certificates.

The use of the TLS protocol also provides protection of data from corruption, a specific Part 96 requirement. Any intervening node that modified content, or any loss of data in the messages, would be detected as a signature error, and the message would be discarded or re-transmitted until corrected.

**Table 14.1** Authentication of Part 96 CBRS ecosystem interactions.

| Certificate | Usage | Protection provided |
| --- | --- | --- |
| Professional Installer | Registration of all Category B (outdoor, or high power), and CBSDs without position detection | Assures that data is provided by a certified professional installer, and that there is non-repudiation for the data in the transaction. Issued to professional installers who successfully complete a compliant installers' curriculum, and have not had the privileges revoked. |
| CBSD | Required for all CBSD interaction with the SAS | Ensure that the SAS is interacting with an approved CBSD and the Part 96 certified CBSD software. |
| SAS | Required for all SAS-to-SAS interaction | Ensure that SAS-to-SAS exchanges are only performed with authorized SASs. Avoids insertion of fraudulent data into the SAS ecosystem through an imposter SAS performing SAS-to-SAS synchronization, or obtaining direct access to SAS database content. |
| | Required for SAS-to-CBSD interactions | Avoid CBSD contacting or being directed by a fraudulent SAS. Precludes unauthorized SASs that might not implement the regulatory protections. |

The FCC did not explicitly discuss the authentication of professional installer certifications in the regulations. However, the National Association of Broadcasters (NAB) pointed out in several regulatory and court filings that there were obvious location errors in the Television White Space (TVWS) databases. These were attributed to a number of causes, including lack of discipline and accountability for self-certified professional installers. This, it was argued, created a weakness in the protection scheme for broadcast transmissions. To avoid this possibility, the CBRS industry (via WinnForum) decided to enforce the cryptographic authentication and consequent non-repudiation provided by having each professional installer demonstrate its qualifications, register with a CA, and obtain a unique certificate.

## 14.4     PA Protection Requirements

Priority Access License (PAL) protection is provided by declaring a Priority Access License Protection Area (PPA) around the deployed node. A PPA is not associated with a census tract license, but only exists after the PAL node receives a grant. Specific features of a PPA include:

- The service area boundary is defined as the region that can receive a signal at a level of $-90\,\mathrm{dBm}/10\,\mathrm{MHz}$ at a height of 1.5 m.

- The registration of each device can declare a PPA, but the SAS must assure it does not exceed the signal level defined dimensions (above).
- The extent of the PPA cannot exceed the dimension of the PAL license.
- The aggregate interference permitted at this boundary, or interior, of a PAL service area is $-80\,\mathrm{dBm}/10\,\mathrm{MHz}$. This is computed at a height above the ground of 1.5 m.

## 14.5 Naval Radar Protection Requirements

Naval radar protection in the US CBRS framework is relatively crude, as it is a fixed geographic exclusion, rather than an actual interference protection. This framework obscures the actual location of the naval radar, at the price of likely overprotection of the radar by excluding more area than necessary to ensure protection.

Each ESC sensor location has an associated protection zone, which represents the region from which nodes could interfere with the radar, assuming the radar could be in any of the possible locations from which it could be detected. The operation of devices in any channel occupied by the radar is suspended. The devices can request another channel from the SAS, or remain suspended.

Suspension is accomplished by waiting for the next heartbeat request from each of the nodes in the protection area, and informing them of the suspension.[1] They must terminate their operation when their current authorization expires. Even if no heartbeat request is received, or replied to, the device will terminate its operation when the existing authorization expires.

The heartbeat authorization time is set to ensure that the device authorization will expire within the time period permitted from radar detection by the ESC to device shutdown. The regulator established time limit is 300 seconds. The device has 60 seconds to perform its shutdown after being suspended, or when not receiving a heartbeat at all. The device could increase its warning time by requesting heartbeats more frequently, since the authorizations are for a fixed period from the reply. Asking for a heartbeat at 150 seconds would provide at least 150 seconds of warning about shutdown, since the authorization that was issued would be good for 300 seconds, and would have 150 seconds remaining, even if a radar was detected on the device's grant frequency.

## 14.6 Secondary Market in CBRS

One of the major additions in the second FCC report and order [2] was inclusion of secondary market rules to enable secondary markets as an optional feature. Some SAS operators may choose to offer support to this market, while others may elect to not support entry of devices under the terms of secondary licenses. In any case, it would not preclude the SAS from providing entry of devices from the primary licensee.

---

[1] A non-response is adequate to implement suspension. The device need not be explicitly informed.

An additional consideration in secondary licensing is the operation of a marketplace for secondary spectrum rights. Although the FCC did not adopt the very dynamic nano-market that the President's Council of Advisors on Science and Technology (PCAST) proposed, the secondary market provides an opportunity to create such markets with whatever spectrum rights PAL holders are willing to submit for secondary market access. This is essentially a two-sided market. The market-maker sells spectrum access to device operators that require protection by transferring rights from unused PALs to temporary holders. Since the market has not yet appeared, we have no sense of how it will emerge; what are time scales in which spectrum will be offered; what premium licensees will receive (if any); whether they will be used for sustained operation, or as "gap fillers"; and other details that will emerge as operators gain experience.

Execution of the SAS responsibilities for devices registered under a secondary license is very similar to the process for a PAL. The secondary registration cannot conflict with a PPA already registered, so the SAS is not only protecting a PAL holder against General Authorized Access (GAA) uses, it is protecting the PAL holder against interference from its own sub-licensee. A PAL holder can waive protection of its own devices from its own devices, as in a dense Long-Term Evolution (LTE), since it is assumed that the interference management functions of the Radio Resource Management (RRM) will manage this contention. In traditional secondary markets, the licensee has the responsibility to validate the interference impact of the secondary usage it is licensing. However, in the case of the CBRS secondary users, the FCC has allocated protection to the SAS, rather than the licensee.

## 14.7    Transparency and Visibility Requirements, and SAS-to-SAS Coordination

Communications providers (TELCOs) are historically very private about their infrastructure, access point hosting arrangements, siting, and vulnerabilities. Users of these services are private about their network choices, and the detailed data regarding its usage. Yet, spectrum sharing is dependent on awareness of all of the location, operation, usage, and sensitivities of the devices in the band. These are inherently conflicting objectives. In implementing the US rules, this subject was highly contentious. Beyond the minimum regulatory requirements, the implementation of the coexistence service requires even more awareness of device position and operating characteristics.

We consider the requirement for transparency as serving several purposes within the ecosystem:

**Protection** One of the key purposes of exchanging registration and grant data is to assure that the protection mission of the SAS is met. Information on all CBRS entities is required in order to compute aggregate interference into protected nodes, and information regarding any additional protected nodes (such as PAL entries) must be added to the list of nodes protected locations.

**Coexistence** Determination of potentially interfering nodes, and plans for mitigating or eliminating this interference, require precise knowledge of node positioning. Vertical placement within a building, at least as it relates to floor levels, is critical in three-dimensional deployments, as well. Not only is the emission Power Spectral Density (PSD) essential, but fairly granular knowledge of the capability of the node to implement coexistence techniques, such as alignment of timing, allocation of sub-channels, sensing of the channel occupancy, or other technology-specific features is also required.

**Diagnostics** Protected and unprotected users encounter system issues often during the operation of a communications or sensor system, and interference is one of the first causes to be considered. In order to meaningfully consider whether there is a possibility of interference, the operator needs to know what devices are present that could be the cause of the disruption. In the case of CBRS, this could be a peer node, a protected CBRS node, a radar, a Fixed Satellite Service (FSS) site, or protected Part 90 (wireless Internet service provider (WISP) or Utility) operator. When competitive SAS operators are present, it is important that all SASs have sufficient visibility of all other SAS-managed nodes, in order to support protected user diagnosis without forcing these users to go to each SAS operator individually.

**Resolution** Knowledge of node positions is critical to resolving interference issues that are shown to arise after whatever diagnostic process in use is executed. At this point, resolution is a matter between the SAS operator, the protected tier user, and the owner of the device causing interference.

To understand the information exchange requirements, the following taxonomy of exchangeable information is useful.

**Public** The regulatory policy mandates that some information regarding licenses be made public. In the case of the CBRS band, the FCC mandates that location, device characteristics, grant information, and related data be public. The release of this public data was required to have personally identifiable information obscured for individual privacy. This information is required to be exchanged without limitations.

**SAS-Enabling** Information that is not required to be public, but is needed by other SAS operators to implement the regulatory requirements for the SAS. Exchange of this information is mandatory, but may be provided with limitations on use for purposes other than SAS operation. For example, the SAS may provide other SAS operators with the name of the device owner, but the other SAS operators would be prohibited from using this contact information for any other purpose than SAS operation, such as marketing their services to these device owners, seeking competitive advantage, or further distributing it.

**Proprietary** SAS operators will have contractual arrangements with device operators. Device operators will have technical operating data on how they have configured their physical and Media Access Layer (MAC) layers, and may have actual usage statistics. This data is not required for the baseline functions of a SAS, and therefore is not required to be exchanged. However, cooperation between SAS operations may be enhanced by the shared access to this information, and SAS operators are free to make arrangements to exchange this information with other SASs under bilateral agreements. The device owner or user may also be a party to these agreements, in addition to the SAS operator. An example of this is information on network usage that is used by the coexistence layer, if present.

## 14.8    Implementation of SAS Synchronization

The information in the previous section is especially significant when one considers a multi-SAS regime. Multiple SAS operators are certainly desirable from a competition point of view. It provides some level of cost control, increases the likelihood of continued availability of SAS services, and creates incentives for innovation and cost reduction on the part of the SAS operator(s). However, it does require some mechanism of coordination between SAS operators authorized to make assignments that could either interfere with each other, or aggregate to interfere with other protected parties.

Several approaches to synchronizing SAS operations emerged in the CBRS SAS design process, but they fit broadly into three classes:

**Truth** SAS operators exchange the identical, interference-related information used in the registration and grant transactions. This includes the exact location of the node (latitude, longitude, and elevation), antenna data, device class, installer identification (if present), and the specifics of the eventual grant. The receiving SAS can therefore apply the same processes it used to admit its own nodes to determine the aggregation impacts of the nodes provided by the interfacing SAS. Nodes synchronized, or nodes admitted, are treated symmetrically.

**Abstracted** It was proposed that SAS operators could "hide" the actual network's deployments by abstracting the aggregate emissions and protection boundaries (the PPA in CBRS terminology) and only provide this abstraction to other SAS operators. The most natural abstraction model is a "heat map" of points outside of the PPA boundaries, supplemented by specific records required to address directionality.

**Allocation** SAS operators agree on an allocation of interference into all protected nodes, and to new protected nodes as they enter. Each SAS would manage within its own allocation of interference, with no coordination necessary. Although this model was not fully developed, one could imagine an extension to the interfaces that created a market for interference rights between the SAS operators as market-makers for their customers. This could essentially form a two-sided market, with SAS operators acting as the middle agent.

There is no question that the approach of exchanging truth data is sufficient to assure synchronization. This is the same data that the original SAS used to perform the analysis itself. The adequacy of the abstraction approach is more complex, largely due to the difficulties introduced by the directionality of both receive and transmit antennas.[2]

---

[2] In reading this, the reader is informed that the author was an opponent of introducing any abstractions into the synchronization model, due to his belief that it unnecessarily complicated the SAS, violated the

Consideration of a fixed allocation between SASs was not pursued due to concerns that such agreements might constitute an allocation of the market, since, on a large scale, there is a monotonic relationship between the number of nodes that can be registered and the allocation of interference. This approach would reduce competitive pressures and advantages in areas of dense spectrum usage, where the interference ceilings were a constraint.

Another driver for the full disclosure of truth data is in the necessity for very detailed data to support the coexistence of peer devices. The FCC established an arbitrary interference limit at the boundary of a PPA, and defined that it was determined at a height of 1.6 m above ground level. This is reasonable from a regulatory perspective, but it is not a sufficient framework for coexistence. Small cell, or Access Point (AP) focused bands will likely have their densest deployment occur vertically, in offices and Multi-Dwelling Unit (MDU) buildings. Interference levels at the street height (where PPAs are determined) and vertically above the clutter may be (and are likely to be) very different. This requires much more precise data than required for protecting the arbitrary contour of a PPA.

While the USA went down a path of full disclosure of registration data, there is no reason that these alternatives could not be pursued in other situations.

The eventual SAS-to-SAS synchronization design is provided in WinnForum documentation. The details of that interface's implementation will not be discussed here. It essentially serializes all of the registration and related data to a set of JavaScript Object Notation (JSON) objects. SAS synchronization provides for exchange of public, private and proprietary data, subject to bilateral agreements between the synchronizing SAS operators, and customers of the providing SAS. The data blocks for exchange are shown in Table 14.2.

## 14.9 Impacts of SAS Differentiation

The logical argument that SAS differentiation is an acceptable and consistent option was provided in Section 8.8. In this section, we consider the impact of SAS differentiation on the implementation of SAS implementations. Some general principles apply:

- Collaborating SAS operators must not be forced to accept computations that are dependent on path loss calculations. For example, this would preclude the *"heat map"* approach of SAS coordination, since it would not provide the details needed by the receiving SAS to apply its own models.
- It is desirable that SASs using more than the least sophisticated model inform other SASs of their path loss determination for nearby nodes that receive protection from aggregate interference for nodes they enter. Other SASs can compute their own path loss estimate and select the highest of the two. This avoids excessively penalizing

principle of shared spectrum transparency, and limited the ability to provide device coexistence as the band becomes more crowded. It is included in this book due to it being an option that was proposed by a number of participants in the standards process, and therefore should be considered by other regimes.

**Table 14.2** SAS-to-SAS synchronization exchange.

| Data Record | Treatment | Usage |
| --- | --- | --- |
| SAS Administrator Record | Public | SAS identification, point of contact, public key(s). |
| ESC Administrator Record | Proprietary | ESC naming and operator identification. Only exchanged if parties have bilateral agreement to share ESC data. |
| ESC Sensor Locations | SAS Essential or Proprietary | ESC Siting and Installation data only exchanged if: (1) parties have bilateral agreement to share ESC data, or (2) SAS operators require that the ESC locations be protected from emissions from CBSD deployments. |
| CBSD Registration Record | Public, SAS Essential, or Proprietary | Data defined by the FCC is public. CBSD installation parameters and grant information (e.g., BW, power, etc.) are SAS Essential. Usage, network specific or extensions beyond Part 96 protections are proprietary. |
| CBSD Operator Record | Proprietary | Data regarding interconnection of CBSDs such as proxy arrangements not related to interference. Exchanged under restrictions, and under mutual agreements, for the purposes of scenarios such as exception handling, etc. Includes Mobile Network Operators (MNO), Mobile System Operator (MSO) or neutral hosts, or enterprise identity, as examples. |
| CBSD Measurement Record | Proprietary | Any measurements of interference, signal or other signal levels. |
| Domain Proxy Registration Record(s) | Proprietary | Similar to CBSD Operator record, providing proxy ownership data on the network operators. Exchanged under restrictions, and under mutual agreements, for the purposes of scenarios such as exception handling, etc. Includes Mobile Network Operators (MNO), Mobile System Operator (MSO) or neutral hosts, or enterprise identity, as examples. |
| Professional Installer Record | Proprietary | Exchanged under restrictions, and mutual agreements, for the purposes of scenarios such as exception handling, etc. Association of professional installer with a given CBSD record is included in the CBSD registration data. |
| PAL Census Tracts Area | Public | All PPAs must be declared to other SASs. |
| Secondary Market License Agreement (SMLA) | Proprietary, licensee can determine | Must be exchanged if secondary licensee might use a different SAS operator than the primary license holder is using so they can process the possible conflict in the PAL PPA. |
| PPA | Public | All PPAs must be declared to other SASs to ensure protection. |

SAS with less sophisticated algorithms. Otherwise, a 10 dB difference in propagation loss estimates could result in admissions by the high performance SAS, effectively precluding less sophisticated SASs from making any assignments at all, even though adequate interference margin might be present.

- The documentation on all non-baseline propagation models should be publicly available for both science validation, and challenges to its application.

A differentiated SAS environment would have to add a record type to Table 14.2 to provide propagation loss estimates to a list of nodes with propagation less than a certain threshold. SAS exchange would have the exchange of this data to be optional, as determined by the receiving SAS.

## 14.10 Suggested Reading

- The FCC Report and Orders [2, 3] for the CBRS band are clearly one of the major references in the subject, and define in detail aspects of how this first admission control system must operate.
- The WinnForum has developed a set of baseline and derived requirements for the SAS required by the FCC regulations. Current versions of these documents are available from the Forum web site [1, 4, 5, 6, 7, 8].
- The process to obtain regulatory authority to operate an approved SAS must disclose many of the design and operating policies of the proposed SAS systems. These documents, as well as the entire certification proceeding, are publicly available, and instructive in how different organizations chose to implement the necessary techniques.

## References

1 Spectrum Sharing Committee (SSC), *CBRS Communications Security Technical Specification*, 1st edn (Wireless Innovation Forum, 2016).
2 Federal Communications Commission, *Amendment of the Commission's Rules with Regard to Commercial Operation in the 3550–3650 MHz Band, GN Docket 12–354* (2016). https://apps. fcc.gov/edocs_public/attachmatch/FCC-16-55A1.pdf.
3 ——, *Amendment of the Commission's Rules with Regard to Commercial Operation in the 3550–3650 MHz Band, GN Docket 12–354*, Order on Reconsideration and Second Report and Order (2016). https://apps.fcc.gov/edocs_public/attachmatch/FCC-16-55A1.pdf.
4 Spectrum Sharing Committee (SSC), *CBRS Operational Security Technical Specification*, WINNF-15-S-0071-V1.0.0 (Wireless Innovation Forum, 2016).
5 ——, *CBRS Operational and Functional Requirements*, WINNF-15-S-0112 (Wireless Innovation Forum, 2016).

6 ——, *SAS–SAS Protocol Technical Specification*, WINNF-16-S-0096-V1.0.0 (Wireless Innovation Forum, 2016).

7 ——, *SAS to CBSD Technical Specification*, WINNF-16-S-0016-V1.0.0 (Wireless Innovation Forum, 2016).

8 ——, *Spectrum Access System Requirements* (Wireless Innovation Forum, 2016).

# Part VI

# Future Bands, Network Services, Business Models, and Technology

# 15 Potential Services and Business Models Enabled by Three-Tier Spectrum

## 15.1 Introduction

Although much of the emphasis in the discussion of three-tier has been focused on spectrum sharing, the innovation will arise from a shared band and a shared technology that is usable by both traditional mobile operators, and non-operators, such as venues, enterprises, and even residences. Wireless Fidelity (Wi-Fi) created a common *"Lingua Franca"* among a wide range of deployers (mobile operators, residences, enterprises, venues, etc.), but the technology partitioned service quality, security, assured connectivity, and other highly desirable, and often necessary features, Assured quality was only available (or at least promised) through mobile operator technology and services. The use of the three-tier common band removes this partitioning between services, and can enable all wireless suppliers to offer identical technology, and equivalent, or differentiated, services.

Spectrum policy discussions tend to focus on the impact of spectrum regimes on band usage and technology. It is perhaps even more important to examine the impact of spectrum policy on the cycle of innovation. Wireless is facing a challenge to double bandwidth every year. Yet the current architecture took from five years to decades to deploy, little new spectrum is available, and the technology is approaching the theoretical link capacity asymptotically. In situations where linear advances are not adequate, innovation is essential.

From an economic perspective, we have learned certain truths. One of them is that domestic markets strongly support, and often lead to, leadership in international markets. Europe led national deployment of Global System for Mobile Communications (GSM), and ended up dominating handsets for a decade. The United States of America (USA) created the first unlicensed band, and US companies dominate much of the Wi-Fi industry.

This section will have an emphasis on mobile wireless services, since this is the major source of new spectrum demand. However, the same rationales as presented here should be applicable to new spectrum uses that emerge in the "post-cellular" world, as that evolves! This includes wireless fixed services, as well as fully integrated wireless fixed/mobile services, such as contemplated in 5G.

## 15.2    Opportunities Introduced by Common Shared Spectrum

The whole effort of developing a three-tier spectrum structure would not have as much value if it did not create new opportunities, both to extend current enterprises and to create new ones. In this chapter, we investigate the new opportunities that can emerge due to the availability of three-tier spectrum. The focus will be on the more disruptive opportunities that may address some of the structural issues with existing wireless services, such as cost, coverage, and competition.

One of the fundamental problems with the wireless ecosystem is the difficulty in entering the industry and offering a unique technological or service model product. Traditionally, the only viable wireless enterprises were those that had dedicated, exclusive spectrum access that in many countries is only obtainable through the spectrum auction process. This process is incredibly venture- or startup-unfriendly. In the USA, the National Broadband Plan [1] provides a history of spectrum auctions that shows the time from identifying spectrum for use, to the ability to actually use the spectrum was a delay of 8.4 years. A delay of 8.4 years hardly supports the "innovation economy" that is heralded as the future.

We need to define what scalable entry means for a spectrum regime. The definition applied here is that the regime enables spectrum access at costs and availability that are near linear with the extent of spectrum access obtained, in terms of frequency range, geographic coverage, and duration of rights. The acquisition of these rights through scalable means should not be at a significant cost disadvantage over the less scalable mechanisms.

An example best demonstrates why spectrum scalability is enabling to innovation in spectrum-dependent services. We consider a startup venture that desires to experiment with the market acceptance of a new technology. In the current exclusive-use paradigm, with long-term licensing, the company would have to purchase perpetually renewable rights throughout its potential deployment area in auctions. Or the company could purchase the perpetual rights to a limited area, but would then have to hope another auction would come along for compatible spectrum, and then bid in that auction in order to deploy their technology. This is an unattractive investment, regardless of the technology:

- At most times, no spectrum is available for auction at any price.
- If an auction was pending, it would typically require massive amounts of capital to purchase exclusive spectrum licenses. This capital investment likely dwarfs the expenses of developing and deploying the technology at small scale.
- Auctioned spectrum typically has to be cleared before use, a process that takes from 5 to 10 years typically.
- If the concept is successful, there is no assured method to expand it in current or new markets with additional spectrum. Expansion of the product would depend on regulatory action, not market acceptance.

In the three-tier regime, the same business plan might require no spectrum acquisition at all (if operated in General Authorized Access (GAA)), but at the most would require only a one year Priority Access License (PAL) auction success in whatever geographic areas the initial trial was to occur in. After that, the footprint could be expanded at will annually, into only those areas where operation would be initiated within each auction interval. If the business plan was not successful, the spectrum cost would likely have been a small consumer of resources. If the plan works, the company is set to acquire protected spectrum in whatever regions it wishes to operate.

Some attributes of a regime that supports scalable entry include:

- There is no necessity to acquire more spectrum rights than necessary because they will not (or might not) be available in the future.
- Spectrum rights can be acquired granularly for the range of time when they are required. An innovation, new market, or new business model that cannot have its deployment footprint expanded has limited value, and acquiring all of the necessary rights in advance will limit consideration to only low risk, and thus less innovative, opportunities. Longer term arrangements, such as perpetual right of renewal, inflate the costs and constitute barriers to entry.
- Spectrum rights can be acquired that are very granular in space. Forcing large areas to be licensed raises costs and is a barrier to entry.
- Spectrum rights can be acquired that are very granular in frequency. Deployments must be able to start with the minimal spectrum needed, but be able to compete with other uses to acquire spectrum rights as market demand dictates.

Scalable entry removed the necessity for regulators to judge the best use of spectrum. Instead of predicting market needs, the markets can be allowed to form, verify the existence of demand and the degree of cost elasticity, and then naturally cannibalize less popular (as measured in ability to maintain spectrum rights) services, displacing them in the spectrum.

With scalable entry, there is less need for us to consider the possible markets that might evolve in a three-tier spectrum regime. However, to justify the disruption that this regime might be considered to cause, it is worthwhile to examine some of the new models that are already being pursued, or seriously contemplated.

## 15.3 Venture Funding Friendly Innovation

Venture capital funding has been a primary driver for many of the disruptive technologies that have arisen in the communications, networking, and applications markets. To see continued innovation, it is important that the spectrum regime enable innovation to occur with a financial pace that matches how new ventures are funded, and roll out innovations to the marketplace. Venture funded efforts do not go full scale at the start. They do not invest billions into purchasing non-expendable resources, such as building a factory, or an office building. Similarly, investing the bulk of the capital into spectrum is not attractive and inherently limits innovation in any product that is spectrum access dependent.

The venture capital funding profile requires a scalable process for initially rolling out and market testing a service. This is the rationale for arguing for a spectrum regime that is itself, scalable. If we want an innovation economy, we want spectrum to not constrain the types of innovation that can be explored. This is not to say that spectrum should be free or not at a market rate, but the access to the spectrum must be flexible and scalable.

## 15.4    Private Network Opportunity

Traditionally, network technology, business models, and deployment methods are partitioned between unlicensed and exclusively licensed models. Unlicensed spectrum currently is dominated by self-deployed Wi-Fi networks that are rarely monetized. Exclusive use spectrum is held by traditional Mobile Network Operators (MNO) service providers, uses Third-Generation Partnership (3GPP) technologies, such as Long-Term Evolution (LTE), and is heavily monetized. While MNOs have deployed some unlicensed Wi-Fi, they have not generally monetized these deployments, or fully integrated them into their service architecture. Spectrum has partitioned the technology and business models arbitrarily, even when the technology objectives are very similar.

One of the unique aspects of three-tier spectrum regimes is that they provide spectrum bands that are accessible to any organization, and under a wide range of licensing terms. Technologies developed by any community in this band become available to all communities that can access the band. In the current mix of technologies, this would provide the ability for a MNO to acquire protection rights for a Wi-Fi network, and for an individual enterprise, venue owner, or even residence owner, to deploy a private LTE network. Perhaps more important, they would be assured of the right to acquire protection, even if this need was unlikely to actually occur.

In the US Citizens Broadband Radio Service (CBRS), one (at least initial) emphasis has been on making Time Division Long Term Evolution (TD-LTE) available for enterprise and venue operation as private networks, supplementing traditional Wi-Fi applications. The author's mantra to small cell suppliers has been to ask *"Would you rather sell ten thousand boxes to a carrier, or ten million to enterprises?*[1] The fact that the CBRS band was already a 3GPP standard[2] band made the rapid development of products possible. The US regulations required that all devices in the band cover the full range of the band, so equipment sold for carrier use would inherently operate with the private networks, so long as authentication was provided.

One could imagine a hospital that used a common network technology to integrate secure medical telemetry and internal paging, and to offer cellular services to visitors, charged back to their MNO or Mobile Virtual Network Operator (MVNO). Within the building, there might be no interference from external uses, due to the spatial offset of the building, and the path loss in the walls and interior, so (assuming the protection

---

[1] As an example, the International Wireless Communications Expo (IWCE) Dynamic Spectrum Summit [2].

[2] The CBRS band overlaps the upper portion of the 3GPP Band 42, and the lower portion of band 43.

models recognized scatter loss), the hospital could use most, if not all, of the sharable spectrum. Because a mobile Access Point (AP) is not permitted, they could ensure that no other devices were on these frequencies, due to their control over the physical space. At higher frequencies, such as above 3 GHz, physical control over a space is a more meaningful consideration than spectrum rights in that space.

The same model can be used for any number of applications where a single entity controls the physical placement of nodes. Even if a protected-tier license exists in the area, it cannot deploy in the spaces to which it has no access. Therefore, the lower-tier spectrum usage in many locations is unconstrained by the existence of the protected license. Sports venues, refineries, factories, shopping centers, race tracks, and other venues all have this attribute.

A final consideration is that the consumer mobile equipment is a mass market, with highly competitive, commodity-priced products. Wi-Fi equipment has similar cost pressure, due to high volume supporting many competitive options, and openness of standards. Cellular infrastructure, on the other hand suffers from lower-volume, closed management protocols, and much less price competition. The creation of a market that both the operator and non-operator communities supports will benefit operators in creating volume, and benefit non-operators in creating opportunities for a much wider range of technology and infrastructure options. Increased volume and more open markets will be more supportive of innovation and the entry of new concepts, again benefiting operators.

## 15.5 Neutral Host Services in Common Spectrum

Neutral host services where one network can serve multiple MNO services can become an attractive opportunity to reduce duplication of network resources, particularly indoors, or in venues with limited accessibility for MNO deployments. In most cases, what are proffered as neutral host systems are really a collection of operator-specific Radio Frequency (RF) systems, operating in exclusive spectrum, and sharing backhaul and some of the infrastructure.

We might consider these multiple RF systems to be *"aggregated host,"* rather than neutral host. True neutral host operation is not possible in spectrum that is exclusively licensed to a single operator, and may have limited coverage in other MNO handsets. True neutral host would not have proprietary MNO or MVNO network control over the access service, and could be flexibly utilized by any operator without explicit licensing of exclusive spectrum from the supported operator. At the same time that spectrum limits the implementation of true neutral host services, the growth in user bandwidth demand has continued unabated, and is particularly focused on indoor environments that are not served well by traditional outdoor operator solutions. To provide the dense, indoor networks needed to address this demand, several conditions must be met:

- Deployments must be simplified and inherently operable on all carriers, without complex carrier relationships.

- The costs of deployment should be recoverable from the carriers whose traffic is offloaded by the systems. It is unreasonable for the carriers to shift the burden for providing solutions to this bandwidth demand onto the owners of interior spaces without compensation, such as Distributed Antenna System (DAS) solutions.
- The integration of neutral host services into an existing Information Technology (IT) structure should be scalable and consistent with the rest of the IT infrastructure, and maximally leverage wiring, backhaul, and layer 3 services. Dark fiber, and unique layer 2 interconnects are cost and complexity drivers.

We can consider this in contrast to the obviously competitive technology DAS. For those not familiar with it, DAS is a network of antennas connected to a common RF section. The DAS system divides its power across the many antennas, so that it creates high-reliability coverage, but has the minimal radiation out of the local environment, and thus minimizes impact on any macrocell usage of the same channels. Some issues with DAS that we will consider include:

1. DAS must operate on the carrier's exclusively licensed spectrum. Therefore, any DAS deployment is carrier-specific,[3] so the DAS installation is effectively the carrier's, not the DAS owner's.
2. Many DAS are deployed to improve the wireless experience in a building, and are installed by building owners or tenants as a service to building occupants. Typically, they are not reimbursed for this cost, and cannot collect revenue from the usage of the DAS. Building owners have little incentive to deploy this capacity. The penetration of DAS indoors is very low.
3. The purpose of a DAS is to improve coverage, not capacity. The multiple antennas all operate as one site, and are not accretive of capacity. No matter how many DAS antenna heads are present, the aggregate capacity is the same, since they are just one cell site, regardless of antenna count.
4. DAS cable signaling from the RF section is unique to the DAS system, and not consistent with the cabling, Ethernet, and Internet Protocol (IP) networking in a typical building or enterprise.

By contrast, the same deployment made using independent small cells resolves many of these issues. By comparison with the previous list:

1. The private network need not operate on spectrum that is licensed to a carrier. As such, the owner of the network is free to operate, manage, and use the network in any manner they desire. The network can be internal to the enterprise security perimeter.
2. Each small cell AP deployed provides additional bandwidth. Assuming the access points are not interfering, the aggregate bandwidth is linear with the number of access points. Independent AP devices provide capacity; DAS provides coverage. Such deployments may have so much physical layer connectivity that they are backaul-limited, even with multi-gigabit fiber connectivity.

---

[3] Some DAS deployments are multi-carrier, but this is essentially multiple RF access systems in one package.

3. The private network is free to make any financial arrangements it wishes to make, and can make them with any of the possible MNO or MVNO that offers to make such arrangements attractive. There is no coupling of spectrum use and network charging.
4. The backhaul of the network operates over conventional IP and ethernet networks, so the private network can be expanded and scaled as needed using the same network installation as used for all of the other enterprise IT services. Deploying additional small cells is similar to deploying additional Wi-Fi AP devices.

There are several models for how neutral host systems might be operated. They might be deployed directly by the MNO, selectively by an MVNO, or through private networks operated by building owners and tenants. The next section discusses this last option.

## 15.6 Third-Party Offload Opportunity

The advent of true neutral host services, and the ability to deploy wireless technology that is scalable and based on low cost, is a recent opportunity. It made sense for wireless carriers to be the providers for outdoor wireless services. There was no existing infrastructure available to support this deployment. New towers, backhaul, power sources, and equipment were required, so no one community was better suited to perform this deployment.

When we consider indoor deployment, most of these conditions do not apply. The owner and tenants of building space have superior access to the premises; they already have had to invest in backhaul and power for their own operations and use, and they have built out cabling plants and internal connectivity solutions throughout their interior spaces. The marginal cost of adding an LTE band to an existing Wi-Fi deployment is low, particularly compared to standalone deployment of an LTE small cell network.

There is reason to believe that a decentralized architecture for providing indoor coverage is an attractive option that can be enabled by three-tier spectrum. The capital cost per unit of capacity for such deployment should be much less than macro services or a DAS alternative. It is true that such a network may not have the same reliability or Quality of Service (QoS) of a carrier standards deployment but there are several reasons why this may be less of a concern:

- This network is an offload network, and there is likely still a macro cell network that can back up the offload network in the case of localized outages. As more traffic is diverted to the offload network, macro capacity may be a surplus, and available as backup to the localized networks Reliability investment can be gauged to the degree of macro and small cell overlap.
- Dense deployment within a given locale will typically have overlapping coverage, since there is sufficient spectrum to not require spacing appropriate to the reuse of one channel.
- The same model can also create abundance of access, and thus localized redundancy between peer providers.

- Traffic growth of wireless systems is tied to Internet access, not conventional voice. This traffic is more tolerant of disruption and lower QoS than the voice application that drove the specifications for current wireless systems.

## 15.7    Potential Enhancements to Three-Tier Concept to Increase Innovation Friendliness

It is worth considering what enhancements could be made to the three-tier regime that would make it even more effective in encouraging innovation in wireless- and spectrum-dependent services, business models, and technology. There has been no meaningful experience with this regime, so no "lessons learned" have appeared, therefore the topics to follow are largely speculative.

One obvious solution is to lower the cost of spectrum, but that would only distort the market towards innovations that could not scale once they were burdened with the full market prince of the spectrum they required. The objective should be to assure that the small unit cost of spectrum is not appreciably greater than "bulk" purchases of spectrum rights.

An effective secondary model for spectrum is not precluded by the three-tier model, but it is not mandated either. Integration of a secondary market could result in even more flexible spectrum access as very short-term dynamics, and localization, could be exploited, and spectrum licenses that were no longer required, or completely utilized, could be monetized. This would enable them to retain an asset value for the duration of their validity.

A secondary market for spectrum would reduce the risk of acquiring licenses that would be excessive if the innovation or market was abandoned. It might provide a level of liquidity that can be used to reduce the cost of deployments by utilizing "distressed" secondary-market spectrum at rates below the market price.

It can be argued that some protected tier 1 user's are not effective uses of the spectrum. In effect, the opportunity cost of their use is greater than the value of the service they are providing. While useful to the operators of the incumbent systems in the spectrum, they fail to fully capture the opportunity cost of that spectrum to other applications. For example, in the USA there are several dozen legacy Fixed Satellite Service (FSS) sites below 3 700 MHz. These are limited by regulation to point-to-point services, so are less than the optimal use of satellite services, as the service they are providing is one that fiber-optic can provide, without the spectrum opportunity cost. However, the spectrum resource is free to them, and a fiber alternative would have cost, so they have no incentive to migrate technologies. Such a situation challenges us to find a spectrum regime that benefits both the incumbent user, and the community sharing the spectrum.

In this case, the FSS operators might benefit from a fiber solution. It would free up considerable square miles of spectrum usage; enough to position thousands of small cells. However if one party could make a deal with the FSS operator to provide them with the fiber alternative, that party would receive no more consideration in the use of the spectrum than any other user of the band. Therefore, no single party has an incentive to make such an arrangement, even though the benefits overall would be significant, and beneficial, for all involved.

It may be that a right of interference protection needs to be made transferable so that legacy uses can be "bought" out of the spectrum. Secondary markets have focused on the transfer of a "Right to Transmit," but an equally important right to monetize and transfer is the "Right to Receive." The three-tier framework provides an exceptional framework to enable this to occur on a case-by-case (rather than band-by-band) basis. The US Incentive Auction is a special case of such a rights transfer, although it was for transmission rights, rather than receive-only rights.

The three-tier framework provides a unique mechanism to clear spectrum on an incumbent-by-incumbent basis. There is no need to clear entire bands. As specific incumbents leave the spectrum, their spectrum footprint can be "returned" to the SAS and then issued to new users. This process can be adjusted to the life cycle of the systems, to economic drivers, private investment, or any other mechanism. Different mechanisms can be even used for different incumbent uses.

## 15.8 Suggested Reading

- The author has done a number of talks in industry seminars in order to describe the opportunity that the new band would create for innovation and new markets [2].
- Although the President's Council of Advisors on Science and Technology (PCAST) report [3] did not want to threaten existing interests, it did hint on the types of innovation that could be encouraged by the three-tier regime, and argued for spectrum policies that were explicitly focused on promoting innovation.

## References

1 Federal Communications Commission, *Connecting America: The National Broadband Plan* (2010). http://download.broadband.gov/plan/national-broadband-plan.pdf.

2 P. F. Marshall, History and opportunities of 3.5 GHz, three tier spectrum. Video in *International Wireless Communications Expo (IWCE) Dynamic Spectrum Summit* (2016). www.youtube.com/watch?v=Ky1_SVDDang.

3 President's Council of Advisors on Science and Technology, *Report to the President: Realizing the Full Potential of Government-Held Spectrum to Spur Economic Growth* (Executive Office of the President (EOP), Office of Science and Technology Policy (OSTP), 2012). http://obamawhitehouse.archives.gov/sites/default/files/microsites/ostp/pcast-stem-ed-final.pdf.

# 16 Candidate Incumbent Bands for Three-Tier Spectrum Sharing

## 16.1 Introduction

This chapter will survey a number of spectrum bands for their applicability to the three-tier spectrum sharing regime developed earlier in this book, and in terms of the incumbent usage. Some of the bands are not apparent candidates for sharing, but are included for purposes of discussion, and to demonstrate methodology.

The complexity of a generalized approach to three-tier spectrum is apparent when one considers the balkanized nature of primary and secondary national spectrum assignments. The National Telecommunications and Information Agency (NTIA) map of these assignments [1] in the United States of America (USA) is shown in Figure 16.1. Although it is hard to read in this book's reduced, gray-scale printing, the complexity of the chart, the division of spectrum into uniquely managed spectrum slivers, and the range of uses of most of the individual frequencies is readily apparent. Undoing this legacy of partitioning and stacking of fixed rights to spectrum access is not a trivial challenge, particularly when, in most cases, the existing structure must continue to operate, even in the presence of the new regime.

Because of the myriad of unique bands, and primary/secondary/tertiary usage permutations, the discussion in this chapter will initially focus mostly on the characteristics of different allocation categories and usage, rather than specific bands. With the discussion of specific sharing and protection constraints completed, the results will then be mapped to the International Telecommunication Union (ITU) international allocations.

Even with this structure, individual National Regulatory Authority (NRA) decisions have established unique national primary and secondary allocations in many bands, so the results of this chapter must be further reconciled with national practices. For example, many countries do not use several of these allocation types at all. International agreement on reallocation may not be possible, but might not be required for action by individual nations, so long as not interfering with international commitments.

## 16.2 Evaluation Criteria for Considering Opportunities for Three-Tier Spectrum

It is necessary to have a scientific, evidence-based process to determine what bands are good candidates for three-tier spectrum, and which ones are really unsuited to such

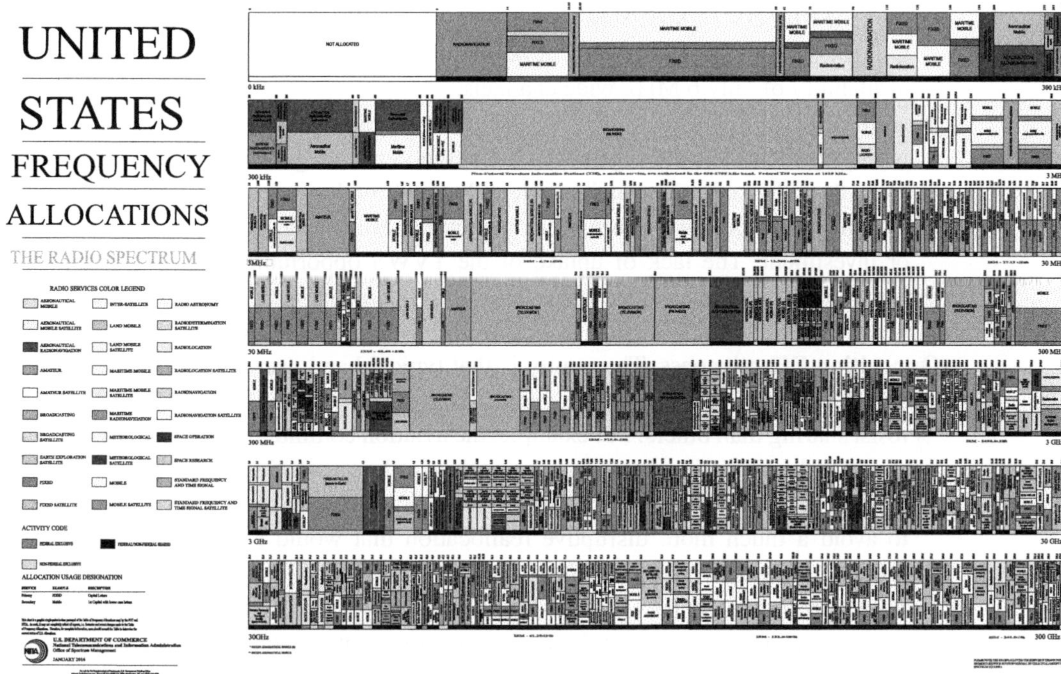


**Figure 16.1** 2016 US spectrum allocation map (from NTIA [1]).

a regime. Spectrum politics often obscures the fundamental technical issues, and the basis and outcomes of decision-making are not always driven by the scientific reality, unfortunately.

The following criteria will be used to consider the entire set of ITU defined primary allocations.

**Incumbent Awareness** The three-tier regime is dependent on accurate awareness of the primary protected incumbents, as well as the participants in the three-tier regime. It is possible to enforce this on the participants in the lower tiers. However, the incumbent tier is what it is, and the characteristics of these systems must be accommodated. The difficulties in the primary awareness could be due to excessive mobility, such as airborne services, location privacy, such as either commercial proprietary or national security concerns, or such dense operation that the information needed cannot practically be collected and managed, among other reasons. A special case in this concern is the existence of passive receivers, such as Television, Receive Only (TVRO) stations, Global Positioning System (GPS), radio astronomy, other navigations services, and similar uni-directional uses.

**Meaningful Bandwidth** Unfortunately, spectrum has been extensively "balkanized" into small, special purpose allocations as small as 5 MHz, or less than 0.05% of the frequency. Although all spectrum is valuable, there is a minimum amount of bandwidth necessary to create a meaningful opportunity for wireless innovation. Unless small allocations can be aggregated into meaningful bands, the small orphaned blocks of spectrum are not a high priority for transition to a three-tier regime. The Television White Space (TVWS) suffered from the likely channel availability of only 6 MHz-wide channels.

**Meaningful Opportunities** The likely spectrum availability in any resulting regime is important in considering a band for three-tier management. A band that might meet the other criteria, but is so densely used, or has such restrictive protection, that there will be few opportunities for lower-tier sharing, is not attractive to users or investment. The lack of uptake of the TVWS opportunity was also due to the perception that the available opportunities in populated areas were not sufficient to create enough usage to justify sufficient investment in the necessary technology.

**Opposition Effectiveness** The reality is that typical band users do not desire to have "their" band shared. However, some band operators may be in a stronger position to resist sharing than others. For example, the aviation community has a record of being successful in resisting spectrum encroachment by presenting possible interference issues and consequences. Other band operators may view sharing as an opportunity to avoid a much more disruptive reallocation that would result in a total loss of spectrum access, or greatly increase its cost.

**National Scope** The impact of three-tier sharing may be limited to the region within the NRA boundaries, minus any border accommodations, or may have impact beyond the borders, and require regional or international agreement. Since nation-by-nation is a more rapid path to adoption, the level of necessary agreement is an important factor. On the other hand, national deployments may not create the scale of market opportunities to exploit the opportunity.

Some other criteria that might be considered would reflect the suitability of the spectrum, and the availability of equipment in the shared band. These obstacles to exploitation of the band are not considered here, as these can be overcome with a suitable opportunity and as technology advances. In fact, implementation in less dense spectrum may be much less contentious.

Regulators should not prejudge the utility of the spectrum. An anecdotal example of the utility of an allocation was in the establishment of the unlicensed spectrum now used for Wireless Fidelity (Wi-Fi). Although this band was considered a "garbage band" [2], it became one of the engines of innovation, initially in the USA, and eventually world wide. One should never prejudge the power of innovation, once you are given an opportunity, is the clear message of this experience.

## 16.3    Introduction to Usage-by-Usage Analysis

The previous section (16.2) provided five criteria for considering a band for management and sharing under a three-tier regime. Tables 16.5–16.7 (at the end of the chapter) provides a list of the ITU allocation categories present between 1 and 6 GHz; the possible candidates for near- and mid-term sharing under a three-tier framework. It provides a summary of the evaluation of each against the criteria of Section 16.2, summarizing the analysis of Sections 16.4–16.11. These allocation can be mapped to the many ITU allocations that permit that usage for a more detailed assessment. A list of the specific allocations is provided in the ITU Radio Regulations Articles [3].

This analysis illustrates several strong candidate bands. It is likely that incumbents might disagree with this analysis, but several of these bands appear to be candidates for national or international consideration. The following sections discuss some of the major categories that have an appearance of viability. The text that follows discusses the specifics of three-tier sharing of many of the ITU allocations. Tables 16.5–16.7 provide a cross reference from the ITU allocation in the table, and the text discussion specific to the specific allocation, if any. In many cases, a single ITU allocation is replicated across a wide range of spectrum, and the range usage and spectrum characteristics are unique enough that sharing is quite different in each of these instances. These details are provided in the following text.

## 16.4　Satellite Uplinks and Downlinks

There is an extensive inventory of spectrum that is allocated and used primarily for satellite operations. In most cases, they are partitioned between uplink and downlink operations, with unique bands for each. Satellite cross-links typically share spectrum with the terrestrial links, but highly directional antennas minimize interference from terrestrial sources.

C-band was once the dominant satellite service, but has lost many of its commercial mission areas to competitive technologies such as fiber, and its consumer market to the Direct Broadcast Satellite (DBS) Ku-band services. Fixed, point-to-point data transfer has largely migrated to lower cost fiber-optic and Internet connectivity. These avoid the bandwidth limits of individual C-band transponders, and are essentially infinitely scalable.[1] Direct-to-customer C-band Television (TV) services have migrated to the increasingly pervasive cable services, or to Ku-band DBS (designated as Broadcasting Satellite Service (BSS) by the ITU). This migration off of C-band has been particularly true in the northern hemisphere. C-band maintains an advantage over Ku- and Ka-band services due to its reduced sensitivity to rain fade; a more significant factor in the southern hemisphere.

The fundamental differences in usage between the northern and southern hemispheres are significant. As an example, in the Intelsat Licensee LLC filing regarding the US position on FCC Docket 04-286, Intelsat advised the FCC that:

*"Intelsat has global operations that utilize C-band differently domestically and abroad, and that rain fade, which degrades Ku- and Ka-band signals, is not an issue within the United States. By contrast, in many areas of the world, such as the tropics, rain fade makes C-band the preferred method of satellite communications. Accordingly, C-band use is much more prevalent – and critical to communications … in other parts of the world … Given the different nature of C-band*

---

[1]　Scalability for video distribution is an increasingly valuable attribute as the number of video resolutions increases, and size of the most demanding one increases, such as the increasing demand for 4K video, or Ultra-High Definition Television (UHD), resolutions.

*satellite use in different parts of the world, protection mechanisms proposed domestically will not necessarily work globally."*

C-band dishes are quite large (typically over 3.3 m in diameter), so their use is largely in fixed sites, such as cable operator antenna farms, temporary uplink trucks used for broadcast backhaul, or industrial facilities where other forms of data service are not available. The fixed, or at least scheduled, nature of C-band usage is a fundamental factor in its suitability for sharing. To some extent, it resembles the TVWS sharing; with the bulk of protected users using fixed installations and a few "pop-up uses," such as wireless microphones that are known and scheduled.

Ku- and Ka-band satellite services are much more diverse than in C-band. This is largely due to the small size of uplink and downlink aperture. These are sufficient to close the links, and to resolve the orbital slots to avoid interference between adjacent satellite orbital positions. The sharing of spectrum with these satellites is a tension between the smaller size of the up and downlink spot sizes, and the very significant differences in the usage characteristics.

### 16.4.1    C-band FSS Uplink

Sharing of the C-band satellite uplink is constrained by the relatively large earth surface footprint of the uplink receive beams of the space segment. They typically are continental or more in their patterns; covering from near the poles in latitude, and up to $\pm 80°$ in longitude. Even if the energy is not directed to the satellite, the aggregation of energy from many devices will raise the noise floor that is received by this uplink. C-band satellites operate on the "bent pipe" principle, so any noise received by the space segment is radiated, along with the intended signal, and therefore impacts the downlink reception across the entire region that is served by any transponder. Uplink interference is additive to the noise inherent in the downlink. With typical $2°$ spacing, omni-directional interference could degrade the performance of up to 80 satellites' slots, over $\pm 80°$ of longitude. For this reason, the sharing of the uplink requires an international action, and may not be possible to accomplish in a shared environment, without reallocation of the Fixed Satellite Service (FSS) operations. Table 16.1 provides additional detail regarding three-tier application to the C-band uplink.

The inability to isolate use of the band from the earth coverage uplink antennas, and the consequential impact worldwide, makes this band a poor choice for spectrum sharing, without a fundamental change in spectrum allocations at the ITU, resolution of the users of the band, and disposition of the space segment assets.

### 16.4.2    C-band FSS Downlink

C-band downlink sharing is an essentially local decision. The operation of each receiver is independent. The downlink band already is allocated for fixed wireless, on the assumption that NRA processes can protect the fixed receivers. Table 16.2 provides additional detail regarding three-tier application to the C-band downlink.

**Table 16.1** C-band FSS uplink sharing analysis.

| Criteria | Analysis |
| --- | --- |
| Incumbent Awareness | FSS sites are registered for protection, and, at least in the USA, under the FCC's second Part 96 order [4] are required to register the specific elevation/azimuth angles in use (equivalent to satellite slot). This information is not very volatile. Similar registration information is available in most nations, or could be obtained. Secondary fixed services have been coordinated, and this information is also available. |
| Meaningful Bandwidth | The band provides from 500 to 800 MHz of contiguous spectrum, with few sharing complexities other than the protection of C-Band operations and some directional fixed wireless links. |
| Meaningful Opportunities | While the density of C-band FSS sites is quite low, in terms of sites per $km^2$, the effective density is the coverage patterns of the C-band uplink antennas in the space segment. These antennas essentially cover the entire occupied earth, so the effective density of usage is 100%. |
| Opposition Effectiveness | Use of the uplink would essentially limit use of C-band throughout the world, at least with the current space segment assets. Although use of C-band is declining in fiber-rich regions, the impact on less fiber-intensive nations, who are more dependent on C-band usage, would be considered to be highly disruptive. |
| National Scope | The international coverage of the uplink antennas would require an international, multi-region consensus at the ITU to effectively terminate the C-band allocations, at least in the spectrum segments that were to be shared. |

While there would likely be opposition from the FSS community, the C-band downlink meets the criteria for being shared in a three-tier regime. It is already allocated for fixed wireless, and the USA will be gaining valuable experience in managing a three-tier regime in downlink spectrum.

It should also be noted that three-tier spectrum solves many of the issues in sharing this band. Less sophisticated approaches would risk interference by devices that could roam into positions that would result in interference to FSS downlinks. The absolute control over device and network admission that is provided by the three-tier regime would enable the sharing regime to preclude entry of Access Point (AP) deployment in any location where the associated client devices could cause interference in any of their antenna configurations. In some locations, the only devices permitted entry might be point-to-point links, with tight control over their directional characteristics. Such a highly situational decision-making structure is ideally suited to the three-tier admission process.

### 16.4.3  Ka- and Ku-band FSS Uplink

Ku- and Ka-band satellite services are much more diverse than in C-band. This flexibility is largely due to the small size of uplink and downlink apertures that are

**Table 16.2** C-Band FSS downlink sharing analysis.

| Criteria | Analysis |
| --- | --- |
| Incumbent Awareness | FSS sites are registered for protection, and, at least in the US, under the FCC's second Part 96 order [4] are required to register the specific elevation/azimuth angles in use (equivalent to satellite slot). This information is not very volatile. The fixed wireless usage, if present, is also regulated and coordinated, so the information on these deployments is available for an admission control system, as well. |
| Meaningful Bandwidth | The band provides from 500 to 800 MHz of contiguous spectrum, with few sharing complexities other than the protection of C-band operations, and some point-to-point usage, which is similar in character. |
| Meaningful Opportunities | The density of C-band FSS sites is quite low, in terms of sites per $km^2$, so relatively short distance devices can share the spaces in between these sites, and can even locate in close proximity to them if appropriate control over beam orientation is provided. |
| Opposition Effectiveness | Some types of sharing of this band has been proposed in many recent World Radio Conference (WRC) agenda items. It has not been adopted, but the band has been increasingly subject to sharing rights, including the national control over the extended C-band frequencies, licensing of fixed services, and the recent US Part 90 and Part 96 Citizens Broadband Radio Service (CBRS) rules, which establish sharing of the extended C-band, and dynamic interference management to incumbents above 3.7 GHz. |
| National Scope | Downlink sharing has no impact outside of the immediate vicinity of the sharing device. There would be no international consequences, other than the typical coordination close to NRA borders. |

sufficient to close the links, and to resolve the signal from a much smaller aperture. This has created a large demand for small aperture services that provide Internet and data communications services for low volume uses, in addition to the traditional fixed uplink stations used for high data rate video and data services.

The mobile, transportable, and dynamic nature of these applications makes Ku- and Ka-band uplink sharing difficult. The benefits of more directional uplink antennas in the space segment are more than counterbalanced by the much greater diversity of the uplink sources.

## 16.4.4     Ka- and Ku-band FSS Downlink

Sharing of the Ka- and Ku-band downlinks is driven by the same considerations as the uplink; a very diverse, and relatively dynamic, ground segment (in terms of location), and a wide range of usage use-cases. These inherently constrain the uplink sharing opportunity. Although Ka- and Ku-band spot beams are quite small, and should enable sharing, they do not always partition along national borders, and many of the applications utilize much wider-coverage antennas.

Although individual Ku- and Ka-band services are increasingly sophisticated, and make greater usage of spectrum sharing across their own footprint, these are a minority

**Table 16.3** Ku- and Ka-band FSS downlink sharing analysis.

| Criteria | Analysis |
| --- | --- |
| Incumbent Awareness | These bands have very significant consumer use for receive-only applications, primarily for DBS TV. These devices are widespread, which reduces opportunity for sharing, and have a significant mobile, or highly transient, component. Creating protection for each of these receivers would be a challenge many orders of magnitude beyond that of the C-Band downlink sharing. |
| Meaningful Bandwidth | The Ku- and Ka-band bandwidth available for sharing is extensive, and would enable many wideband (gigabit level) service models. |
| Meaningful Opportunities | The extensive consumer use of devices likely reduces the extent of sharing in this band. This would be a significant impediment to any business case that was not a limited use product. |
| Opposition Effectiveness | This sharing would likely be opposed by the satellite industry, but would also be presented as a threat to individual consumers. |
| National Scope | This bandwidth would be available on a nationwide, and international scale. |

of the satellites on orbit. With $2°$ spacing, the worst case (in terms of interference) conditions constrain the spectrum sharing opportunity, rather than the most advanced one.

Table 16.3 provides additional detail regarding three-tier application to the Ku- and Ka-band downlink.

## 16.5　Ground-Based Radar Bands

Ground-Based Radar (GBR) deployments are largely for weather, aviation positioning, or military uses. Radars vary in frequency depending on their mission. Weather radars are typically lower in frequency, as are long-range surveillance. Doppler weather radars are higher in frequency, as are military tracking and fire control. For example, the high-accuracy tracing and fire-control radars within the US Terminal High Altitude Area Defense (THAAD) and Ballistic Missile Defense (BMD) GBR missile defense radars, operate in the X-Band allocation. For North Atlantic Treaty Organization (NATO), this is 8.5–10 GHz.

The considerations of sharing spectrum with this mix of uses are highly varied. The air-traffic control, approach, and related radars are relatively large, fixed installations. Similarly, weather radars are also typically large and fixed enough that their location should not be considered as sensitive. Large, fixed military radars may have their location hard to disguise. The portable or transportable radars, such as used in air defense, can be rapidly positioned, and their operational status may itself be

**Table 16.4** Ground-based radar (radio location) sharing analysis.

| Criteria | Analysis |
| --- | --- |
| Incumbent Awareness | Depends on radar type. Weather and air traffic control radars should be readily available. Some mobile military radars may not have location data available. Sensing may be acceptable in these cases, as in the US CBRS Environmental Sensing Capability (ESC) framework, or the Dynamic Frequency Selection (DFS) approach. |
| Meaningful Bandwidth | The radar bands typically have large contiguous bandwidth available for use. |
| Meaningful Opportunities | High altitude radars are typically spaced at large distances, and have a limited view of ground level once beyond (below) their line of sight. There will be large areas of exclusion on occupied frequencies, but meaningful opportunities should exist in most places and channels. |
| Opposition Effectiveness | The aviation and military communities have been highly resistant to any sharing of their spectrum.[a] |
| National Scope | This sharing can be best accomplished at the national level, since each country has a different mix of systems deployed, and therefore less resistance to sharing may occur. |

[a] There are major exceptions in the case of militaries, some of which seek a new relationship with civil users. The US CBRS effort in 3.5 GHz is an example of this movement. The same is not true of the aviation community.

considered sensitive. Therefore, sharing with military radars will be highly country- and band-specific.

Table 16.4 provides additional detail regarding three-tier application to ground-based radar bands.

## 16.6 Non-Communications Satellite Space Allocations

This section provides a more abbreviated discussion of several of the non-communications satellite allocations that support space science and operations.

### 16.6.1 Space Research

Few countries in the world perform space research, although some others do host ground stations for sensing, control stations, and tracking services for other nations. This is another case where sharing of the band nationally may be consistent with the ITU treaty terms, if used on the uplink or sensing frequencies, but not if performed on the frequencies that are listened to by the space segment.

**Table 16.5** Analysis of selected ITU allocations in terms of three-tier spectrum criteria (page 1).

| Usage | Prim | Sec | Text Para | Incumbent Awareness | Meaningful Bandwidth | Meaningful Opportunities | Opposition Effectiveness | National Scope |
|---|---|---|---|---|---|---|---|---|
| Satellite Services | | | | | | | | |
| Aeronautical Mobile | X | | 16.8 | Poor for air platforms | Minimal | Significant | Extremely High | International |
| Broadcasting | X | X | 16.4.3, 16.4.4 | Difficult | Yes | No | High | International |
| Earth Exploration (passive) | X | | | Orbital | Yes | Yes | High | International |
| Earth Exploration (active) | X | X | | Orbital | Yes | Yes | High | International |
| Earth Exploration (Uplink, Crosslink) | X | | 16.4.1, 16.4.3 | Orbital | Yes | Yes | High | International |
| Earth Exploration (Downlink, Crosslink) | X | | 16.4.2, 16.4.4 | Yes | Yes | Yes | High | National |
| Fixed (Uplink) | | | | | | | | |
| C-band | X | | 16.4.2 | Available | Yes | Yes | Significant | International |
| Ku- and Ka-band | X | | 16.4.4 | Poor for mobile | Yes | Yes | Significant | International |
| Fixed (Downlink) | | | | | | | | |
| C-band | X | | 16.4.2 | Available | Yes | Yes | Significant | Localized |
| Ku- and Ka-band | X | | 16.4.4 | Poor for mobile | Yes | Yes | Significant | Localized |
| Meteorological (Downlink) | X | | 16.4.2, 16.4.4 | Yes | Yes | Yes | H_gh | National |
| Mobile (Downlink) | X | X | 16.4.2, 16.4.4 | Not available | Yes | Limited | Moderate | National |
| Mobile (Uplink) | X | X | 16.4.3, 16.4.4 | Not available | Yes | Limited | Moderate | International |
| Radio Navigation (Uplink) | X | | 16.4.1, 16.4.3 | High | Minimal | Yes | High | National |
| Radio Navigation (Downlink) | X | | | Not available | No | No | High | International |
| Radio Navigation (Crosslink) | X | | | High | Yes | None on Ground | High | International |
| Radiodetermination | X | X | | Not available | No | No | High | International |

**Table 16.6** Analysis of selected ITU allocations in terms of three-tier spectrum criteria (page 2).

| Usage | Prim | Sec | Text Para | Incumbent Awareness | Meaningful Bandwidth | Meaningful Opportunities | Opposition Effectiveness | National Scope |
|---|---|---|---|---|---|---|---|---|
| Space Operations | | | | | | | | |
| Uplink, Space-Space | X | | 16.6.2 | Yes, for uplink | No | Yes | High | International |
| Downlink | X | | 16.6.2 | Yes | No | Yes | Variable | National |
| Downlink, Space-Space | X | | 16.6.2 | Yes, for downlink | No | Yes | Depends | National |
| Space Research | | | | | | | | |
| Active | X | X | 16.6.1 | Yes | No | No | High | International |
| Deep Space, Uplink | X | | 16.6.1 | Yes | No | Yes | High | International |
| Deep Space, Downlink | X | | 16.6.1 | Yes | No | Yes | Variable | National |
| Deep Space, Uplink, Space-Space | X | | 16.6.1 | Yes | No | Yes | High | International |
| Deep Space, Downlink, Space -Space | X | | 16.6.1 | Yes | No | Yes | Varies | National |
| Passive | X | X | 16.6.1 | Yes | No | No | High | International |
| Aeronautical | | | | | | | | |
| Unspecified | X | | 16.8 | No | No | No | Very high | International |
| Mobile | X | | 16.8 | No | No | No | Very high | International |
| Radio Navigation | X | | 16.8 | No | No | No | Very high | International |

**Table 16.7** Analysis of selected ITU allocations in terms of three-tier spectrum criteria (page 3).

| Usage | Prim | Sec | Text Para | Incumbent Awareness | Meaningful Bandwidth | Meaningful Opportunities | Opposition Effectiveness | National Scope |
|---|---|---|---|---|---|---|---|---|
| Location/Navigation | | | | | | | | |
| Radionavigation | X | | 16.11 | No | No | No | Very high | International |
| Radiolocation | X | X | 16.5 | Military an issue | Yes | Yes | Military | National |
| Maritime Navigation | X | | | No | No | No | High | International |
| **Miscellaneous** | | | | | | | | |
| Amateur | | X | 16.7 | None | Yes | Poor | High | Yes |
| Aeronautical | X | | 16.8 | Poor for air platforms | Minimal | Significant | Extremely high | International |
| Fixed | X | X | 16.8.1 | High | High | High | Technical | National |
| Mobile | X | X | 16.9 | Proprietary issues | Yes | Yes | Very high | National |
| Radio Astronomy | X | X | | High | Limited | Yes | High | International |
| Meteorological Aide | X | | | No | No | No | High | International |
| Broadcasting | X | | 16.10 | Yes | Yes, typically 6 MHz | High in some areas | Significant | National |

For spacecraft-focused activity, the range of distance to the spacecraft is critical to establishing the potential for interference. If the spacecraft is at low or medium altitude, then many earth locations do not have simultaneous visibility to the spacecraft control or payload functions. Therefore, there is no practical interference to the craft. At higher orbits, such as geosynchronous orbits or deep space, the spacecraft antenna is likely to have a large spot size on the earth, and the distance separation required to isolate interference becomes continental to hemispheric; therefore sharing of bands used for these systems is not practical, except with tight coordination with the using organizations.

Most space research facilities are publicly identified, and are non-military and non-proprietary. The research spacecraft orbital or deep space positions are generally well known, and highly predictable on reasonable time frames.

## 16.6.2    Space Operations

Space operations has a similar usage footprint to space research. Few countries in the world perform space operations, although some that do not perform space operations do host ground stations for sensing or tracking. These provide control stations, tracking services, and downlink for other nations. This is another case where sharing of the band nationally may be consistent with the ITU treaty terms, if used on the uplink or sensing frequencies, but not if performed on the frequencies that are listened to by the space segment. Another consideration is the possible use of mobile control terminals that might be on ships or temporarily hosted.

This determination is a case-by-case one. Well-designed spacecraft should not respond to transmissions that are not authenticated through the command authenticator. Therefore, emissions on the control or downlink frequencies should not hinder the operation of the satellites when they are over regions sharing the spectrum, with no visibility to spacecraft that also have visibility to their control or payload ground sites or other control resources. A special case that must be considered is the space operations, control, and communications functions that might be relayed through other space vehicles, such as the USA Tracking and Data Relay Satellite System (TDRSS) system.

## 16.7    **Amateur**

Amateur usage is essentially uncoordinated within the bands it is authorized in. There is essentially no way to be aware of this incumbent class's usage. There are no enforced emission types, or equipment classes, so there is no way to recognize signals as specific to amateur operation. It is a shared service, but it would be difficult to implement a protected tier in a band also assigned an amateur allocation. Amateur usage does not receive protection, so it need not be protected from the three-tier usage. The amateur community is very consistent in defending the rights to meaningful spectrum access. So, despite their secondary status, they would probably strongly (reasonably and successfully) resist any attempt to locate a high density application within their spectrum allocations.

## 16.8 Aeronautical Services

Sharing of aeronautical spectrum (in any form) has three fundamental challenges:

- The platforms (at least one end of most links) are highly mobile, and therefore provide little incumbent awareness. While the fixed side of these links is typically well known (radar sites, air traffic control transmitter and control facilities, navigation stations, such as Instrument Landing System (ILS), Microwave Landing System (MLS) VHF Omnirange/Tactical Area Navigation (VORTAC), or Distance Measuring Equipment (DME) sites) the airborne components are typically rapidly moving platforms.
- The operators of the aeronautical systems (typically national regulatory organizations and aircraft operators) claim that the systems are extremely fragile to interference, and that any additional energy in these allocations would have safety-of-life impacts.
- Public concern over the sensitivity of aeronautical systems to interference would make sharing of these, and adjacent bands, highly unpopular. This was one of the reasons that the LightSquared/GPS issue [5] was decided against the interests of LightSquared, even though LightSquared was not operating in an aviation band!
- There are additional treaty considerations in the United Nations (UN) treaty obligations under the International Civil Aviation Organization (ICAO), which generally assure international operation of aviation-related electronics and non-interference. These are in addition to the ITU obligations.

### 16.8.1 Fixed Services

The fixed wireless community has managed spectrum through coordination processes for an extensive period. This process is not essentially different than the processes in an admission control system, except it is executed by humans through a regulatory mandate or arbitration, rather than automated as in the three-tier model. It has many of the elements of a sharing candidate. A problem with sharing with fixed bands is that many of them also have allocations for mobile and other uses, which includes incumbents that are less practical sharing partners.

There is extensive spectrum that is used for fixed wireless, and in many locations, it is not fully utilized and could provide spectrum access to uses that were deconflicted with the incumbent usage.[2] Of course, any attempt to share spectrum creates a reaction from the incumbent users and band managers, but the arguments in this case would have to be much more technical, as the framework of the three-tier model closely resembles that of the current approach to deconfliction.

---

[2] Fixed wireless usage is as much in angle space (direction) as it is in location. Deconfliction is typically a matter of frequency and direction, rather than assuring separation distance.

## 16.9    Mobile Services

Mobile uses span a wide range; from high frequency radios that can operate on a continental scale, to short range cellular devices. To provide consistency, and focus on possible spectrum sharing candidates, we will only consider the applications above 1 GHZ, which are dominated by cellular (or International Mobile Telecommunications (IMT) in ITU terms).

Cellular mobiles do not have their locations typically (or at least publicly) tracked. The base stations are a different case. Since a cellular device cannot operate outside of a base station service area, the base stations are a proxy for the possible locations of the handsets.

Obtaining awareness of base stations, or small cells, or any type of access point is not a technical issue, but a policy one. ITU obligations are quiet on the specifics of national licensing, so some nations may treat the siting of IMT capacity as public, while others might treat it as proprietary. Particularly in exclusive-use spectrum, the presumption has favored proprietary. Previous chapters have discussed the issues regarding this disclosure even in shared bands, such as the United States (US) CBRS band.

Objections to this sharing might arise from three considerations:

- Existing mobile operators would resist the sharing of their deployment data with competitors or the public. Without this data, it would be difficult to implement the range of protections to User Equipment (UE) and AP devices.
- The likely users of the spectrum might/would be competitive to the primary user's interests and business.
- No matter how rigorous the sharing regime, there is always the possibility of it generating interference, for which the primary user received no premium for risk acceptance.
- There is legal ambiguity regarding the scope of the rights obtained by cellular operators in their spectrum acquisition, and any forced sharing could be resisted in the appropriate courts.

The use of cellular spectrum is driven largely by population, or subscriber, density. Therefore, there is a shortage of spectrum in densely populated regions, and likely, a surplus of spectrum in less dense areas. Therefore, there are opportunities for sharing spectrum in rural and possibly suburban regions, if there is a business case for deployments that are counter-correlated with population density.

## 16.10    Broadcast Services

One of the first examples of database driven spectrum sharing is the opportunity to share unused TV spectrum; an option that is available in many countries throughout the globe is TVWS. TVWS represents the largest block of spectrum currently suitable for multi-tier sharing. Other broadcast spectrum allocations would only be suitable for relatively narrowband uses.[3]

---

[3] Satellite broadcast is not considered here, it is addressed in the Ku- and Ka-band up and downlink analysis section.

The nature of broadcast regulation and commercialization assures a high level of incumbent awareness; this is the basis of the TVWS database solution. The consequences of inadvertent interference are inconvenient, but not on the scale of aeronautical or space control. The opportunities for sharing are quite extensive in less developed regions (even within the developed world). The future of sharing in this band might be challenged or significantly reduced by government desire to monetize the unused spectrum through the "digital dividend," or incentive auction, such as in the USA.

## 16.11 Radio Navigation

Radio navigation includes the non-radar positioning services, such as LORAN and GPS. The criticality, and political support for these types of services was apparent in the USA when the proposed LightSquared use of INMARSAT spectrum was effectively denied by the Federal Communications Commission (FCC) [5]. Older narrowband systems, such as LORAN are seeing decreasing use and support. Aviation services, such as MLS, ILS, DME, and the like, are necessary to the continued operation of the international airspace system, and will be difficult to share spectrum with. The spectrum used by these systems is not extensive. The issue they raise is more on compatibility (such as LightSquared), rather than direct band sharing.

## 16.12 Suggested Reading

- There is a good summary of how to read the ITU allocation tables available at the SpectrumWiki site. The pdf of this document is available at: www.spectrumwiki.com/wp/allocations101.pdf.
- The last several ITU WRC agendas have had items addressing mobile use of C-Band spectrum. This has developed a considerable literature about the relative benefits of C-Band spectrum, from both its supporters and detractors. Some of the proceedings filed to influence the US position on the WRC agenda formulation include submissions by Intelsat LLC (already quoted).
- The ITU Radio Regulations [3] provide all of the specific allocations to which the analysis in this chapter is applicable, as well as the "fine print" that define these usages, both as primary and secondary usage.
- The issues that were raised in the LightSquared [5] proceeding in the USA are instructive of impact of popular support on spectrum decisions. Regardless of the merit of the positions on those in favor, and those opposed, the result reflected a number of non-technical concerns and issues, along with the technical ones.

## References

1 National Telecommunications and Information Administration, *United States Frequency Allocations: The Radio Spectrum Chart*. Consultation (2016). www.ntia.doc.gov/page/2011/united-states-frequency-allocation-chart.

2 *The Economist*, A brief history of Wi-Fi. *Technology Quarterly*, **2Q** (2004). www.economist. com/node/2724397.

3 International Telecommunications Union, *Radio Regulations International Telecommunications Union Radio Regulations, Volume 1* (2012).

4 Federal Communications Commission, *Amendment of the Commission's Rules with Regard to Commercial Operation in the 3550–3650 MHz Band, GN Docket 12–354* (2016). https://apps. fcc.gov/edocs_public/attachmatch/FCC-16-55A1.pdf.

5 National Space-Based Positioning, Navigation, and Timing Systems Engineering Forum (NPEF), *Assessment of LightSquared Terrestrial Broadband System Effects on GPS Receivers and GPS-dependent Applications* (2011). www.gps.gov/spectrum/lightsquared/docs/2011-06-NPEF-lightsquared-report.pdf.

# 17 From Shared Spectrum, to Shared Infrastructure, to a New Model of 5G

## 17.1 Introduction

This chapter will explore the changes in the deployment model for the continued deployment of 4G technology, and to meet the unique challenges of Fifth-Generation Wireless Systems (5G) deployment. While the specifics of 5G deployment are not yet finalized, there is general agreement on many aspects of this transition, and these areas of agreement are sufficiently converged to provide a solid foundation for considering how three-tier spectrum could be a significant factor in this evolution,.

Previous chapters in this book have emphasized shared, three-tier spectrum as a potential disruptive technology in spectrum management. In this chapter, we extend this to consider one of the consequences of that disruption, the viability of an extensive shared 5G infrastructure, and new business models in the wireless communications industry. If these paradigm changes do emerge, they will require new technology to link less vertically integrated telecommunications services, new methods to differentiate operator offerings, and a new industry to deploy these solutions, similar, but more expansive than the current Distributed Antenna System (DAS) industry.

As discussed in the introduction to Chapter 15, this opportunity arises from spectrum sharing between carriers and non-carriers, and the creation of a common band. The question is whether this becomes an interesting aspect of one band, or the first in a series of fundamental adaptations of the wireless service model.

## 17.2 The Traditional TELCO Model of Infrastructure

Although the spectrum community focuses on exclusive spectrum as one of the core tenants of TELCO approaches, spectrum is not the sole resource that is managed primarily as an exclusive asset. It would be a mistake to approach the concept of exclusive-use spectrum without considering it in the context of all of the other resources that make up a TELCO network.

Traditionally, the entire TELCO infrastructure has been owned and operated exclusively by one or more TELCOs. Exclusive, absolute control over the components of the network has evolved as the standard practice for a number of reasons:

- It was the sole means to control the network and assure the Quality of Service (QoS) provided under various Service Level Agreement (SLA) terms.

- Many TELCOs were national monopolies, or quasi-monopolies, and therefore had no natural sharing partner for most resources.
- The specifications for TELCO standard interconnections are hostile to shared technologies. For example, the backhaul for Long-Term Evolution (LTE) eNodeB's is specified, at layer 2, and is likely to be dedicated to a single use, and is typically incompatible with the Internet Protocol (IP) networks that dominate commercial networking.
- Until recently, few industries were potential partners in sharing resources, or creating third party suppliers, because TELCO technology was not shared by other enterprises, unlike today's extensive enterprise networking, cloud computing, and unified networking.

Business realities have made some impact on the "exclusive-everything" model. In the USA, several operators divested their tower holdings, and space on these towers is now leased by a third party to all-comers. The fiber trunks used for backhaul to/from macrocell sites may not be carrier owned; but the individual strands used for the dark fiber connections are typically exclusively dedicated to a single operator. Sharing of the macrocell, antennas and other air interfaces has typically not been present.

## 17.3    The Internet Model of Infrastructure

The contrast to the TELCO model is the one inherent in the Internet. The Internet is shared between billions of devices, servers, services, providers, and applications. Cloud computing has extended that sharing into the compute fabric, where even competitive products share the same cloud computing resources.[1]

The basic infrastructure standards of the Internet have remained quite stable over the four decades of its operation. In less time, the wireless industry has gone through several incompatible generations of infrastructure design and standardization.[2] It is a reasonable question to consider why one set of standards has been massively resilient, even in the face of changes in bandwidth capability of over $10^6$ (from kilobits per second to tens of gigabits), and the other has been redesigned under the pressure of less than an order of magnitude increase between technology generations.

Perhaps the reason that the Internet has been so resilient to technology change, while the carrier systems have not, is the level of abstraction that is inherent in the Internet design. Devices on the Internet do not know, or care, if they are connected by a slow

---

[1] For example, competitive video streaming services Netflix™ and Amazon Prime™ both run on the same Amazon Advanced Wireless Services (AWS) infrastructure.

[2] In both cases, edge equipment technology transitions are essentially ignored. The edge itself is quite technology-sensitive, and advances rapidly. The core technology of the Internet infrastructure (Boundary Gateway Protocol (BGP), Exterior Gateway Protocol (EGP), Open Shortest Path First (OSPF), IP, User Datagram Protocol (UDP), Address Resolution Protocol (ARP), and Transport Control Protocol (TCP) have advanced, but evolutionarily. On the other hand, the cellular network infrastructure for the various generations, such as Advanced Mobile Phone System (AMPS), Global System for Mobile Communications (GSM) and now LTE, are fundamentally different from each other.

speed Digital Subscriber Line (DSL) line, or a 10 Gbps fiber optic strand. The layers are isolated by very generalized abstractions of the pipes and protocols that operate over the network. Products such as Transport Layer Security (TLS), Hypertext Transfer Protocol (HTTP), Secure Hypertext Transfer Protocol (HTTPS), and even Network Address Translation (NAT), have been overlaid on the foundational design with no disruption of the underlying systems. The abstraction framework has enabled the Internet design to meet emerging challenges almost entirely by enhancements in individual layers, with no disruption to other layers. Not only has this reduced cost, it has enabled the network to upgrades asynchronously, as each layer can be enhanced independently.

Most important, the level of abstraction in the Internet has allowed for the emergence of widely different implementations of its core functions. Google search can be used by users connected with dialup modems, or with the emerging gigabit Fiber to the Home (FTTH) connections. A laptop that works on one Internet Service Provider (ISP) is highly likely to work on any other ISP. Applications developed for ten-year-old devices run on the new devices (often). Few Internet companies are as vertically integrated as TELCOs; they pick only the portion of the network where they have unique skills or technology, and share the rest of the network, leaving it to other specialists.

In the following sections, we will consider how three-tier spectrum can be instrumental in providing new opportunities for wireless networks to achieve the benefits of shared infrastructure.

Abstractions that "hide" the specifics of network operators, infrastructure suppliers, spectrum usage, and other temporal or situational aspects are the key to evolution to shared networks. These abstractions must provide the flexibility required to be flexibly integrated into personal, enterprise, or mobile operator networks with flexibility and scalability. In Section 17.9, a potentially significant new TELCO network abstraction will be discussed.

## 17.4     Issues in Deploying 5G Above 6 GHz

5G is proposed to provide deployment both below and above 6 GHz as options. The proposals for 5G below 6 GHz are a more incremental extension of the current LTE technology already in use in that frequency range, with a significant increase in antenna processing to increase bits per Hertz and provide significantly more bandwidth. The disruptive opportunity in 5G lies in the use of frequencies above 6 GHz; most likely above 20 GHz. These bands offer the wide spectrum extents that can support gigabit speeds. The short wavelength, highly directional beam patterns reduce energy demands, provide high data rates, and enable these next generation antennas to fit into a form factor consistent with current User Equipment (UE) concepts.

The rest of this chapter's discussion will therefore concentrate on the use case above 20 GHz. This is not to say that a three-tier, or neutral, host, has no role below 6 GHz. The previous chapters, and particularly Section 15.5, described the contribution three-tier spectrum can make to the deployment of the current generation of wireless technologies, and these points are equally applicable to 5G. However, it is when we consider systems above 20 GHz that the three-tier model becomes an enabler of fundamentally different

**Table 17.1** Propagation loss at the millimeter wave frequencies.

| Material | Approx. 20 GHz | 60 GHz |
| --- | --- | --- |
| Wall Board | 6.8 [1] | 0.7 [4] |
| Brick Masonry Wall | 28.3 [1] | 25.3 [3], 135 [4] |
| Cinderblock Wall | | 218 [4] |
| Wooden Cabinet | | 35.3 [3] |
| Untreated Glass Pane | 4.0 [1] | 5.5 [4] |
| Tinted Glass | 24.5–40.1 [1] | |
| 1/8″ Aluminum Sheet | 46.0 [2] | |
| Plywood. 0.75″, Dry | 4.0 [2] | |
| Plywood. 0.75″, Wet | 32.0 [2] | |

models of wireless deployment, in the cases where traditional models of mobile operator deployment appear to be highly unattractive to both users and operators.

The impacts of the propagation loss, and material attenuation, both of which increase with frequency, is an important consideration in 5G above 20 GHz. Not only does it impact the specifics of the network design, it has implications that go far beyond the technology; the challenges it poses significantly changes the deployment strategy that has remained essentially static through the first four generations of commercial wireless systems.

Table 17.1 provides some reported signal attenuation of common materials in the range contemplated for the high-bandwidth 5G millimeter wave band deployments [1, 2, 3, 4]. Note that there are some materials that have a wide range of reported attenuation, but they are all consistently high loss.

This attenuation loss is an addition to the increase in the path loss of at least 20 dB. The difference between dry and wet plywood in the 20 GHz band is particularly indicative of the challenges in creating high reliability, predictable services in an outdoor environment, when operating above 20 GHz. Unfortunately, there is no consistent source of millimeter wave measurements, such as was cited [5] in support of Table 12.1, so the values in this table are from multiple sources, and have significant differences in methodology. Not surprisingly, there are significant inconsistencies when they are aggregated, but the trends are clear despite these methodological differences.

Some of this path loss can be reclaimed by antenna beam-forming at either end of the path, but the very high material scattering losses have a fundamental impact on the nature of 5G deployments. Millimeter wave signals will typically be restricted to open areas. Passage through walls, even human bodies, will so reduce signal strength that each space within a building will often require several access points to ensure coverage. We can see the scope of access point deployment in Table 17.2, which depicts some typical deployment scales for various categories of macro and small cell. Actual deployments vary widely, but the trends in Table 17.2 are indicative.

The likely coverage scope of an above 20 GHz Access Point (AP) has implications for its deployment. One or more AP installations per room is a significant challenge;

**Table 17.2** Geographic scope of cellular frequencies and access point category.

| Deployment | Typical Freq. | Scope | Area |
|---|---|---|---|
| Macrocell | 700 MHz | 10 km radius | 300 km$^2$ |
| Macrocell | 1 700 MHz | 3 km radius | 27 km$^2$ |
| Small Cell | 1 700 MHz | An outdoor urban block | < 1 km$^2$ |
| DAS | 1 700 MHz | A single building | 0.1 km$^2$ |
| Femtocell | 1 700 MHz | A single floor and region within a building | 0.01 km$^2$ |
| Millimeter Wave AP | 30 GHz | Some, or all, of a single room | 0.001 km$^2$ |

multiplying this by the number of competitive operators in an area makes a difficult problem become impractical and uneconomic, if not impossible.

## 17.5 Neutral-Host/Shared-Infrastructure Model for 5G Deployment

Section 15.5 introduced the concept of private, carrier-grade networks offering connectivity services to mobile operators on a neutral-host basis. This model is attractive for 4G, or more specifically, for below 6 GHz deployments, but is much more so when contemplating deployments above 20 GHz. The discussion in this section was focused around the progress in commercializing the first three-tier band. In this chapter, we now consider if this model can be translated to the millimeter wave region above 20 GHz, with economic benefits to operators, deployers, and users.

If the above-20 GHz deployment is to be a room-by-room, or space-by-space endeavor, it is worth considering what organizations are in the best positions to perform this deployment. Traditional carrier infrastructure was deployed by each carrier, in parallel. It is the premise of this section that this model is not applicable to many of the deployment situations that will arise in 5G deployment. Some of the advantages that a premises owner/tenant has in performing this deployment include:

- The venue owner is often in the best position to install the necessary infrastructure as the building is constructed or renovated, when integrated into other Information Technology (IT) infrastructure.
- The networking backhaul and other resources can be integrated into the same infrastructure that is shared by users in the space. Enterprise networks now provide bandwidth on a par with that of the carrier networks, and are capable of supplying support to the devices.
- The building facility can provide many of the support utilities, such as uninterrupted power, from on-site capabilities that will be required to support other building and enterprise functions, such as servers, emergency lighting, and the like.
- The deployment of the devices can be performed integrally with other enterprise services, such as enterprise Wireless Fidelity (Wi-Fi), telephony, and Internet of

Things (IoT), and even provided in the same deployed edge equipment. The effective cost of this deployment would be the marginal cost of the additional 5G circuitry in the AP node, rather than the complete cost of a stand-alone deployment.

- The building or enterprise can use the same wireless network for internal uses, such as Wireless Local Area Network (WLAN), IoT, telephony, and other internal uses.
- Building operators would find this more attractive than DAS. DAS requires unique spectrum arrangement with each carrier, typically, requires multiple bands, and requires unique fiber-optic connectivity that cannot be shared with the other IP-based services.

These are significant advantages for a premise-based deployment. However, without operator involvement, there is no financial incentive for the building operator to cooperate in the deployment. If operators do not deploy their own networks, then there is no reason for UE manufacturers to include the band coverage, or even for chip suppliers to offer it. However, there are good reasons to believe that operators would find the use of a shared infrastructure attractive, once they have overcome any reluctance to operate exclusively over infrastructure that they did not own or operate exclusively.

- The carrier has no equivalent opportunity to deploy 5G within a building. The "outside-in" strategy of lower bands is not practical, due to the high loss in the building structure.
- The operator obtains capacity with no requirement for capital investment.
- Presumably the operator can acquire access to the interior at a much lower cost per unit of capacity than other deployment alternatives. Even if there are no cost savings by having the deployment integrated with other building construction and operations, the cost of deployment is being shared by multiple operators. So long as 5G provides excess capacity, there is no cost disadvantage in the relationship, and considerable upside.
- The cost of the network capacity is linear with use, proportional to the actual use, with limited fixed cost. The otherwise fixed depreciation and maintenance cost, which is insensitive to traffic or revenue, is not present on the balance sheet.

## 17.6    Barriers to Shared Infrastructure Model

The transition to significant adoption and use of a shared infrastructure will face significant challenges. Some of these are technical and economic barriers to adoption, and others are cultural.

1. Building operators may not find the economic advantages of this model sufficiently attractive to deploy sufficient 5G capacity to be economic for inclusion in UE and other devices.
2. Operators may not want to lose the ability to differentiate their product due to much of their indoor service being offered through a common infrastructure. Many industries have created differentiation even though using common elements (trucking

companies use the same roads, internet applications use the same Internet, airlines use the same airports).

3. Operators may not be comfortable in adopting a new model that reduces their ownership of the critical infrastructure for their service. It significantly reduces the distinction between a Mobile Virtual Network Operator (MVNO) and a Mobile Network Operators (MNO), opening up the traditional operators to direct competition from non-operators with strong consumer or business marketing, sourcing or financial capabilities.

4. New mechanisms for integrating the dedicated operator core networks with the emerging shared infrastructure will be required. These networks must be usable by multiple operators, and will be under the physical control and operation of a third party.

The question is the extent to which these barriers might impede, or even negate, a shared infrastructure deployment.

The first issue (economics from the perspective of a building owner) is driven by the level of carrier payment for traffic carriage. There are still considerable unknowns in this question, but there are good reasons to believe that a reasonable charge-back rate is feasible from both the venue and MNO, MVNO, or even Mobile System Operator (MSO) perspectives.

As previously mentioned, the installation of 4G or 5G infrastructure may be less costly for a building or enterprise that can integrate it into other build-out costs. Additionally, the cost of deployment is shared among the operators; so the cost of this capacity to an operator should be significantly less than exclusive use deployment. The barrier to operator adoption of the model may be driven by the indoor demand level. If demand is not present, than this model will not succeed, but that is really saying that operator-driven, indoor 5G is not viable; an assumption few have suggested.

The second concern is certainly fundamental. It is necessary to determine how operators will differentiate services in the 5G timeframe. The advertising the operators are using may be highly instructive in their private perceptions of buying behavior and differentiation. In the United States of America (USA), one of the major wireless carriers had a motto *"Can You Hear Me Now?,"* intended to emphasize its voice quality as a service differentiator. This motto is no longer used, perhaps reflecting that data is a commodity product. Another advertisement by a different operator, with less network coverage than the dominant operator, states that they are only 1% poorer than the dominant carrier in a metric of quality, and claimed that they had price and capacity advantages. This campaign would imply that they perceive customers are less interested in a highly reliable dial tone, but instead are interested in video streaming quality access and price.

If bandwidth is a commodity, and customer interest is in capacity and price, then the opportunity to acquire high bandwidth indoors at lower cost per bit in Operational Cost (OPEX), and with no Capital Expenditure (CAPEX) may be an attractive proposition, even though it reflects the loss of some product differentiation.

The third issue, that operators will not be willing to utilize an infrastructure that they do not own and operate, is certainly both a cultural and technical barrier. Yet, this model does have precedent in the operators' use of independent shared DAS suppliers in locations that they cannot physically or economically deploy. With the right economics, the cultural barrier appears to be capable of being overcome. The issue of lack of technical control over the service quality and security of the link is an important consideration, but one that can be overcome by sufficient economic advantage. The Internet approach of over-provisioning bandwidth and using abundance instead of complex management has not been applicable to highly spectrum- and bandwidth-constrained wireless networks in the past, but could become effective in 5G, simplifying the service architectures.

Security from the AP to the operator can be provided through secure tunnels to operator infrastructure, but that does create a point of vulnerability at the AP site itself. With increased use of higher-level encryption, such as Secure Socket Layer (SSL) and TLS, this vulnerability is increasingly being mitigated, but it certainly is real. Behaviorally, the widespread use of "Open" Wi-Fi hubs would support the argument that it is not a significant behavioral concern by consumers, as operators deploy offload services, and users search for open access points seemingly without regard for security.

The loss of control over network service quality is a factor that is very likely to concern operators. The TELCO culture is still very rooted in the issues of voice traffic, with very high standards for latency, reliability, drop out periods, etc. In a digital world, it is not clear that these metrics are as important to the user experience as they were with voice. Certainly, the trend has been for application designers to address the shortfalls in internet data services. This has been accomplished through techniques such as buffering content, channel adaptive resolution and compression, more extensive edge device processing, and IP address insensitivity, to reduce impact on the user experience caused by latency, packet loss, drop outs, and other imperfections in the end-to-end service. The willingness of users to use Wi-Fi in dense, congested, and high packet loss conditions indicate that these measures have generally been successful. The tight control over network service quality may be a diminished concern as voice services recede as a major application of these emerging networks.

## 17.7 Abstraction of Heterogeneous Network Technologies and Deployments

The last of the potential barriers raises the issue of developing a new model for the relationship of these edge networks to the operator's core network. A new definition of how the premise-based network deployment is presented to the operator infrastructure, and integrated into the functionality of the operators core networks is required. Certainly, one option is to provide access to devices similarly to how they are represented within the exclusive use infrastructure today. In the case of LTE, this is a complex system within the Extended Packet Core (EPC), including Mobility Management Entity (MME) and Serving Gateway (SGW) components that manage the mobility of UE devices, and routing of traffic. While LTE is the first TELCO system that is IP-based, it has a unique protocol to move data from the UE to the exit point of the TELCO network. By contrast,

Wi-Fi operates entirely in the IP space, and does not require that traffic be managed by unique protocols as it transits the intermediate IP network(s).

Architecturally, there are significant differences in the delivery architecture of LTE and Wi-Fi. In the LTE model, traffic from an endpoint transits the interior of the EPC (carrier network) in a star network to the core, and then is interconnected to internet exchange points, or peering arrangements. These points concentrate the traffic from all users to one or more central interchange points. By contrast, Wi-Fi networks typically have no core, and "breakout" traffic at the nearest exchange, interconnect, or peering location. The internet application has no visibility into the network design, topology or performance. Each element of the delivery path is independent, and the interaction between them follows a common abstraction, defined by the internet standards.

Although the shared infrastructure is a significant change in the TELCO business model, it has a parallel, but lesser impact on the technical interactions between the TELCO operator and the shared infrastructure provider.

- Devices must serve multiple operators, so no operator can have sole control over the device.
- Enterprises may operate the access devices within fire-walled networks in order to host high security applications over their wireless networks. This implies that the device control functions are likely to be unreachable through a strict fire-wall policy. The TELCO will receive an IP stream, but may have to delegate decisions such as mobility to the shared network, rather than the central management of the EPC functions. Fortunately, levels of this abstracted delegation of authority to local networks is evolving under the general framework of LTE Self-Organizing Network (SON) concepts.
- Device and network identity and authentication may become more complex. Is the enterprise device "owned" by the enterprise, and then roaming onto an operator network, or the reverse? Are there tens of thousands of operators, or are they aggregated? There are an entirely new set of challenges in scaling the existing authentication framework into the shared infrastructure regime.

There are some exciting concepts that could evolve in this model. With significant levels of abstraction, aggregators mobile operators could essentially operate two-sided models, when multiple shared infrastructure has overlaps in coverage. They are selling an integrated network service to their subscribers, but can acquire the capacity through any number of unique market models, timeframes, and QoS levels. Price transparency has the advantage that it would encourage deployment in underserved locations due to the visible premium that these locations achieve.

## 17.8    Implications for the 5G Spectrum Policy

There have been some interest in communities that have advocated for increased spectrum sharing, either as unrestricted unlicensed, such as used in Wi-Fi, or as the three-tier framework, as advocated by President's Council of Advisors on Science and

Technology (PCAST), and implemented by the Federal Communications Commission (FCC). On the other hand, the US FCC has proposed [6] an allocation of spectrum that has approximately 4 GHz in the 27.5–28.35 GHz and 37–40 GHz bands for exclusive use through licensing (presumably) by auction, and an additional 7 GHz for unlicensed usage in the 57–71 GHz band [6]. This is a considerable additional allocation of unlicensed spectrum, but the approach does not appear to reflect the emergence of shared infrastructure, neutral host services, and the very progressive ideas in their own Citizens Broadband Radio Service (CBRS) rules.

The issues with this allocation are numerous, including:

- Although a considerable extent of spectrum is made available for unlicensed usage, it is at a very high frequency band, and affordable user equipment technology will likely lag the availability of the 20 and 40 GHz technology. Although all of these bands have technology challenges, achieving operation at 70 GHz will likely be delayed until long after the 20 and 40 GHz technology has matured.
- The lesson from the early CBRS thinking is that carrier-grade technology should be deployable by both operators and by third parties. It is in all parties' interests that the capacity deployment be possible by the party in the best position (control of backhaul bandwidth, access, power, etc.) to perform this deployment economically. The innovation in the CBRS band is the access to common spectrum by the widest range of users. Common user- and customer-premises equipment is best achieved when the bands are in common usage. Exclusive-use spectrum essentially precludes any shared infrastructure, since such infrastructure could not be shared in a predictable manner. And yet, deployments that are on the scale of individual rooms and spaces is not likely to be performed exclusively by any one operator, nor used exclusively by one operator.

Traditionally, operators have argued to regulators that they required exclusive-use spectrum in order to protect their expansive networks from interference, to protect their investment, and to incentivize them to deploy capacity. It is worth examining how the emergence of shared infrastructure, and neutral host, impacts these rationale, and how that might impact both political considerations and spectrum buying behaviors.

The exclusive licensing model was well suited to towers that typically had up to $100\,\text{km}^2$ of coverage. Elevated towers made them vulnerable to interference from any emissions in the band, and the propagation on these frequencies was favorable. This meant that exclusive use over large areas was required to have practical deployment engineering. While there are certainly policy arguments regarding exclusive spectrum, there is no question that it has significant benefits to the operators in terms of reliability and consistency. However, above 3 GHz, and certainly above 20 GHz, it is unlikely that macrocells will be the dominant deployment, and the argument for exclusive spectrum becomes moot.

The massive Multiple Input/Multiple Output (MIMO) and beam-forming planned for these frequency ranges may be highly beneficial, but the high scattering losses will likely limit this use of macrocells in these bands. If true, then the deployment model may be more like that planned for 3.5 GHz. The same cost and deployment drivers of 3.5 GHz band will apply to 5G, probably more so. These strongly favor the shared-infrastructure model.

If shared infrastructure becomes an important aspect of the 5G deployment, then the role of shared spectrum is an obvious issue that requires re-examination. The issues we have to consider are:

1. Is the exclusive-use model still appropriate to networks that have a typical service area of under $0.001$ km$^2$, and may also be deployed vertically?
2. If the exclusive model is not appropriate, what model is appropriate?
3. If there is a better model, will it be politically, financially (from a government perspective) and commercially sufficiently acceptable to the stakeholders to be adopted?

First, it is unclear that an exclusive-spectrum regime is optimal for a shared-infrastructure deployment. It would have several impacts.

- Making extensive regions of the spectrum available only to single operators would block shared-infrastructure suppliers from deploying the capacity that the operators would find useful to utilize. It would force deployment of dedicated infrastructure, often contrary to operator best interests.
- Reduced sharable spectrum will similarly reduce the opportunity for third parties to deploy sharable infrastructure. In turn, this reduces the capacity available to operators, and likely reduces the competition in supplying this sharable infrastructure. Loss of competition will lead to higher marginal costs to the operators because they tied up the spectrum exclusively.
- Exclusive licensing causes the handset ecosystem to fragment. One of the reasons Wi-Fi is successful is that all devices can share the spectrum, and therefore all devices include the full band coverage. In contrast, typical LTE UE devices support only a small selection of bands. This limits the devices mobility to other operators, or to shared infrastructure. If one operator has exclusive use of a band segment, it is likely that other operators phones will not support it. This increases cost, and reduces flexibility for both operators and users.[3]

Second, there are only three models of spectrum commonly made available and well-understood: exclusive, shared three-tier, and unlicensed. We have already considered the exclusive model, whether applied at a primary or secondary level (such as in Authorized Shared Access (ASA)/Licensed Shared Access (LSA). The other alternative to three-tier is the unlicensed, unprotected model. Even in the three-tier model, it is likely that most of the spectrum usage will be in the unprotected tier. In these cases, there is no substantive difference between unlicensed and three-tier General Authorized Access (GAA). However, the three-tier model has several advantages over the pure unlicensed.

---

[3] In the US CBRS band, the regulation requires devices to cover the entire band in order to preclude this very situation, but this is unique in this band.

- The spectrum need not be occupied solely by the 5G applications, and some degree of legacy usage can continue in the band, and receive whatever protections the regulators define. That reduces cost, and increases the availability of spectrum for 5G.
- The ability to acquire protection, whether actually required or not, is important to many investors in the band. Certain high-cost investments may not be possible without recourse to protected spectrum.
- There likely will be situations where usage demand is so dense that the protection provided by the protected tier is necessary to assure good order in the band, and enable operators to assure QoS levels.

Considering the last question, there is definitely a trend by governments to exploit spectrum licensing revenues as revenue sources, rather than as marketplace arbitrators. Some have referred to this reliance on spectrum auction revenue as *"Auction Addiction."* The three-tier model offers a revenue stream, but it is not yet known how significant the revenue will be, and it does not provide the "instant gratification" that an auction provides.[4]

However, the revenue from spectrum was driven by two spectrum auction bidder perceptions. The first was that there was some linear relationship between spectrum holdings and bandwidth capability. Therefore, operators were perceived to believe (and might have believed) that there was no alternative but to pay whatever was necessary to acquire rights to use a suitable portion of the available spectrum. Since densifying low-band spectrum is difficult, due to the high cost of the infrastructure and the long range of the signals, this perception had considerable truth.

The second perception was of spectrum scarcity. At least initially, the significant amount of spectrum being proposed for 5G above 20 GHz would not appear to be "scarce." The amount of spectrum proposed to be offered for 5G is an order of magnitude more than all of the spectrum auctioned to date in the USA. Considering that the possible coverage in the 5G bands is $10^5$ (10 to the fifth) that of the same spectrum in the traditional 4G bands, and more of it is available, it is quite abundant in terms of the aggregated capacity that can be offered.

This linear relationship between spectrum bandwidth and digital service bandwidth does not apply to deployments in the higher frequencies. Clutter losses, not free-space losses, dominate, and infrastructure deployment is often limited by backhaul, rather than spectral density. Spectrum bandwidth is required to achieve peak capacity, but not to create aggregate capacity. Therefore, despite the wishes of governments, the windfall from the "spectrum addiction" may be coming to an end due to physics, regardless of policy. Further accelerating this loss of revenue is that the operators might be de-appetized because of the inapplicability of exclusive spectrum to support a shared-infrastructure model.

The answer is not to believe that any policy-maker or regulator can determine the right mix of shared and exclusive spectrum. Neither the advocates for unlicensed, nor the ones advocating exclusive licensing, have sufficient knowledge of how this band

---

[4] For example, in the USA, spectrum revenue is estimated and "scored" when auctions are authorized. These funds are available to Congress for spending immediately, even if the auction is years away, and independent of the actual revenue raised.

will evolve or deploy. The uncertainties in the technology, use cases, cost and service models, and customer demand, are all massive. No informed decision is possible at this time. However, the three-tier structure provides a mechanism to balance the demand for both categories of spectrum dynamically and locally.

Dense urban areas with high clutter losses may be dominated by small cells that rely on scattering loss to isolate their operations, and can readily operate in shared spectrum. Operators in less urban low-band areas may utilize regulatory protection through second-tier rights to assure protection. Three-tier spectrum management enables a market-driven solution to spectrum allocation. There is no need for regulators to make arbitrary and uninformed decisions. In fact, there is every reason to not make them.

Policy-driven allocations of spectrum must be performed many years ahead of usage, in order to provide the ecosystem stability required to develop the equipment. The use of the three-tier (or the lower tiers in unused spectrum) provides a mechanism to both provide industry guidance on the spectrum that will be available, and to defer the binding of that spectrum to specific regulatory categories. The US 5G policy statement for above 20 GHz spectrum could have instead stated that:

*"We will commit to 11 GHz of additional spectrum to be available for protected and unprotected tiers, with some amount across all bands, to be available as non-exclusive, with annual auctions for limited period protection rights, and some spectrum assured for unprotected, general access."*

This policy would fully abandon the command-and-control mentality of spectrum regulations, and enable the transition to a number of different spectrum rights models, driven by market realities, rather than rapidly obsolete regulatory decisions.

## 17.9    Impacts on Infrastructure and Virtualization

One issue that might constrain the growth of access points from the tens of thousands to the tens of millions is the inherent inflexibility of the TELCOM infrastructure, and practical ability to extend it these three orders of magnitude required to integrate the access networks into the existing TELCO networks. Even without the consideration of neutral-host networking, many participants in the TELCO world had recognized the need to re-imagine the central office, and how equipment is integrated into networks.

One result of this recognition is the emergence of a large-scale project to develop a virtualized, commodity, open-source framework for managing central offices with the same level of abstraction and virtualization as is present in datacenter management. This industry program, Central Office Re-architected as a Datacenter (CORD), is attempting to develop such a framework. Its objectives are stated to be:

*"Our mission is to bring datacenter economies and cloud agility to service providers for their residential, enterprise, and mobile customers using an open reference implementation of CORD with an active participation of the community. The reference implementation of CORD will be built from commodity servers, white-box switches, disaggregated access technologies (e.g., vOLT, vBBU, vDOCSIS),[5] and open source software (e.g., OpenStack, ONOS, XOS)."*

---

[5] Virtual Optical Line Termination (vOLT), Virtual Baseband Unit (vBBU), and Virtual Data Over Cable Service Interface Specification (vDOCSIS)

This development of a community-accepted, scalable, and broadly-based infrastructure model, and even implementation, is key to scaling the number of access points supporting individual TELCO networks upwards by two to three orders of magnitude. CORD may lead to that answer, or at least indicates that there is recognition of the need for such scalable architectures. That recognition should lead to a solution, whether it is CORD or not. It would appear to follow the same path as the scalable computing fabric projects, such as OpenStack have successfully followed, and has a similar scale of technical complexity.

## 17.10    Suggested Reading

The emergence of a shared-infrastructure model is fairly recent. However, many of the basic concepts are well developed in more general literature, and many can be directly extrapolated to wireless applications.

- There is extensive literature regarding 5G wireless. By nature, much of the specifics are speculative, or even wishful thinking on the part of advocates, for one or more of the technologies, but there is a common consensus that is useful and well documented in these works. One recent work is by Osseiran [7].
- There has not yet been extensive peer-reviewed studies of propagation above 20 GHz to the same extent as below 3 or 6 GHz. Some data has been published, and is instructive in considering realistic options of the use of this band in cellular systems [1, 2, 3, 4].
- This chapter raises questions regarding the US FCC action on 5G allocation above 20 GHz. This policy is provided in the FCC Report and Order [6].
- Introduction of two-sided markets is a logical consequence of the transition to shared infrastructure. There is considerable literature on this subject in scholarly work, including one book by Zhao [8].
- There is an extensive set of material describing, and implementing the CORD project, including a fairly extensive wiki [9].

## References


1 H. Zhao, R. Mayzus, S. Sun, M. Samimi, J. K. Schulz, Y. Azar, K. Wang, G. N. Wong, F. Gutierrez, and T. S. Rappaport, 28 GHz millimeter wave cellular communication measurements for reflection and penetration loss in and around buildings in New York city. *2013 IEEE International Conference on Communications (ICC)* (2013), 5163–5167.

2 E. Violette, R. Espeland, and G. Hand, *Millimeter-wave Urban and Suburban Propagation Measurements Using Narrow and Wide Bandwidth Channel Probes*. National Telecommunications and Information Agency, Technical report. NTIA Report 85–184 (1985).

3 F. Fuschini, S. Häfner, M. Zoli, R. MÃijller, E. M. Vitucci, D. Dupleich, M. Barbiroli, J. Luo, E. Schulz, V. Degli-Esposti, and R. S. Thomä, Item level characterization of mm-wave indoor

propagation. *EURASIP Journal on Wireless Communications and Networking* **1** (2016), 1–12. doi: 10.1186/s13638-015-0502-3.

4 J. Lu, D. Steinbach, P. Cabrol, P. Pietraski, and R. V. Pragada, Propagation characterization of an office building in the 60 GHz band. *The 8th European Conference on Antennas and Propagation (EuCAP 2014)* (2014), 809–813.

5 National Institute of Standards and Technology, *NIST Construction Automation Program Report No. 3 Electromagnetic Signal Attenuation in Construction Materials*, NISTIR 6055 (Gaithersburg, MD, 1997). fire.nist.gov/bfrlpubs/build97/PDF/b97123.pdf.

6 Federal Communications Commission, *Use of Spectrum Bands Above 24 GHz for Mobile Radio Services, GN Docket No. 14–177; Establishing a More Flexible Framework to Facilitate Satellite Operations in the 27.5–28.35 GHz and 37.5–40 GHz Bands, IB Docket No. 15–256; Petition for Rulemaking of the Fixed Wireless Communications Coalition to Create Service Rules for the 42–43.5 GHz Band, RM–11664; Amendment of Parts 1, 22, 24, 27, 74, 80, 90, 95, and 101 to Establish Uniform License Renewal, Discontinuance of Operation, and Geographic Partitioning and Spectrum Disaggregation Rules and Policies for Certain Wireless Radio Services, WT Docket No. 10–112; Allocation and Designation of Spectrum for Fixed – Satellite Services in the 37.5–38.5 GHz, 40.5–41.5 GHz and 48.2–50.2 GHz Frequency Bands; Allocation of Spectrum to Upgrade Fixed and Mobile Allocations in the 40.5–42.5 GHz Frequency Band; Allocation of Spectrum in the 46.9–47.0 GHz Frequency Band for Wireless Services; and Allocation of Spectrum in the 37.0–38.0 GHz and 40.0–40.5 GHz for Government Operations, IB Docket No. 97–95*, (2016). http://apps.fcc.gov/edocs_public/attachmatch/FCC-16-89A1.pdf.

7 A. Osseiran, J. F. Monserrat, and P. Marsch, eds., *5G Mobile and Wireless Communications Technology* (Cambridge University Press, 2016).

8 W. Zhao, Two-sided markets model and its applications. Unpublished Ph.D. thesis, Stanford University (2011).

9 Central Office Re-architected as a Datacenter (CORD), Wiki Home. wiki.opencord.org/display/CORD.

# 18 Future Actions to Deploy More Three-Tier Spectrum Bands and Nations

## 18.1 Issues for International Expansion

It might be considered that spectrum policies are developed through a common process that is similar throughout the world, and standardized. However, spectrum policies have generally followed one of two alternative paths in their development and implementation:

**ITU Down** An International Telecommunication Union (ITU)-Down strategy focuses on achieving international, or at least ITU regional, acceptance of establishing three-tier spectrum policies. The process of achieving agreement at the periodic World Radio Conference (WRC) events is at least a long process. It typically addresses band-specific policies, rather than policies that are band-independent. And as will be discussed later, such international harmonization is not necessary for the adoption of three-tier spectrum.

**Nations-Up** Nation-by-nation approaches focus on developing support in typically early-adopter nations as a secondary allocation. The regulations would need to protect ITU-compliant uses, but these are typically incumbents in the band that are protected under the three-tier concept in any case. For example, in the USA, the Citizens Broadband Radio Service (CBRS) band is the extended C-Band satellite downlink. The rules protect the international activity in the band, and therefore comply with ITU treaties.

The strategy for introducing three-tier into National Regulatory Authority (NRA) regimes is significantly aided by a factor the author has referred to as *"regulatory virtualization."* This provides the flexibility to adopt these regulations on a band-by-band, and nation-by-nation process that requires no coordination across borders, and can be adopted to suit localized, rather than international, needs.

## 18.2 Regulatory Virtualization

One of the constraints of spectrum policy has been that device manufacturers have "baked in" most of the regulatory limits imposed by national and international regimes. This has limited the flexibility of nations to manage spectrum in unique ways, since they would be forcing a fragmented market, and higher costs. They would be vulnerable to an inflow of non-conforming devices due to device purchase in other countries, as roam the world. Similarly, the rules, once established, cannot be readily updated to either restrict or open up the rights regime.

To the extent that band rules are implemented in the access control system, national administrations are free to establish unique rules and bands without any impact on, or fragmentation of, the device ecosystem. Devices need not even be aware of these national differences, as they are inherently reflected in the acquisition of spectrum rights through the admission control process. Examples of spectrum regime features that can be addressed and adjusted through the admission process include:

- Adjust the types of devices that receive protection.
- Adjust allowable power limits, bandwidth, channelization, and modulation types.
- Create markets and structures for rights acquisition.
- Notification methods for incumbent protection.
- Selection of specific bands and frequency ranges.
- Limitations on type of usage (fixed, mobile, etc.).

We consider this virtualization because the individual devices and operators are in fact, ignorant, or at least could be ignorant, of the regime. A device can be sold anywhere in the world, and be ignorant of the rules in any specific country, and yet operate in accordance with any regime that is imposed in the location in which it registers with the admission control system. Not only can a device roam geographically, it can roam in terms of the regulatory regime in which it operates.

The virtualization of regulation should enable worldwide adoption of the three-tier model relatively unsynchronized, reducing the necessity for an "ITU-Down" approach. The ITU is not necessary to a nation-by-nation rollout, and is not necessary if worldwide, or regional, band, usage and interference criteria protects the primary allocation in the three-tier regimes. This approach is a significant departure from traditional spectrum policy making, but is uniquely enabled by the addition of the admission control component to the policy tool set.

## 18.3    Device Commonality

The economics of modern consumer wireless is in use of high volume to reduce consumer and infrastructure costs. This has been achieved through international cooperation in terms of broadband standards, and international, harmonized standardized band assignments. An example of this is the rapid near-universal adoption of the Third-Generation Partnership (3GPP) standards for Long-Term Evolution (LTE) [1, 2, 3] for standard band definitions.

An objective of any implementing NRA should be to avoid impacting the device ecosystem, and limit the unique aspects of their regime to the admission control system wherever possible. Unique devices, and unique device software, partition markets, limit entry and competition, and reduce benefits to both operators and consumers. In the first implementation of the principles in the US CBRS band, it was obviously necessary to make changes to the Access Point (AP) software to add the Spectrum Access System (SAS) interface and interaction. Significant effort went into assuring that unmodified LTE User Equipment (UE) devices developed for 3GPP band 42 and 43 could operate without modification. This preserved the economics of the high volume UE ecosystem, even if it forced some unique development of the less numerous AP equipment in the band.

## 18.4    Admission Control System Commonality

Similar principles apply to the software and hosting that make up the admission control system. There are a significant number of algorithms and methods that make up the system. While large, developed markets might be able to support extensive development of unique system software, smaller markets would find such development a burden, and might slow, or even block, the use of the regime.

The United States of America (USA) has defined one set of regulations. It is likely that the next administration that adopts three-tier will require changes in algorithms as well as rules. However, as the number of administrations and bands increases, the viability of the regime will be optimized in the basic framework of the protections and limitations that are consistent with a template of national rules and a band-by-band basis. The number of variants should converge, and not grow linearly with adoption rates.

Non-recurring development costs for each incremental national or band-specific addition to the regime should decrease over time. Eventually, the number of permutations should converge to where the introduction of additional bands and regions would equate to more of a template filling-out operation, rather than a unique software development.

## 18.5    A Model for International Three-Tier Multi-Band Roaming

The following discussion is not intended to provide, or advocate, for any specific design of a device-to-admission-control-system interface, but to show how one such design could enable worldwide mobility, support multiple numerous regulatory regimes, and operate across a wide range of non-harmonized frequencies.

There are a number of aspects that impact a device's ability to utilize spectrum that is available through a three-tier regime. These include:

**Locally Available Bands** Three-tier spectrum is unlikely to be available on an internationally harmonized basis, so the expectation is that different local NRA may elect to make different bands available on a nation-by-nation basis. These rules may be more dynamic than traditional bands, so it is best that the admission control system inform the devices about spectrum availability and the regulatory requirements for use.

**Required Certifications** Each NRA will likely require that devices operating in their spectrum have some level of regulatory approval. Presumably, the admission control system will inform the device of the specific certification(s) that are required for each spectrum segment. These could be national, reciprocal, or collateral with other countries. Devices will be capable of holding multiple certificates, potentially for different bands, and different countries, and would offer these in the registration of, the device, instead of the fixed certificate in the CBRS regime.

**Device Capability** Three-tier spectrum principles can be applied to a wide range of spectrum with fairly localized propagation, from the region from 1 GHz, up to the millimeter wavelength bands. Single devices will likely not be able to operate across the full range of possible spectrum. Devices must have algorithms to select among the available bands, associate them with the certifications required to operate in them, and hardware limitations, and negotiate with the appropriate admission control system to reflect the intersection of the opportunities and the device capability.

**Protection Licenses** Devices may hold rights to utilize mid-tier spectrum, which is afforded protection or usage rights. The admission control system must inform devices of what licenses are required to access which spectrum channels, and devices need to assert their rights under these licenses.

**Marketplace Identity** The initial US CBRS band has no dynamic markets, but the President's Council of Advisors on Science and Technology (PCAST) proposal provided a very dynamic, nano-marketplace, and the US CBRS regulations do provide for secondary markets, from which dynamic markets could emerge. Admission control systems may offer such services, or they could point to the marketplace. Similarly, devices would need to retain identity in order to enter into transactions in such markets.

Figure 18.1 illustrates a generalized interaction with an admission control system and a device when it first contacts the system and performs an inquiry. Interaction with an admission control system would not be in clear text, obviously, and would use Extensible Markup Language (XML) or JavaScript Object Notation (JSON), and would be processed automatically by the device with no, or limited, human-decision interaction.

In this first interaction, the admission control system provides an inventory of the spectrum that is available. Just for completeness, other unlicensed, but not three-tier spectrum is included in the menu to reflect the differences that are emerging in even unlicensed spectrum.

In this example, we assume some of the existing unlicensed and lightly licensed methods into the admission-control inquiry functionality. This provides the ability to not only virtualize the three-tier regime, but to utilize the same mechanism for informing devices of opportunities in other bands, as well. In this case, the opportunity to share Television White Space (TVWS), and to apply Dynamic Frequency Selection (DFS) in the 5 GHz band [4]. If no radars required protection within a region, the DFS requirement could be removed from one or all of the channels in which radar protection was no longer required.

| | | | Spectrum Access System Inquiry | | | |
|---|---|---|---|---|---|---|
| **Inquiry Range** | | 500 MHz to 6,000 GHz | | | | |
| Location: USA | | | Administration: USA | | | |
| **Band** | **From** | **To: Nomenclature** | **Regulation** | **Certificate** | **Access System** | |
| 1 | 614 | 698 TVWS | US:Part15 | N/A | tvws.google.com:80 | |
| 2 | 2 400 | 2 464 ISM | USA:Part15 | N/A | N.A | |
| 3 | 3 550 | 3 650 CBRS | USA:Part96 | USA:CBSD | self | |
| 4 | 3 650 | 3 700 CBRS | USA:Part96 | USA:CBSD | self | |
| 5 | 5 170 | 5 250 ISM | USA:Part15 | N/A | N/A | |
| 6 | 5 250 | 5 730 ISM:DFS | USA:Part15:DFS | N/A | N/A | |
| 7 | 5 735 | 5 835 ISM | USA:Part15 | N/A | N/A | |

**Figure 18.1** Notional display of a notional spectrum admission control system inventory.

| | | CBRS Channel Occupancy and Availability Inquiry | | | |
|---|---|---|---|---|---|
| **Channel** | **From** | **To** | **From** | **PAL Holder** | **Location Specifics** |
| 1 | 3 550 | 3 560 | PAL | Verizon | In PPA |
| 2 | 3 560 | 3 570 | PAL | Verizon | In PPA |
| 3 | 3 570 | 3 580 | PAL | AT&T | PPA Near |
| 4 | 3 580 | 3 590 | PAL | Sprint | |
| 5 | 3 590 | 3 600 | PAL | Google | |
| 6 | 3 600 | 3 610 | PAL | | |
| 7 | 3 610 | 3 620 | PAL | | |
| 8 | 3 620 | 3 630 | GAA | N/A | Naval Radar |
| 9 | 3 630 | 3 640 | GAA | N/A | Naval Radar |
| 10 | 3 640 | 3 650 | GAA | | |
| 11 | 3 650 | 3 660 | GAA | | |
| 12 | 3 660 | 3 670 | GAA | Unavail | |
| 13 | 3 670 | 3 680 | GAA | Unavail | |
| 14 | 3 680 | 3 690 | GAA | Unavail | WISP Protection |
| 15 | 3 690 | 3 700 | GAA | Unavail | WISP Protection |

**Figure 18.2** Notional display of a notional spectrum admission control system band detail.

The device can scan this list (depicted in Figure 18.1), and then select a specific band to perform a detailed availability inquiry. In this case, the graphic depicts the response to an inquiry in the US CBRS band where there are several Priority Access License (PAL) licensed, but with only several occupied: a naval radar in two channels, and several non-interfering General Authorized Access (GAA) channels. The logical contents of this interaction are shown in Figure 18.2.

In this example, the device is informed about the channels that are available for PAL use. If it requests channels 1 or 2, the request will be rejected because the device is within a PAL Priority Access License Protection Area (PPA), and must be rejected. In channel 3, it is near a PPA. This is a hint that this request may be rejected, but that decision depends on what the request specifics are. Channels 4 and 5 are PAL channels that have PAL licenses, but no PPA is present in the vicinity. These channels can probably be used by a GAA device, but they may be forced to relocate if the PAL holder registers a device and associated PPA.

Channels 6 and 7 are available in future auctions for PAL licenses, and so a device might be relocated after these auctions occur, but no PAL holder could displace the node at this time. Channels 9 and 10 are not available due to naval radar operation that were detected. Similarly, channels 14 and 15 are unavailable due to the requirement to protect WISP operation over the five-year grandfathering period.

This framework avoids the requirement for internally recognized bands or regulations. Commonality has advantages in lower cost through reduced fragmentation of markets, but it is not essential to the regime, and market forces will inherently correct against an overly fragmented international regime through the cost advantages of high-volume commodity parts, and code-base reuse across bands and national regulations.

## 18.6   Dynamic Markets

The PCAST report proposed a nano-market that would enable escalated spectrum rights to be exchanges on a dynamic marketplace that would be flexible in terms of the rights that were sold, the time duration, and the geographic extent [5]. The market established by the CBRS rules did not include this within the PAL structure [6]. However, the second order generally described the regulations for a secondary market for the PAL spectrum rights that were acquired at the more static, annual PAL auction [7].

Other nations may adopt a more PCAST-like market framework; but even in the absence of such government-executed markets, the secondary markets may create a dynamic exchange for spot-market acquisition of spectrum rights from holders of the longer-term licenses. At an extreme, the annual markets for government rights could become a wholesale market, with the secondary market supporting end users in managing the dynamic nature of their traffic delivery, without a planning or investment in full time, prediction-based commitments.

## 18.7   US Experience as Confidence Building

The US CBRS will be the first implementation of a three-tier regime, even though several features were not incorporated in the initial band regulations. For this regime to be attractive to the rest of the world, or to advocates for additional bands, the band must demonstrate its attributes to three communities:

**Consumers and Users** There should be demonstrable increases in available bandwidth or a reduction in the cost of bandwidth that can be explicitly traced to the capacity introduced by the regime. Alternatively, it may turn out that the band introduces a new modality that could not be supported through existing spectrum models, such as for Internet of Things (IoT).

**Industry and Operators** One of the core arguments of the regime is that it will create new investments and innovative products. The USA experience should demonstrate that this investment occurs, even though its scope is initially constrained by the single market and single band.

**Incumbents** A core tenant of the three-tier regime is that incumbents are protected from interference from the secondary users managed by the three-tier regime. The US experience must demonstrate that this commitment is met. There will likely be claims of disrupted operation, but these must be either shown to not be due to the three-tier regime, or to have been rapidly resolved, and the regime updated, to avoid recurrence.

At this point in the deployment process, it is premature to evaluate the first and third of these criteria. However, some initial indications are apparent in regards to satisfaction of the second criteria. The Wireless Innovation Forum (WinnForum) process has attracted over 50 organizations and companies who perceive an interest in the band, including the major chip and system manufacturers, and all four of the US major cellular operators.[1]

To address the technology-specific application of Time Division Long Term Evolution (TD-LTE), the CBRS Alliance industry organization was formed by Qualcomm, Nokia, Alphabet Access, Intel, Ruckas, and Federated Wireless to promote the use of TD-LTE in the band. A more important object, from the innovation and capacity perspective is the development of neutral-host networks in partnership with the Mobile Network Operators (MNO) community.

None of these events by themselves satisfies the second criterion. However, they do show the investment is widely perceived to have value, and that a business opportunity is very possible. At this point in the evolution of the band, it is a positive sign.

## 18.8    Suggested Reading

- Many of the techniques discussed in Cognitive Radio (CR) research would be highly useful when a range of spectrum is made available to devices. Device decision-making would be able to trade range, power limits, bandwidth, propagation and interference conditions dynamically. Work in this field includes two books by the author [8, 9], the first of which focuses on algorithms to perform this kind of decision-making. A number of works with material that may be relevant to strategic management of spectrum decisions include a number of works in CR [10, 11, 12, 13].

- Dynamic secondary markets may emerge as a natural consequence of the diversity of spectrum that becomes available in three-tier regimes. These should encourage the dynamics and transparency that would maximize the benefit to all of the users. These

---

[1] This includes Qualcomm and Intel on the chip side, Nokia, Ericsson, Ruckas, and Huawei on the system-infrastructure side.

holdings are not standalone, as in multi-billion investments, but can be highly linked, as described in terms of a portfolio [14], and with dynamic pricing that minimizes the incentives to overly restrict spectrum access [15].

- The industry progress is an indication of the investment in the band, and therefore industrial estimates of its economic opportunity the band affords. Progress in both the WinnForum and CBRS Alliance is indicative of this. These organizations have web pages accessible at www.wirelessinnovation.org and www.CBRSAlliance.org, respectively.

## References


1 E. Dahlman, S. Parkvall, and J. Skold, *4G: LTE/LTE-Advanced for Mobile Broadband* (Amsterdam: Academic Press/Elsevier, 2011).

2 A. Ghosh, J. Zhang, J. G. Andrews, and R. Muhamed, *Fundamentals of LTE* (NJ: Prentice Hall, 2011).

3 A. Ghosh and R. Ratasuk, *Essentials of LTE and LTE-A* (Cambridge University Press, 2011).

4 The Wi-Fi Alliance Spectrum and Regulatory Committee, Spectrum Sharing Task Group Regulatory Task Group, *Spectrum Sharing in the 5 GHz Band – DFS Best Practices*, (2007). www.ieee802.org/18/Meeting_documents/2007_Nov/WFA-DFS-Best-Practices.pdf.

5 President's Council of Advisors on Science and Technology, *Report to the President: Realizing the Full Potential of Government-Held Spectrum to Spur Economic Growth* (Executive Office of the President (EOP), Office of Science and Technology Policy (OSTP), 2012). http://obamawhitehouse.archives.gov/sites/default/files/microsites/ostp/pcast-stem-ed-final.pdf.

6 Federal Communications Commission, *Amendment of the Commission's Rules with Regard to Commercial Operation in the 3550–3650 MHz Band, GN Docket 12–354*, Report and Order and Second Further Notice of Proposed Rulemaking and Order (2015). http://apps.fcc.gov/edocs_public/attachmatch/FCC-15-47A1.pdf.

7 ——, *Amendment of the Commission's Rules with Regard to Commercial Operation in the 3550–3650 MHz Band, GN Docket 12–354* (2016). https://apps.fcc.gov/edocs_public/attachmatch/FCC-16-55A1.pdf.

8 P. F. Marshall, *Scaling, Density, and Decision-Making in Cognitive Wireless Networks* (Cambridge University Press, 2012).

9 ——, *Quantitative Analysis of Cognitive Radio and Network Performance* (Norwood, MA: Artech House, 2010).

10 L. Doyle, *Essentials of Cognitive Radio* (Cambridge University Press, 2009).

11 Federal Communications Commission, *Facilitating Opportunities for Flexible, Efficient, and Reliable Spectrum Use Employing Cognitive Radio Technologies*, ET Docket No. 03–108 (2003).

12 M. P. Fitz, T. R. Halford, I. Hossain, and S. W. Enserink, Towards simultaneous radar and spectral sensing. *2014 IEEE International Symposium on Dynamic Spectrum Access Networks* (2014), 15–19.

13 QinetiQ, *Cognitive Radio Technology: A Study for OFCOM – Summary Report* (2007). www.ofcom.org.uk/__data/assets/pdf_file/0017/40364/cograd_summary.pdf.

14 P. K. Muthuswamy, K. Kar, A. Gupta, S. Sarkar, and G. Kasbekar, Portfolio optimization in secondary spectrum markets. *2011 International Symposium on Modeling and Optimization in Mobile, Ad Hoc and Wireless Networks (WiOpt)* (2011), 249–256.

15 H. Mutlu, M. Alanyali, and D. Starobinski, Spot pricing of secondary spectrum access in wireless cellular networks. *IEEE/ACM Transactions on Networking*, **17**/6 (2009), 1794–1804. doi: 10.1109/TNET.2009.2019959.

# 19 Alternatives to Three-Tier Operation

## 19.1 Introduction

This chapter considers some of the alternative future concepts and approaches that can serve as alternatives to the three-tier approach. They are considered in the context of the evaluation criteria used in Chapters 3 and 4. Where there are shortfalls in our three-tier criteria, methods of resolving them without the implementation of a three-tier regime will be explored.

## 19.2 Cognitive Radio

Cognitive Radio (CR) has been a concept that has had considerable research performed in the academic community, and has achieved some interest in the commercial world, as well. There are significant overlaps between the three-tier concept and CR. The fundamental difference in the academic CR research is that the CR device makes spectrum decisions based on its perception of conditions, and in three-tier, it is centralized in a database-driven process, with very little spectrum decision-making performed by the radio.

It should be noted that many of the participants in the President's Council of Advisors on Science and Technology (PCAST) study were active researchers in the field of CR. It is instructive to consider why the PCAST recommendations did not include more of the CR fundamental concepts. Some of the reticence to advocate for CR principles is undoubtedly to create a lower risk set of proposals, and thus increase the likelihood that the recommendations could not be effectively challenged. But there were also challenges that current CR technology could not address, as well.

**Sensing** As discussed in the section on Television White Space (TVWS), there are difficulties in relying on sensing as a mechanism to create awareness of other devices operating in the band, unless the uses are consistent with sensing, which many existing uses are not.

**Tier Detection** Even if the sensing problem is solved, the ability to determine the status of the tier they are operating in is difficult, if not impossible, if multiple tiers have common technologies. In the band considered by the United States (US) PCAST study, the primary incumbent is a radar with a single set of operating characteristics and is always the primary, so there is no difficulty in detecting the radar, and determining its tier. However, the second and third tiers have no technology limitations, and cannot be differentiated through sensing methods.

**Receive Only** The assumption in most CR discussion (including by this author) is that protected users can be detected by their emissions. While the item above raises questions about that premise, there is no doubt that receive only devices, such as satellite and broadcast receivers cannot be detected. A similar argument can be made for systems that operate in some variant of Frequency Division Duplex (FDD) Media Access Layer (MAC) layers.

**Rights Acquisition** Rights acquisition through some intermediary, or even one-on-one, implies common brokers or other interconnection that may involve nodes that are in range, or have common protocols, cryptographic material, and policies that can enable a self-forming market, that is not reliant on non-automated intermediary facilitators.

It is a fair question if one or more of these constraints can be remediated in order that the centralized functions in the three-tier framework can be devolved to execution by the devices themselves. A distributed, autonomous approach would appear to benefit a network that was evolving towards an edge-focus, such as is being implemented in small cells using Long-Term Evolution (LTE) Self-Organizing Network (SON) technology. Certainly some of these individual issues may not be constraining in specific band applications.

The objectives of the three-tier regime can be at least partially achieved by mitigating the issues described above:

**Sensing** The general framework of the three-tier regime was not based on the use of self-sensing due to the number of special and corner cases that could create, or be perceived to create, vulnerabilities in the ability of the admission control system to be fully aware of protected nodes. The United States of America (USA) Citizens Broadband Radio Service (CBRS) has a hybrid approach, that uses sensing for the naval radars, promulgated by a database system, and declared locations for Fixed Satellite Service (FSS) and Part 90 users. It is possible that a specific implementation of three-tier could depend on sensing for some, or even all, of the protected node detection.

**Tier Detection** A more subtle spectrum sensing question is the ability to determine if a node that is sensed has superior, equal, or inferior rights to spectrum access. If the use of each tier has unique and distinguishable technology to a sensor, then it might be possible for all of the devices in an area to organize themselves in a manner that is appropriate to the rights structure. A radar is distinguishable from a communications device. However, within a class of devices, it will be difficult for a sensor to tell the rights of the devices it detects. PCAST and the FCC proposed that the use of Tier 2 and Tier 3 had no differences in the emission rights, and therefore could not only be similar technologies, they could be identical equipment. Therefore, a sensor could not determine which tier a signal was associated with. However, if the technology was detectably unique in each tier, the sensing of a CR could be effective in managing the hierarchy of tiers.

**Receive Only** The premise of cognitive radio is the awareness of the environment, so detecting devices that have no signature is probably not an option. However, for bands with no FDD or receive-only operating modes, this would not be an issue.

**Rights Acquisition** There may be methods for devices to negotiate some rights between themselves if they share a common channel, waveform, and crypto, but the number of conditions in which this could be performed is limited. It is certainly not a general approach, and is not technology neutral. Decisions about rights can be made through cognitive radio, but the coordination of this is likely to be performed through some out of band mechanism.

The introduction of self-sensing in a three-tier device hybridizes these approaches. In a hybrid regime, the admission control system provides a range of "legal" spectrum choices, and the device uses CR techniques to select the best option among these choices. This approach may be adopted to address some of the peer coexistence requirements that will emerge as the band becomes fully occupied.

## 19.3  Market-based Alternative

The three-tier proposal presented in the early chapters implicitly was an offer by a National Regulatory Authority (NRA) to share spectrum that was unused by the primary licensee of the spectrum, and an offer to acquire privileges over other users that have accepted this offer. It is certainly possible that this same offer could be made by a licensee that has sufficiently flexible rights under its license with the NRA.

There may be objections to letting a primary licensee generate revenue from secondary licensing. This use was presumably not contemplated when the license was issued. If it was purchased at auction, the valuation likely did not reflect its secondary-use value. If the spectrum was licensed for a specific use without charge, such as satellite and broadcast spectrum, there may be issues with private income being derived from this public grant. This last consideration was certainly an issue when the USA proposed to allow broadcasters to benefit from the repurposing of their spectrum in the incentive auction. In this case, the decision essentially shares the revenue from reduced broadcast spectrum between the stations relinquishing the spectrum, and the Federal government.

However, if these objections can be overcome, these are advantages to enabling incumbents to form such markets:

- One of the major obstacles to spectrum sharing is resistance to sharing by the incumbents. However, if the incumbent has a vested interest in the success of the sharing regime (such as income), it is reasonable to assume that their support will be more forthcoming.
- The incumbent best understands their operating concept, actual link margin, weather conditions, etc. Use of this insight into the incumbent operation is invaluable in maximizing the degree of sharing that is possible.
- The incumbent may have greater flexibility to set up and operate the sharing regime without the more extensive processes that are required when similar regulations are developed and implemented by governments.

Markets established by current spectrum incumbents would operate similarly to those established by NRA organizations. The enforcement capability of a privately operated three-tier market may require some recognition in national regulations. Assertions made about device operation and compliance with licensing conditions is typically a legal violation (of variable significance) when these assertions are made to governments, and the conditions issued by governments. However, the status of the same violations when linked to a private sharing arrangement is less clear, and may not provide sufficient disincentives to malicious or fraudulent behaviors.

The PCAST and CBRS proposals implied that General Authorized Access (GAA) (unprotected) access was free of any charge, and that the protected middle-tier Priority Access (PA) or Priority Access License (PAL), respectively, had a cost that was established by auction. However, to incentivize commercial offering of sharing, there is no reason that a three-tier system must offer a "free" tier that mimics traditional unlicensed spectrum. For example, it would not be unreasonable that there be a larger fee for protection, and a lesser fee for unprotected access. This is the spectrum equivalent of "reserved seating" and "general admission."

It may be that the private sector opportunity in three-tier management is so attractive that this is the most effective way to recapture the opportunity cost of underused spectrum. It may not have been the intent of these licenses, but it may be a pragmatic means to overcome the typical incumbent resistance to any spectrum.

## 19.4    Management and Allocation by Regulators

In this implementation, we consider the same spectrum policies that were presented in the three-tier spectrum model, but that they were instead implemented through a manually executed, regulatory process. This process is presumably similar to licensing in the current regime, except permitting the third-tier usage. Therefore, the spectrum management benefits would be essentially the same as our three-tier baseline, with the exception of an automated admission control process. In practice, techniques such as aggregation into thousands of nodes would be hard to implement, unless it was so automated as to be the functional equivalent of a SAS admission control system.

It is possible that the same framework that allocates spectrum could allocate any number of tiers manually through a similar process to that used by the automated process. Consider the possibility of errors. In the case of the USA, it is estimated that ten million access points could be deployed over the first three years, with half of this deployment occurring in the third year. Assuming it took one hour for a human to process a request completely, and 1 860 work hours a year, it would take a staff 2 688 employees in the third year. Even if the volume enabled a four times increase in speed of processing the staff needed would still be 672 employees. Anyone who has worked with regulators would find the idea of an answer in one hour to be quite unrealistically optimistic, much less the 15 minute estimate!

This option becomes even more unrealistic when we consider any level of higher-tier spectrum reclaiming activity. If half of the transactions are for protected tier entry, and each entry impacts three existing nodes, the volume of transactions increases from five million to over twelve million, with a linear increase in workload.

The conclusion is that a three-tier regime is dependent on an automated admission control system that is empowered to execute interference analysis and to perform the admission decisions independently of direct human intervention.

## 19.5 Removal of PA Rights Tier

Removal of the PA layer creates effectively an exclusive tier, and a tier resembling unlicensed. The question this approach would raise is whether suitable investment will be made if the option to acquire protection is removed. The PCAST argument was that protected access was an essential aspect of a regime that desired to attract infrastructure investment. While it was not clear that protection was, or is, a superior approach assured operation for performance-sensitive users, the right to obtain it, if needed, assured that the effectiveness of their investments in infrastructure could be protected.

Experience with the three-tier regime is required before the necessity of a protected tier can be definitely determined. However, communications services providers are inherently conservative, due to their traditional emphasis on reliability and predictable performance. The goal of the admission control system is to minimize the extent to which such protections deny useful access to other users. If successful, this minimizes the imposition on access by other users. And, so long as the protected tiers are populated with devices to the carrying capacity of the band, the objectives of the regime are still fulfilled.

## 19.6 Removal of GAA Rights Tier

Removal of the GAA tier essentially reverts the three-tier framework to a set of primary and secondary exclusive licenses, similar in effect to Authorized Shared Access (ASA)/Licensed Shared Access (LSA). It undercuts the fundamental principle that all spectrum should be usable by someone at all times and places. As such, it fails to be disruptive to the existing structure of wireless services. Although such regimes do have value, they are not relevant to the innovation objectives of three-tier spectrum management.

## 19.7 Reallocation as an Alternative

The PCAST study proposed the three-tier framework because it viewed the option of clearing and reallocation of many of the (even lightly used) Federal spectrum bands was not practical. Therefore, three-tier spectrum management appeared to be a more effective use of the spectrum than the exclusive use by the less densely deployed Federal agencies. The significant interest in developing systems for this band would appear to support this belief, although it is too early to know how the band will evolve over the next decade.

Whether this assumption was right or wrong, it is a different question than whether three-tier can produce greater value from the spectrum than relocating the incumbents, in those situations when reallocation is a practical option. A fair question is: *"should two- or three-tier models[1] replace exclusive, effectively one-tier, licensing?"*

Technically, it is hard to argue that a single tier model can achieve the density of use that the three-tier model achieves. Any spectrum not used by the sole licensee would be effectively wasted. So, if any area was not fully deployed, the case for three-tier is valid. Effective 100% coverage might be possible at lower frequencies, but it is not likely in the higher frequency spectrum that is typically used in small cell deployments. Even if the spectrum is used extensively in two dimensions, it is likely that it will not be fully exploited in all three dimensions.

The question of the value of this lower-tier use could be open to question. However, if the three-tier interference protection is effective, then whatever benefit is achieved is at least a societal benefit, and comes at no cost except the complexity of the regime. Establishing the value of this spectrum access is a question for economists, not engineers!

## 19.8    Summary

Table 19.1 summarizes the degree to which these alternative mechanisms achieve the objectives established in Chapters 1 and 3. The objectives of three-tier can be achieved to a significant extent with different approaches, depending on the specific characteristics of the incumbents of the band.

Perhaps the most important, and differentiating, feature in the multi-stakeholder framework is the use of an automated admission control system. This enables the use of complex algorithms, fine-resolution spectrum and spatial extents, with the corresponding high number of unique transactions.

The mix of protected and unprotected users (first and third tier) may make the middle tier meaningless if they are effective in the spectrum. In this case, the intent is met. However, it is more likely that different locations will have different usage mixes, with a PA tier being maximally occupied in some regions, and the lower tier the dominant usage in others.

## 19.9    Suggested Reading

Most of the appropriate readings have already been suggested in previous chapters:

---

[1] Since the spectrum is being reallocated, it implies that the first tier is being relocated, leaving the three-tier model operating only the lower two tiers.

**Table 19.1** Satisfaction of three-tier objectives by alternative approaches to its implementation.

| Approach | Multiple Protected Tiers | Lower Tier Shared | Lowest Tier Without Protection | Small Allocations (Frequency and Spatial) | Automated Transactions |
| --- | --- | --- | --- | --- | --- |
| Cognitive Radio | Difficult to isolate tier users, unless there is a forced differentiation in the air interface appearance to receivers. | Tiering difficult to establish among homogeneous users. | Possible. | No allocations used. | A central admission control system is not needed due to autonomous nature of decision-making. |
| Market-based | Could be provided by the market maker. | Assumed that the rights owner would require compensation. | Could be provided by the market maker. | Could be provided by the market maker. | Could be provided by the market maker. |
| Regulator Allocation | YES, manually. | Possible, but may be impractical due to volume limit of regulatory transaction rate and cost. | Possible, but may be impractical due to volume limit of regulatory transaction rate and cost. | Small allocations would drive transaction volume beyond practical limits. | No, relies on regulatory interaction for allocation and assignment. |
| Remove PA | No non-incumbent protection available. | YES | YES | YES | YES |
| Remove GAA Tier | YES | YES | There is no lower tier that is available for unprotected use, therefore all spectrum carries some exclusive rights. | YES | YES |
| Reallocation | Would only establish a single tier. | Only if no protected tier was created. | Only if no protected tier was created. | Possible, but traditional exclusive licenses have been for large regions. | Not applicable. |

- There is extensive literature investigating cognitive radio technology that has emerged over the last two decades. Clearly, some fusion of the principles of CR and of three-tier spectrum will occur in the future, and they will no longer be distinct alternatives. In addition to two books by the author [1, 2], other readings include books focusing on cognitive networks [3], and broader surveys of the field of cognitive radio [4, 5].
- The more traditional approaches discussed here have close analogs in traditional spectrum management. Understanding of these is certainly useful in interpreting more deeply on how the three-tier regime can coexist in these regimes. The work by Cave and Webb is a good resource on the current practices and issues in spectrum management practice and issues [6].

## References


1 P. F. Marshall, *Quantitative Analysis of Cognitive Radio and Network Performance* (Norwood, MA: Artech House, 2010).

2 ——, *Scaling, Density, and Decision-Making in Cognitive Wireless Networks* (Cambridge University Press, 2012).

3 Q. Mahmoud, *Cognitive Networks: Towards Self-Aware Networks* (New York: John Wiley & Sons, 2007).

4 S. Haykin, Fundamental issues in cognitive radio. In *Cognitive Radio Networks* (New York: Springer-Verlag, 2007).

5 L. Doyle, *Essentials of Cognitive Radio* (Cambridge University Press, 2009).

6 M. Cave and W. Webb, *Spectrum Management Using the Airwaves for Maximum Social and Economic Benefit* (Cambridge University Press, 2016).

# 20 Conclusions and a Look Ahead

## 20.1 Benefits and Roadmap of a Transition to Three-Tier

The future of three-tier spectrum is likely in the hands of the industry and operator communities that are engaged in its deployment. The regulators have created an opportunity for a new ecosystem; its success will be evaluated from the outcome of several deployment thrusts, such as the neutral-host model, enterprise, and Internet of Things (IoT), among others. It is not necessary that all succeed; one "killer app" is probably sufficient for an ecosystem to expand beyond the shores of the United States (US), and the 3.55 GHz band. It is not just that the US Citizens Broadband Radio Service (CBRS) is particularly important by itself; but, that a failure in the first deployment would certainly support and empower the opponents of any adoption anywhere, regardless of the reason for failure in the USA.

The rest of this section will address the scenario in which the US deployment is at least a partial success, and has sufficient economic and societal benefit that other nations have a desire to achieve at least these benefits, and maybe to tailor the regime to make it even more effective than the case in the USA.

On the other hand, if the US deployment is perceived to be a failure, then much of this material is irrelevant. The material in Section 20.5 may be of some use in resolving the perceived flaws in the three-tier regime.

## 20.2 Technology and Policy

Much of the discussion of this book has centered on spectrum-sharing technology; technology that is not specific to three-tier spectrum management regimes. The difference between general spectrum sharing and the specific three-tier model is that the three-tier model uses spectrum-sharing technology as a necessary enabler of a new, innovative, and potentially transformative regime, rather than an augmentation of current spectrum policies into a two-party sharing regime. Spectrum sharing is an admirable goal, but it is not equivalent to three-tier in terms of enabling ease of entry, of innovation, scalable deployment, and the maximum utilization of the spectrum.

There are many technical obstacles to freeing up spectrum access, and incumbents deploy them against any threats to either their use of the spectrum, or the business model which they practice, and depend on. It is the job of the policy community to balance these very immediate interests of vocal critics of any increased access to spectrum against the opportunity cost of precluding not-yet-existing, not-yet-vocally-supported, and somewhat vague opportunities.

However, it can be argued that these are the same innovations, technologies, and new business models that produce economic growth, and value to consumers and societies. At least, spectrum policy should be neutral to the forces of creative destruction, rather than a protector of the status quo. The forces of creative destruction have shown themselves to be inevitable; but, it has also shown that they benefit the early adopters, and punish those that cling to old models, in the belief that these models can be saved.

## 20.3    National Activities

The previous chapters showed that opportunities to establish three-tier spectrum sharing through national policy had the most potential for immediate action. Such action would be secondary to international treaty obligations. Further, it had the greatest extent of spectrum that could be repurposed or shared, in the near- and mid-term timeframe.

There are two paths that nations may consider:

- Alliances with spectrum holders that may find their current allocations vulnerable to reallocation, and might view sharing with a broad constituency as providing the support needed to continued retention of their access rights. To some extent, this is the case with the US CBRS band, in that the Department of Defense (DoD) had only very low-density usage of the band. Sharing the band would create a community that has a vested interest in maintaining the Federal allocation for this band, and resist any reallocation of the band.
- Forcible establishment of three-tier regimes. This is most likely in the exclusive usage bands that have been allocated to a single use, but not to a single user. This is the case in bands allocated to broadcast, many of the satellite services, government, and general commercial use. The extended C-band satellite incumbents in 3 600–3 700 MHz in the United States of America (USA) would consider this the case in the CBRS implementation.

## 20.4    International Activities

International activity should have two goals: one focused on achieving the maximum amount of common three-tier spectrum, and one aiming to maximize the harmonization of the sharing principles (even if the bands themselves are not harmonized) in order to achieve the benefits of high-volume devices and access control software. Creating success in making specific allocations to three-tier spectrum across a wide range of interests, as are involved in the International Telecommunication Union (ITU) and World Radio Conference (WRC) processes would be a significant, and unnecessary challenge. Since the premise of three-tier spectrum management is to protect the rights of the incumbent, primary tier, explicit WRC or ITU action is not required for individual nations to proceed. That said, establishing realistic and balanced protection expectations for the services protected under ITU regulations would be highly useful to all nations, and even to incumbents. The harmonization of many of the three-tier spectrum access principles would benefit both the device and spectrum admission control system communities.

The international approach should focus on principles for both the general framework of three-tier rules, and on establishing balanced protection expectations for incumbent systems. Specifically:

- ITU interference recommendations are developed by working groups with common interests in the protected service. The process for their development is expensive to the participants in terms of time and travel, and therefore the participation tends to consist of organizations with a vested interest in the protection of their specific spectrum allocations. If there was no tension between the uses of the spectrum, this would be a reasonable process. However, it has no balancing force to represent the interests of a future in which this spectrum might have to be shared. The protection criteria should be a balance, or compromise, between many competing interests, not just the incumbent's. All of the interference protection recommendations should be revalidated, with a process that provides a balance between the incumbent interests, and the new uses that can be provided in the same spectrum.
- Harmonization of spectrum has been a focus of the ITU. New devices are much more flexible, and strict harmonization is no longer as critical as it once was. Harmonizing the rules (rather than specific band allocations) for devices to operate worldwide is critical to economically viable markets. Harmonization of core spectrum sharing principles and methods will be highly beneficial. Examples include extensions to propagation models to account for clutter loss, methods of interference aggregation, registration, privacy expectations, and other aspects that transcend the individual band allocations, but would comprise the "virtual regulations" discussed previously.

These objectives, while not essential to three-tier spectrum management would significantly benefit all parties. It would reduce regulatory complexity for national adoption, reduce costs to all participants though the largest compatible markets, and provide smaller National Regulatory Authority (NRA) organizations with a practical path to adoption.

As an example, the Third-Generation Partnership (3GPP) Long-Term Evolution (LTE) specifications are far from harmonized in terms of frequency and band plans (43 are defined as of this writing). But they provide a common control system, physical and Media Access Layer (MAC) layer design, and other band-independent aspects of the design. This harmonized technology framework has been readily extended into many internationally harmonized, and nation-specific spectrum bands, as needed. There are massive cost savings from replicating this functional core, even if the Radio Frequency (RF) elements of the design are unique to each band. The same principles can be applied to three-tier spectrum. Some broad international principles can be developed on common, international platforms, and then localized through development of frequency-specific implementations.

## 20.5    What Might Come After the Three-Tier Spectrum Regimes

This book has largely looked to a transition from single-tier spectrum management to three-tier. However, it is worthwhile to begin the process of looking beyond the adoption of three-tier in the bands in which it is appropriate, to spectrum regimes that might supplant it, and create even greater spectrum utilization, and flexibility. Two thrusts present themselves:

- Section 16.2 introduced a number of constraints on the application of three-tier spectrum. A regime that did not inherit these restrictions could be applied more broadly, and would have benefits beyond the current three-tier model.
- The three-tier regime continues an implicit mission to avoid interference to devices. There are alternative approaches that can assure high-confidence communications without the massive loss of spectrum utility that is required to assure interference-free operation on specific and fixed channels.

The next two subsections explore each of these alternatives.

### 20.5.1    Removal of Three-Tier Spectrum Management Application Limitations

We will review the constrains to adoption of three-tier that were provided in Section 16.2, and consider how these limitations could be eliminated if the framework of the three-tier regime was modified, or abandoned.

**Incumbent Awareness** The question is whether an enhanced regime could avoid the necessity to have detailed awareness of protected incumbents. This implies that either the awareness is not needed, or can be obtained through other means. The next section addresses eliminating the need for proactive protection of incumbents through interference tolerance or avoidance. The other alternative is to rely on sensing of these incumbents. The Dynamic Spectrum Access (DSA) community has developed a number of concepts, prototypes, and executed experiments on the "sense and avoid" principle. There are some fundamental limits in the relative EIRP of the incumbent and sharing device [1, 2].

**Meaningful Bandwidth** The question of bandwidth availability is driven by the technology that will utilize the band. The emphasis today is on mobile wireless, which, even with Long-Term Evolution Carrier Aggregation (LTE-CA), requires meaningful segments of contiguous spectrum. Other applications may emerge that need less bandwidth, such as for the hypothesized IoT networks, or systems that can use a large number of independent carriers effectively. The requirement for contiguous spectrum may be much less important in future years if the spectrum is made available on reasonable terms. Contiguous spectrum is important today, but need not be an obstacle in the near future. History has shown that when spectrum is available on advantageous terms, innovation occurs.

**Opposition Effectiveness** The question would be if an enhanced version of three-tier spectrum could be more acceptable to incumbents, and could this model therefore be more successful than the three-tier model, as it is framed today? Resistance to the implementation of spectrum sharing would require that the risks to incumbents be further reduced below the level of risk of the three-tier regime. The three-tier regime is quite conservative, and is likely as low risk as any follow-on approach. The risk to incumbents can only be reduced by changes in the incumbent system's vulnerability to interference, as discussed in the next section.

**National Scope** The question here is if a follow-on to the three-tier regime could be more rapidly adopted on an international basis, and therefore avoid achieving independent decisions by each nation. This question may be independent of the regime that is adopted, whether it is three-tier, or another technology.

## 20.5.2 Transition to a New Set of Objectives for Spectrum Management

The advocates of DSA spectrum management have generally accepted the premise that the goal of the alternative regimes they propose is to emulate the existing processes, in holding the avoidance of interference as a key organizing principle. This is true for the US President's Council of Advisors on Science and Technology (PCAST) and CBRS proposals, as well as the vast majority of material presented in academic conferences on DSA, such as the IEEE Dynamic Spectrum Access Networks Conference (DYSPAN) and Global Communications Conference (GLOBECOM).

The author has proposed two alternative approaches to the enablement of highly dynamic wireless systems. Both of these proposals challenge the concept that spectrum management should focus on assuring clear channels, and the essentially isolated operation of protected devices. In both of them, the spectrum management regime works with devices cooperatively to meet their needs, or supports their independent recourse actions. In both cases, devices have to be sufficiently *"smart,"* or more appropriately, *"cognitive."*

The first proposed principle addresses the issue from the perspective of receiver protection. The author has proposed this approach in several publications and papers [3, 4, 5], including *Dynamic Spectrum Management of Front-End Linearity and Dynamic Range* that appeared in IEEE Communications magazine.

This analysis demonstrated that with typical spectrum densities, devices with even poor front-end linearity and reasonable tunable/selectable filters could locate spectrum and operate at the same front-end overload probability as devices with 20 dB more front-end linearity,[1] with the consequent reduction in power required by the stage, thermal load, and more flexibility in device substrate and fabrication. Reducing the requirement for front-end protection can enable the spectrum to be used much more densely.

---

[1] As measured by the Third-Order Input Intercept Point (IIP3) of the first stage.

The previous concept removed the necessity for the spectrum management regime to manage the environment of the front-end outside of the signaling channel. The second concept addresses the management of the signaling channel.

The author proposed [6] in a paper titled *"Dynamic Spectrum Access as a Mechanism for Transition to Interference-Tolerant Systems"* that the DSA objective of using sensing to avoid interference to other nodes had inherent limitations on the density it could achieve (similar to those of an admission control system). The very conservative assumptions of propagation loss inherently reduces the effective density far below the actual density that could be achieved with the closest possible non-interfering spacing.

Instead of the prediction-based interference avoidance process, the author proposed that the sensing mission should focus on self-protection and channel selection by the device itself. In this framework, channel selection should be based on the device's own sensing. The devices would be interference tolerant because they would have the flexibility, and the right, to select other channels that were superior by whatever measure they chose to apply. No protection would be offered. This approach achieves significantly more density that prediction-based interference avoidance. However, like any spectrum allocation scheme, it eventually could reach a point of saturation. However, it would reach this true interference-limited condition after admitting many more nodes than could be admitted in a risk-averse interference-avoidance-driven process. Devices with the technical capability, and regulatory authorization to make these decisions would be considered to have *"Recourse,"* and would be allowed to operate flexibly in bands shared with other similarly qualified devices.

This approach requires that the devices in the band are sufficiently cognitive and have sufficient flexibility to not be dependent on an external agent to select and protect a channel for them. The benefits in spectrum density arise from eliminating many of the uncertainties that an admission control system must address with the most conservative assumptions. The admission control system is still needed to inform devices of the range of frequencies that they were permitted to utilize. A spectrum regime could position devices that have these recourse features in a shared band, while segregating devices that required an external agent to prevent them from interference into separate, and tiered spectrum.

Some of these uncertainties that self-protection through sensing can minimize include the following:

- Is the device even transmitting? A sensor will detect this dynamically. Users of devices are unlikely to inform an admission control system of such dynamic information. Devices can decide the timescale in which they are willing to move frequencies.
- Propagation loss is always a guess, at best. For any path, there is generally a significant (tens of dB) difference between the median path loss, mean path loss, and the low-risk path loss estimates. However, the low-risk one is the one that must be utilized in spectrum management. The device sensing can determine if a signal is actually present at a level that would cause interference. The ecosystem will operate at the density permitted by the average path loss, not the worst-case one.

- Placement within a location has a major impact on path loss. A firewall in a building might have 50 dB of loss, which an admission control system could never be aware of, or assume is present, without undue risk of permitting interference. Particularly when considering indoor deployments, these scatter loss impacts probably drive the achievable density, yet cannot be included in the estimates by an admission control system.
- The interference level that is tolerable is highly variable. An indoor system may have strong signal levels in its free-space propagation conditions, and could operate effectively, even if there was a significant level of interference power. An outdoor system might be attempting to maximize range in a free-space environment, in which case even low levels of noise-floor elevation would severely degrade the coverage area. Different applications have different Signal to Interference and Noise Ratio (SINR) tolerance. There is no absolute standard for this metric.

In this model, we fundamentally shift the responsibility for channel quality from the admission control system to the device. The admission control system is still critical to this regime as it retains several missions, including:

- It segregates the devices that do not have the technical, or operational, flexibility to dynamically select frequencies. The spectrum that they do not block are then available for the "free for all" usage by the highly dynamic devices this section contemplates.
- It protects receive-only users, such as frequency-translated services, radio astronomy and other passive sensing, satellite uplinks and downlinks, and other uses that cannot be sensed.
- The admission control system would still have to have a high level of interaction with the dynamic device. They would still require awareness of what spectrum they would be authorized to use at one time and place. This could be dynamic, as the admission-control-system-managed assignments to non-recourse-protected devices, and relocated the partition between the two ecosystems (interference-tolerant and non-interference-tolerant) dynamically.

Although this section proposes these methods as alternatives to three-tier regimes, in fact they would be complementary, and would likely coexist for many years, if not forever. The three-tier regime is an essential precursor for the dynamic regime, as it cannot exist without the partitioning provided through an admission control system.

## 20.6    Summary

The three-tier spectrum model is a significant advance in the management of spectrum, enabled by recently available tools, such as cloud computing, embedded processing, and much more flexible RF and baseband systems. It addresses many of the structural weakness of the current one- or two-tier spectrum allocation models, and is fundamentally unique in its ability to foster innovation in wireless technology and markets.

However, it should be recognized as a milestone, not an endpoint. Abandoning the "right to exclude" is fundamental to any advanced spectrum regime, and essential to any of the possible successor regimes. As it stands, the three-tier model is the most conservative of the approaches to eliminate this necessity for this right, so is a necessary first step in whatever ultimate regime is desired.

We only need to look at the dynamic, competitive, and creative ecosystem that has formed in the Internet space to see the benefits of an ecosystem that fosters innovation, enables flexible and scalable entry, and is based on a premise of abundance, rather than scarcity.

## 20.7    Suggested Reading

Most of the appropriate readings have already been suggested in previous chapters:

- Radars are a major, and highly inflexible, user of spectrum. The approach at the end of this chapter could be applied to radar systems, similarly to the discussion of communications systems. There is very limited literature in the area of cognitive radar, but at least one book has appeared [7].
- The previous chapter discussed ways in which Cognitive Radio (CR) approaches did not address many of the needs of an alternative framework for spectrum management. In this chapter we considered a hybrid approach that leverages many of the concepts in CR, but utilizes them in different ways than classical DSA. The readings from that chapter are very applicable to this topic, as well as the last chapter's. This includes the books by the author [3, 4], and others focusing on cognitive networks [8], and broader surveys of the field of cognitive radio [9, 10].

## References


1 R. Chen, J. M. Park, Y. T. Hou, and J. H. Reed, Toward secure distributed spectrum sensing in cognitive radio networks. *IEEE Communications Magazine*, **46**/4 (2008), 50–55.

2 D. A. Roberson, C. S. Hood, J. L. LoCicero, and J. T. MacDonald, Spectral occupancy and interference studies in support of cognitive radio technology deployment. *1st IEEE Workshop on Networking Technologies for Software Defined Radio Networks*, 2006 (2006), 26–35.

3 P. F. Marshall, *Quantitative Analysis of Cognitive Radio and Network Performance* (Norwood, MA: Artech House, 2010).

4 ——, *Scaling, Density, and Decision-Making in Cognitive Wireless Networks* (Cambridge University Press, 2012).

5 ——, Cognitive Radio as a mechanism to manage front-end linearity and dynamic range. *IEEE Communications Magazine*, **47**/3 (2008) 81–87.

6 ——, Dynamic Spectrum Access as a Mechanism for Transition to Interference Tolerant Systems. *IEEE 4th International Symposium on New Frontiers in Dynamic Spectrum Access Networks* (Singapore, 2010).

7 J. Guerci, *Cognitive Radar: The Knowledge-Aided Fully Adaptive Approach* (Norwood, MA: Artech House, 2010).

8 Q. Mahmoud, *Cognitive Networks: Towards Self-Aware Networks* (New York: John Wiley & Sons, 2007).

9 S. Haykin, Fundamental issues in cognitive radio. In *Cognitive Radio Networks* (New York: Springer-Verlag, 2007).

10 L. Doyle, *Essentials of Cognitive Radio* (Cambridge University Press, 2009).

# Part VII

## Appendices

# Appendix A  Terms and Acronyms

| | |
|---|---|
| **3GPP** | Third-Generation Partnership |
| **4G** | 4th-Generation Wireless |
| **5G** | Fifth-Generation Wireless Systems |
| **AGL** | Above Ground Level |
| **AMPS** | Advanced Mobile Phone System |
| **AMSL** | Above Mean Sea Level |
| **AP** | Access Point |
| **ARP** | Address Resolution Protocol |
| **ASA** | Authorized Shared Access |
| **AWGN** | Additive White Gaussian Noise |
| **AWS** | Advanced Wireless Services |
| **BGP** | Boundary Gateway Protocol |
| **BMD** | Ballistic Missile Defense |
| **BSS** | Broadcasting Satellite Service |
| **CA** | Certificate Authority |
| **CAPEX** | Capital Expenditure |
| **CMBR** | Cosmic Microwave Background Radiation |
| **CBRS** | Citizens Broadband Radio Service |
| **CBSD** | Citizens Broadband Radio Service Devices |
| **CDF** | Cumulative Distribution Function |
| **CDMA** | Code Division Multiple Access |
| **CEA** | Council of Economic Advisors |
| **CIO** | Chief Information Officer |
| **CNR** | Combat Net Radio |
| **COMSEC** | Communications Security |
| **CONUS** | Continental United States of America |
| **CORD** | Central Office Re-architected as a Datacenter |
| **CPE** | Customer Premises Equipment |
| **CR** | Cognitive Radio |
| **CTIA** | Cellular Telecommunications Industry Association |
| **CTS** | Clear to Send |
| **CW** | Continuous Wave |
| **DARPA** | Defense Advanced Research Projects Agency |
| **DAS** | Distributed Antenna System |

| | |
|---|---|
| **DBS** | Direct Broadcast Satellite |
| **DFS** | Dynamic Frequency Selection |
| **DHCP** | Dynamic Host Control Protocol |
| **DME** | Distance Measuring Equipment |
| **DNS** | Domain Name Service |
| **DoC** | Department of Commerce |
| **DoD** | Department of Defense |
| **DSA** | Dynamic Spectrum Access |
| **DSL** | Digital Subscriber Line |
| **DYSPAN** | IEEE Dynamic Spectrum Access Networks Conference |
| **EC2** | Elastic Computing Cloud |
| **ECFS** | Electronic Comment Filing System |
| **EDAC** | Error Detection and Control |
| **EGP** | Exterior Gateway Protocol |
| **EIRP** | Equivalent Isotropic Radiated Power |
| **ENG** | Electronic New Gathering |
| **EPC** | Extended Packet Core |
| **ESA** | Electronically Steerable Array/Antenna |
| **ESC** | Environmental Sensing Capability |
| **ETSI** | European Telecommunications Standards Institute |
| **EU** | European Union |
| **FCC** | Federal Communications Commission |
| **FDD** | Frequency Division Duplex |
| **FFT** | Fast Fourier Transform |
| **FSS** | Fixed Satellite Service |
| **FTTH** | Fiber to the Home |
| **GAA** | General Authorized Access |
| **GBR** | Ground-Based Radar |
| **GCP** | Google Cloud Platform |
| **GLOBECOM** | Global Communications Conference |
| **GPS** | Global Positioning System |
| **GSM** | Global System for Mobile Communications |
| **HAAT** | High Above Average Terrain |
| **HTTP** | Hypertext Transfer Protocol |
| **HTTPS** | Secure Hypertext Transfer Protocol |
| **I/N** | Interference to Noise |
| **ICAO** | International Civil Aviation Organization |
| **IETF** | Internet Engineering Task Force |
| **IF** | Intermediate Frequency |
| **IFT** | Federal Telecommunications Institute |
| **IIP2** | Second-Order Input Intercept Point |
| **IIP3** | Third-Order Input Intercept Point |
| **IIT** | Illinois Institute of Technology |
| **ILS** | Instrument Landing System |

| | |
|---|---|
| **IMT** | International Mobile Telecommunications |
| **IoT** | Internet of Things |
| **IP** | Internet Protocol |
| **ISART** | International Symposium on Advanced Radio Technology |
| **ISP** | Internet Service Provider |
| **IT** | Information Technology |
| **ITIF** | Information Technology and Innovation Foundation |
| **ITU** | International Telecommunication Union |
| **JSON** | JavaScript Object Notation |
| **LAA** | Licensed Assisted Access |
| **LAN** | Local Area Networks |
| **LBT** | Listen Before Talk |
| **LC** | Licensed Shared Access Controller |
| **LMR** | Land Mobile Radio |
| **LNA** | Low Noise Amplifier |
| **LNB** | Low Noise Block Converter |
| **LR** | Licensed Shared Access Repository |
| **LSA** | Licensed Shared Access |
| **LTE-A** | Long-Term Evolution Advanced |
| **LTE-CA** | Long-Term Evolution Carrier Aggregation |
| **LTE-U** | Long-Term Evolution Unlicensed |
| **LTE** | Long-Term Evolution |
| **MAC** | Media Access Control Layer |
| **MDU** | Multi-Dwelling Unit |
| **MFC** | Microwave Filter Corporation |
| **MIMO** | Multiple Input/Multiple Output |
| **MLS** | Microwave Landing System |
| **MME** | Mobility Management Entity |
| **MNO** | Mobile Network Operators |
| **MSO** | Mobile System Operator |
| **MVNO** | Mobile Virtual Network Operator |
| **NAB** | National Association of Broadcasters |
| **NAT** | Network Address Translation |
| **NATO** | North Atlantic Treaty Organization |
| **NEC** | National Economic Council |
| **NIST** | National Institute of Standards and Technology |
| **NOI** | Notice of Inquiry |
| **NRA** | National Regulatory Authority |
| **NSF** | National Science Foundation |
| **NTIA** | National Telecommunications and Information Agency |
| **OFCOM** | Office of Communications |
| **OSPF** | Open Shortest Path First Protocol |
| **OOBE** | Out of Band Emissions |
| **OPEX** | Operational Cost |

| | |
|---|---|
| **OSTP** | Office of Science and Technology Policy |
| **PA** | Priority Access |
| **PAL** | Priority Access License |
| **PAN** | Personal Area Networks |
| **PCAST** | President's Council of Advisors on Science and Technology |
| **PCS** | Personal Communications Service |
| **PDF** | Probability Density Function |
| **PN** | Public Notice |
| **PNSS** | Pseudo Noise Spread Spectrum |
| **PPA** | Priority Access License Protection Area |
| **PPP** | Public Private Partnership |
| **PRF** | Pulse Repetition Frequency |
| **PSD** | Power Spectral Density |
| **PW** | Pulse Width |
| **QoS** | Quality of Service |
| **RAN** | Radio Access Network |
| **RF** | Radio Frequency |
| **RMS** | Root Mean Square |
| **ROC** | Receiver Operating Characteristics |
| **RRM** | Radio Resource Management |
| **RSSI** | Received Signal Strength Indication |
| **RTS** | Request to Send |
| **SAS** | Spectrum Access System |
| **SGW** | Serving Gateway |
| **SIA** | Satellite Industry Association |
| **SINR** | Signal to Interference and Noise Ratio |
| **SLA** | Service Level Agreement |
| **SMLA** | Secondary Market License Agreement |
| **SNR** | Signal to Noise Ratio |
| **SON** | Self-Organizing Network |
| **SPTF** | Spectrum Policy Task Force |
| **SSC** | Spectrum Sharing Committee |
| **SSID** | Service Set Identifier |
| **SSL** | Secure Socket Layer |
| **SSPARC** | Shared Spectrum Access for Radar and Communications |
| **SUR** | Spectrum Usage Rights |
| **TCD** | Trinity College, Dublin |
| **TCP** | Transport Control Protocol |
| **TD-LTE** | Time Division Long-Term Evolution |
| **TDD** | Time Division Duplex |
| **TDRSS** | Tracking and Data Relay Satellite System |
| **THAAD** | Terminal High Altitude Area Defense |
| **TLS** | Transport Layer Security |
| **TT and C** | Tracking, Telemetry and Control |

| | |
|---|---|
| **TV** | Television |
| **TVRO** | Television, Receive Only |
| **TVWS** | Television White Space |
| **UDP** | User Datagram Protocol |
| **UE** | User Equipment |
| **UHD** | Ultra-High Definition Television |
| **UK** | United Kingdom |
| **UN** | United Nations |
| **USA** | United States of America |
| **US** | United States |
| **UWB** | Ultra Wideband |
| **VORTAC** | VHF Omnirange/Tactical Area Navigation |
| **WAN** | Wide Area Network |
| **Wi-Fi** | Wireless Fidelity |
| **WiMax** | Worldwide Interoperability for Microwave Access |
| **WinnForum** | Wireless Innovation Forum |
| **WISP** | Wireless Internet service provider |
| **WLAN** | Wireless Local Area Network |
| **WRC** | World Radio Conference |
| **XG** | Next Generation Communications Program |
| **XML** | Extensible Markup Language |

# Appendix B  Symbols

| Variable | Meaning |
| --- | --- |
| $\alpha$ | Propagation exponent (as in $r^\alpha$). |
| $\lambda$ | Signal wavelength. |
| $\sigma$ | Radar cross section (in same units as $\lambda$ and $R_{\max}$). |
| $\theta$ | Angle from earth station to antenna boresight. |
| admission | Logical variable indicating whether an admission control system issues a grant to a device, or not. |
| $B_{\text{radar}}$ | Radar emission bandwidth. |
| $B_{\text{sensed}}$ | Instantaneous sensed bandwidth. |
| $B$ | Bandwidth in Hz. |
| bearing | View angle from earth station to interference source. |
| $C_{\text{net}}$ | Capacity of the network to deliver information. |
| $C_0$ | Mean transmission capacity. |
| $dB$ | $dB(x) = 10\log_{10}(x)$. |
| $D(n)$ | Quantity of data delivered for user $n$. |
| $d_{\text{AP-C}}$ | Distance from an access point to the most distant client that can be serviced. |
| $d_{\text{AP-V}}$ | Distance from an access point to the closest interference victim that might be impacted by the AP or one of its clients. |
| $d_{\text{C-C}}$ | Distance from an access point's client to the closest interference victim that might be impacted by the AP or one of its clients. |
| $\text{Eff}_{\text{Architecture}}$ | Architecture effectiveness metric, in terms of data capacity, over the spectrum and area whose usage was precluded. |
| $\text{Eff}_{\text{Spectrum}}$ | Spectrum effectiveness, in terms of data delivered across a range, over the spectrum, area, and time whose usage is precluded. |
| elevation | Angle to satellite above the horizon. |

| | |
|---|---|
| $f$ | Operating frequency (in MHz). |
| $G_{ant}$ | Antenna gain over isotropic (in dB units). |
| $G_{LNA}$ | Gain of the amplifier stage in volts. |
| $I$ | Interference power. |
| $I_0$ | Worst-case interference range, out to where other uses of the spectrum are precluded. |
| $I(n)$ | User $n$'s interference range, out to a range where other users of the spectrum are precluded. |
| $IIP2_{volts}$ | Second-order intercept point in volts. |
| $IIP3_{volts}$ | Third-order intercept point in volts. |
| interference | Logical variable indicating whether a grant issued by an admission control system issues will actually cause interference to a protected node or network. |
| $k$ | Boltzmann constant. |
| $k$ | Range ratio, namely the mean ratio of actual to maximum communications range. |
| $L_P$ | Path loss over a general distance ($d$), or between two specific points ($x$, $y$). |
| $lat$ | Earth station latitude. |
| $long$ | Earth station longitude. |
| $N$ | Noise power. |
| $n$ | Number of nodes, indicating network scalability. |
| $n_d$ | Node density ($n$ over area). |
| $N_{bandwidth}$ | Sensed bandwidth induced noise elevation. |
| $P_{Dscan}$ | Probability of sensor scan occurring during a radar pulse permitting detection. |
| $N_{integ}$ | Noise energy from integration over time. |
| $P(C)$ | Network delivery reliability. |
| $P_{AP}$ | Access Point EIRP (in dBm). |
| $P_D$ | Probability of Detection. |
| $P_{Emin}$ | Radar detection energy threshold. |
| $P_{FA}$ | Probability of False Alarm. |
| $P_{prot}$ | Maximum interference permitted at an AP or client node (in dBm). |
| $P_S$ | Radar pules (peak) power. |
| $P_{UE}$ | Client EIRP (in dBm). |
| $R_0$ | Maximum possible communications range. |

| | |
|---|---|
| $R_{\max}$ | Radar maximum range. |
| $R(n)$ | User $n$'s actual communications range. |
| $R_{\text{AP-AP}}$ | Range from an Access Point (AP) to its closest protected AP (in meters). |
| $R_{\text{AP-loss}}$ | Distance for the client power to be reduced from $P_{\text{UE}}$ to the value of $P_{\text{AP}}$. |
| $R_{\text{AP-UE}}$ | Shortest distance from AP to its farthest client. This is the AP service area in this model (in meters). |
| $R_{\text{client-loss}}$ | Distance for the AP power to be reduced from $P_{\text{AP}}$ to the value of $P_{\text{AP}}$. |
| $R_{\text{UE-UE}}$ | Range from a client on one AP to a client on an adjoining one (in meters). |
| *slot* | Satellite equatorial (geosynchronous) orbital slot (degrees longitude). |
| $S$ | Signal Power. |
| $S_0$ | Mean amount of spectrum denied to other users. |
| $S(n)$ | Actual spectrum precluded to other users by user $n$'s activity. |
| $T$ | Temperature in Kelvin. |
| $T(n)$ | Quantity of data actually provided to user $n$. |
| $t_{\text{pulse}}$ | Radar pulse duration. |
| $t_{\text{sense}}$ | Sensor scan duration on any given frequency. |
| $v_{\text{out}}$ | Output voltage of the amplifier stage. |
| $v_{\text{in}}$ | Input voltage to the amplifier. |

# Appendix C  Opportunity Cost Model of Spectrum Utilization

The following material is abridged and slightly modified content from Chapter 4 of a previous book (*Scaling, Density, and Decision-Making in Cognitive Wireless Networks*) by the author [1]. It was published in 2012, also by Cambridge University Press. One of the book's premises was that spectrum utilization should be considered to be the quantification of the lost opportunity cost resulting from one user's application on all other users of the same spectrum, and it treated receiver protection and emitter impacts symmetrically. This model was adopted in the President's Council of Advisors on Science and Technology (PCAST) report [2], Appendix B (Proposed New Metric for Spectrum Use) to explain the focus on reducing default receiver protection, or at least reflecting it in the cost charged to spectrum users for the protection.

## C.1  Key Objectives for Introducing Cognitive Processes to Wireless Systems

When communications engineering and operations was dominated by individual links, the evaluation and optimization of these links was relatively scalar; maximize the link performance, or minimize the resources required, to achieve a given link performance. As wireless has moved from wireless access to wireless networking, the evaluation process is much less scalar, and the objectives of optimization are often competing. We will examine four metrics of the aggregate effect of network decision-making. We will see that these metrics are not independent or orthogonal, but approaching them with this assumption is convenient to the early analysis steps, before they are "entangled."

**Capability** $(C_{\mathrm{net}})$**:** We will consider capability to be from the demand to the source of information for the mean request. This metric is therefore not just a metric of the packet network, but also includes the consequences of content location.

**Reliability** $(P(C))$**:** The network reliability is considered in the context of the other three metrics. It is the probability of achieving a given *capacity* in a network of given total membership (*scalability*) and mean *density*. We express this as a function of the capacity.

**Scalability** $(n)$**:** The total number of nodes forming a single address space and for which all other nodes must be sufficiently aware of their location, reachability, or address to communicate.

**Density** $(n_d)$: The physical density of network nodes. This is measured in nodes per unit area, or $n_d = \dfrac{n}{a}$, where $a$ is the area over which the $n$ nodes are deployed.

The purpose of this book is not to be a design text, so our interest is not to determine specific values for each of the variables, but to understand the effect of network decisions on each of the variables. Therefore, the focus will typically be on the derivatives of each variable. That will lead us to understand the sensitivity of the network's performance to each possible operational parameter and condition, and thus each possible decision that can be made about its operation.

## C.2     Metrics for Wireless Network Effectiveness

### C.2.1     Spectrum Usage Effectiveness Metric

There is no shortage of metrics for reflecting different aspects of wireless network effectiveness. Some of these assess physical layer efficiency, as in bits/Hertz, others assess the range achieved, and still others assess capacity, as examples. Unfortunately, none of these reflect the fundamental trades that must be made to operate wireless systems in conditions of high density, spectrum scarcity, and inadequate capacity. Measures of spectrum efficiency fail to reflect the complex trade space of wireless design, and very different mission needs. Additionally, since there is no standard for 100% effective use of spectrum, or most aspects of wireless capacity, the concept of efficiency cannot be applied, since a ratio of actual to perfect is not a meaningful comparison. Instead, we will develop a concept of effectiveness, which is an unbounded metric.

For example, it will be shown later that increasing bits/Hertz is desirable for a single user in the spectrum, but is highly detrimental when users must be densely packed into spectrum, with adjacent interference regions. Similarly, capacity is readily increased with additional spectrum or power, but this mechanism implies that the network is not in any sense more effective or efficient in its use of these resources. In this work, we will consider spectrum not from the perspective of what is used, but from the degree to which its use is precluded to other uses of the spectrum. It therefore reflects the opportunity cost of the spectrum. It makes no difference if the user of the spectrum considers his range to be distance $r$, when the spectral power density precludes meaningful operation out to a range many times that.

Also, we wish to see a metric that *"rewards"* systems that recognize and leverage the actual range of their users from the closest network node. Shouting at a great volume to a user that is ten times closer than the limit of the communications is wasteful of spectrum resources, and denies other uses of the spectrum.

From the network layer perspective, we consider a link from user $B$ to $k$ other users, denoted as $a_1, a_2, \ldots a_k$. Transmissions from and to each user $a_k$ utilize spectrum $S(k)$, occur at a range of $R(k)$, and preclude other users of the spectrum out to a range of $I(k)$. The amount of time taken for a block of data is $T(k)$, and provides a set of data of $D(k)$ bits.

The benefit from this metric system is the distance over which the communication is provided ($R$), and the volume of data ($D$). The cost of the delivery operation is the time (T) over which the spectrum is utilized, the distance over which the spectrum is precluded ($I^2$), and, of course, the amount of spectrum ($S$) that is not available to other users. It is important to recognize that this metric applies to systems of transmitter and receivers. A poor receiver that has significant response, or susceptibility, to adjacent channel signals effectively increases $S$ (lowers the Effectiveness metric) because the receiver requires some form of protection through reduced usage of these guard bands. Spectrum is "*consumed*" by receivers, as well as by transmitters.

$$\text{Eff}_{\text{Spectrum}} = \sum_{n=1}^{k} \frac{R(n)\, D(n)}{I^2(n)\, T(n)\, S(n)} \tag{C.1}$$

where $\text{Eff}_{\text{Spectrum}}$ is the spectrum effectiveness in terms of data delivered across a range, over the spectrum, area, and time whose usage is precluded. $R(n)$ is user $n$'s actual communications range; $D(n)$ is the quantity of data delivered for user $n$; $I(n)$ is user $n$'s interference range, out to a range where other users of the spectrum are precluded; $T(n)$ is the quantity of data actually provided to user $n$; and $S(n)$ is the actual spectrum precluded to other users by user $n$'s activity.

The units of this measure are in $\dfrac{bits}{distance\ Hertz\ seconds}$, which is intuitively pleasing, but reflective of a much more complex set of relationships than the modulation order of the signal. Importantly, it is not the standard licensing metric of spectrum alone, but reflects the possible sharing of time, and the extent over which the spectrum cannot be used by others. This is the metric needed for shared spectrum.

For static systems,[1] such as some cellular downlinks, we can simplify the equation to the steady state value by:

- Replacing the "time and data delivered" term with a general expression of peak capacity, $C_0$.
- Considering the denied area to be constant at $I_0^2$.
- Considering that users are distributed across a range of 0 to $R_0$, with a mean of $k\, R_0$.

The effectiveness measure of Equation (C.1) becomes:

$$\text{Eff}_{\text{Spectrum}} = \frac{k\, R_0\, C_0}{I_0^2\, S_0} \tag{C.2}$$

where $k$ is the range ratio, namely the mean ratio of actual to maximum communications range; $R_0$ is the maximum possible communications range; $C_0$ is the mean transmission capacity; $I_0$ is the worst-case interference range, out to where other uses of the spectrum are precluded, and $S_0$ is the mean amount of spectrum denied to other users.

We can examine how this metric behaves with the introduction of various system-architecture features:

---

[1] Meaning systems with fixed coverage, bandwidth, modulation characteristics, etc.

**Interference Tolerance:** Interference tolerance is strongly rewarded. Any measure that reduces the interference range has a square relationship to the metric. If a given approach reduces the interference range by 44%, the value of the metric is doubled. Similarly, if a given system design can tolerate interference and enables it to operate closer to another system by the same 44%, the metric is doubled.

**Matching Range and Usage:** Communications is most effective if the range of the system is matched to the range of the users. If many downlinks are in range of a handset, then it is likely that the range being provided exceeds that is needed, and the mean range will be far less than a $k$ of 0.5. The metric is optimized when the communications is tailored to the actual user range dynamically.

**Power Management with Dynamic Usage:** Power control is certainly useful, and would be reflected in the dynamic values of $I(n)$, but only if there are provisions to either coordinate the opportunities with other systems in the spectrum, or where the interference caused by high-power operation can be tolerated by other systems, and thus not cause the value of $I_0$ to be established at the maximal, or worst case range.

**Higher Frequencies:** Propagation loss typically is increased for communications links at higher frequencies. However, the increase in loss is generally even higher for the path to possible victims of interference. Therefore, use of higher frequencies often is highly beneficial in terms of the possible density, even if it increases the power requirement for the transmissions. Dense spectrum operation benefits from the availability of a wide range of frequencies, in order to maximize the ratio of $\dfrac{R_0}{I_0}$.

**Receiver Performance:** Poor receiver performance reflects in this metric by forcing more distance between nodes (and thus increasing $I$, and in increasing the amount of spectrum that is required for a service, using that receiver; increasing $S$). For example, the Global Positioning System (GPS) adjacent-channel interference issue with the proposed LightSquared service [3, 4] reflects that the true value of $S$ for GPS is much higher than the GPS waveform bandwidth, as the design of the receivers significantly precludes meaningful usage of additional spectrum, beyond that of the emitted signal. The spectrum needed to provide guard bands to protect the receivers effectively adds to the $S$ value.

## C.2.2    Architecture Effectiveness Metric

A similar measure can be developed to consider the density effects of different system architectures. In this case, we ignore the range term, since we would be considering comparisons of alternative methods to accomplish identical communications tasks. If we examine the effectiveness of different architectures to deliver service to the same set of nodes.

If the serviced set of nodes is equivalent in two architectures, the effectiveness of this architecture ($\text{Eff}_{\text{Architecture}}$) reduces Equation (C.2) further to:

$$\text{Eff}_{\text{Architecture}} = \frac{C_0}{I_0^2 \, S_0} \tag{C.3}$$

where $\text{Eff}_{\text{Architecture}}$ is the architecture effectiveness metric, in terms of data capacity, over the spectrum and area whose usage was precluded. This metric has units of: $\dfrac{bits}{area\ Hertz}$.

For example, this measure is appropriate to compare the effectiveness of a cellular-tower-provided service, or a Wireless Fidelity (Wi-Fi)-provided one (of equivalent character) to the same set of candidate nodes. This measure may explain why Wi-Fi is so effective, and that this small slice of spectrum was reported to offload over 40% of one carrier's smartphone traffic [5], despite its limited coverage or availability. Given the fundamental limits of communication theory, the only viable method to achieve density of information, is to assure locality of communications.

## References


1 P. F. Marshall, *Scaling, Density, and Decision-Making in Cognitive Wireless Networks* (Cambridge University Press, 2012).

2 President's Council of Advisors on Science and Technology, *Report to the President: Realizing the Full Potential of Government-Held Spectrum to Spur Economic Growth* (Executive Office of the President (EOP), Office of Science and Technology Policy (OSTP), 2012. http://obamawhitehouse.archives.gov/sites/default/files/microsites/ostp/pcast-stem-ed-final.pdf.

3 National Space-Based Positioning, Navigation, and Timing Systems Engineering Forum (NPEF), *Assessment of LightSquared Terrestrial Broadband System Effects on GPS Receivers and GPS-dependent Applications* (2011). www.gps.gov/spectrum/lightsquared/docs/2011-06-NPEF-lightsquared-report.pdf.

4 Federal Communications Commission, *Comment Deadlines Established Regarding the LightSquared Technical Working Group Report*. Technical report, DA 11–1133 (2011). http://apps.fcc.gov/edocs_public/attachmatch/DA-11-1133A1.txt.

5 AT&T Inc., AT&T Wi-Fi network usage soars to more than 53 million connections in the first Quarter. Press release (2010).

# Index